Identification Guide
to the Ant Genera of the World

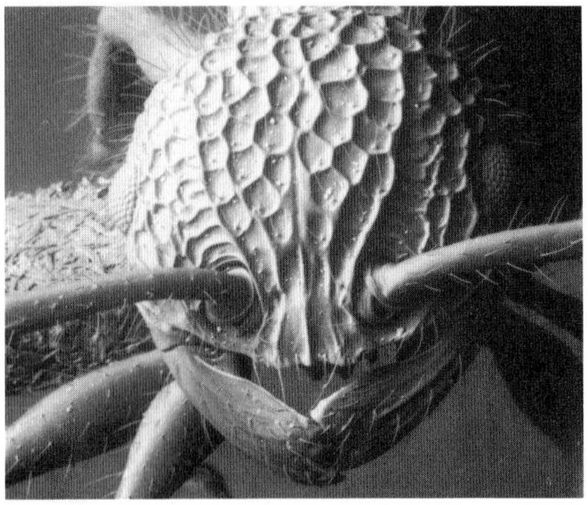

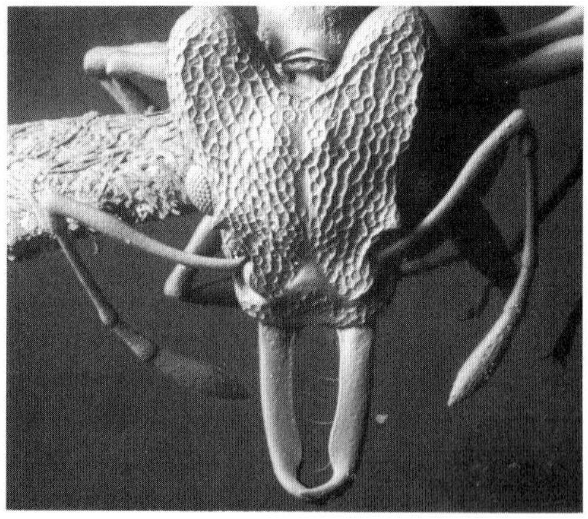

Identification Guide to the Ant Genera of the World

BARRY BOLTON

Harvard University Press
Cambridge, Massachusetts
London, England · 1994

Scanning Electron Microscope photography
by Laraine Ficken.

Printed in the United States of America

This book is printed on acid-free paper, and its binding
materials have been chosen for strength and durability.

Library of Congress Cataloging-in-Publication Data

Bolton, Barry.
Identification guide to the ant genera of the world / Barry Bolton.
p. cm.
Includes bibliographical references (p) and index.
ISBN 0-674-44280-6 (acid-free paper)
1. Ants—Identification. 2. Ants—Classification. I. Title.
QL568.F7B58 1994
595.79′6′012—dc20

93-41270
CIP

Contents

Identification Guide to the Ant Genera of the World

Introduction

Among all the wide variety of insect life on the planet, ants are one of the few forms universally recognized. This is because as a group they are truly ubiquitous, and usually quite conspicuous. They are found in all terrestrial habitats from subarctic tundra to equatorial rainforest, from swamp to harsh desert, from sea coast to great altitude, and from deep in the soil to the tips of the highest trees. Their morphology is as varied as their habitat preference and their range of lifeways is enormous. In size they range from less than 1 millimeter to about 40 millimeters, in temperament from docile to extremely pugnacious, in colony size from a few dozen to many millions. For food various species utilize plant seeds, nectar, honeydew secreted by sap-sucking insects, and fungi, but most are general or specialized carnivores that can exert an enormous control on other invertebrate populations in their vicinity. Some are accomplished scavengers or necrophages, and a few even specialize in preying on other ants. Their nest sites vary from a simple cavity in the soil to extremely complex subterranean excavations, from the space beneath a chip of bark to purposely built leaf shelters stitched together with larval silk, from an abandoned beetle boring in rotten wood to vastly ramifying tunnel systems in timber, and from bare cavities to elegant spaces lined with silk or fine carton. Some species even form symbiotic associations with particular groups of plants, which produce suitable pre-formed nest sites to attract the ants to take up residence. Together with *Homo sapiens* the ants are one of the few animal groups that commonly manipulate and modify their immediate surroundings to suit their needs, and it is a truism that they occupy a position among the terrestrial invertebrates equivalent to that occupied by our species among the vertebrates.

There are about 15,000 living ant species, of which some 9,000–10,000 have been described. All of these fall into a single family, the Formicidae, which is included in the superfamily Vespoidea of the insect order Hymenoptera. The literature dealing with all aspects of myrmecology, the scientific study of ants, is vast. Some specialized areas of myrmecological study have developed into discrete disciplines, with their own methodologies and terminologies. The studies of ant-plant interactions, social parasitism and commensalism, sociobiology, exocrinology, navigation, and communication are all currently receiving a great deal of attention. Strangely, despite their ubiquity, perennial presence, and enormous numbers, ants do not figure strongly in the ecological literature. The reason for this may be that ants are traditionally a taxonomically "difficult" group; a problem that this volume attempts to address.

Two recent books are strongly recommended for introductory reading. The first of these, by Gauld and Bolton (1988), sets the Formicidae in the context of the order Hymenoptera, without giving undue weight to the family. It provides a broad overview of the order, including the morphology, classification, and biology of its families. The second, by Hölldobler and Wilson (1990), deals only with the ants and gives a superb synopsis of the enormous amount of myrmecological knowledge currently available. It is eminently readable and covers all aspects of ant life in considerable detail. Other fairly recent useful introductions to myrmecology, or some of its many aspects, include Beattie (1985), Blum (1985), Dlussky (1981), Dumpert (1978), Gösswald (1985), Hermann (1979–1982, 1984), Jolivet (1986), Kipyatkov (1991), Passera (1984), Schneirla (1971), Sudd and Franks (1987), Wilson (1971), and their included references.

Catalogues of described taxa are available for some regions. The old catalogues of living world ants by Dalla Torre (1893) and Emery (1910b, 1911, 1912, 1921, 1922a,b, 1925) are now very much out of date, but the present author is currently completing a new one which will include not only the extant species but also the fossil forms. Relatively recent and trustworthy catalogues include Kempf (1972a), with additions by Brandão (1991) for the Neotropical region; M. R. Smith (1951, 1958) and D. R. Smith (1979) for the Nearctic; Taylor and Brown (1985) and Taylor (1987a,b) for Australasia; Baltazar (1966) for the Philippines; Shattuck (1993) for the subfamilies Aneuretinae and Dolichoderinae. Catalogues which are mostly out of date but contain some useful information include W. M. Wheeler (1922) for the Afrotropical and Malagasy regions, and Chapman and Capco (1951) for the Oriental and Indo-Australian regions, though the last is rather inaccurate.

The principal aim of the present volume is to provide a series of identification keys to the living ant subfamilies and genera of the world. A major problem in formulating such a survey is the fact that taxonomy, contrary to common belief, is a dynamic rather than a static branch of bioscience. Its nature must be dynamic for the following simple reasons.

(1) Improvements in collecting and ecological sampling techniques continue to unearth previously unknown taxa. These must be fitted into the classification or the system must undergo modification in the light of the new information. Newly discovered taxa may be placeable directly into the current classification or they may demand that modifications to the system must be made. For instance, newly acquired material may indicate that two taxa previously considered distinct are in fact synonymous, or the reverse, that two taxa previously considered synonymous should be separated. They may also indicate that the present suprageneric classification is unacceptable, that supposed relationships between taxa are incorrect, or that zoogeographical knowledge of the group is deficient.

(2) Collections from small areas and the subsequent description of their faunas, as if isolated from, and without regard for, the fauna of the rest of the world, frequently produce a welter of unnecessary names based on uncritical splitting, unrecognized identity, and misinterpretation of generic limits. This was particularly true in the nineteenth and early twentieth centuries, when the amassing of names seems to have been an end in itself, with authors apparently racing each other to produce as many self-attributed names as possible. Of necessity, modern taxonomists must spend much of their time undoing these ancient taxonomic tangles and setting the nomenclature and classification to rights. This is an ongoing process, still far from complete.

(3) The basic structure of, and any modifications made to, the classification must be well founded upon scientific fact and must improve its predictive properties as well as indicate firmer groupings among the various taxa.

These considerations should dispel the frequently made assertion that taxonomists merely change names around to upset other, nontaxonomic, zoologists. On the contrary, every alteration of status, every change of name, every shift in classification, must be in response to newly acquired information, must bring clarity where previously there was turbidity, must improve the classification, and must be done with an aim to increasing our understanding of the natural world.

Problems of Classification

As currently constituted the family Formicidae contains 296 extant genera in 16 subfamilies. As well as this the ants have an extensive fossil record, reaching back to the Cretaceous. This embraces 61 extinct genera referable to living subfamilies, and a further 14 genera that are grouped into four extinct subfamilies. The taxonomy of the fossil forms remains problematical and the status of many taxa dubious. No synthesizing survey or analysis of the whole fossil ant fauna at genus level has been undertaken since the time of Handlirsch (1906, 1907). Carpenter (1992) gives a useful list of genera containing fossil taxa, and recently a number of checklists of the ant fauna of amber deposits have appeared, e.g., Burnham (1979), Keilbach (1982) and Spahr (1987), but these are not analytical in any way. A detailed taxonomic survey, both of forms in amber and those in rock, is most definitely needed.

The classification of the living forms is also extremely decrepit and is untrustworthy in parts. The recent production of a subfamily-level phylogeny by Baroni Urbani, Bolton, and Ward (1992) has helped to pin down the subfamilies rather more accurately, but even here some weak spots still remain. Classification at tribe rank, within several subfamilies, is to a large extent chaotic and awaits the attentions of phylogenetic researchers. In subfamilies containing more than one tribe, those of Ponerinae, Cerapachyinae, Ecitoninae, and Leptanillinae are reasonably stable. The long-established and little-changed tribal classification of Formicinae (e.g., Emery, 1925a; Hölldobler and Wilson, 1990) has recently been challenged by Agosti (1991), on the basis of an analysis of some recently discovered characters, and is in need of a thorough study. The rank of tribe has recently been abolished in Dolichoderinae by Shattuck (1992c), after his research proved that the rank held no value in this subfamily. The tribal organization of Myrmicinae is in an appalling state and a complete overhaul of the classification is long overdue.

Once we turn to the genera a rather more encouraging state of affairs is encountered. Taxonomists over the last 30 years or so have revolutionized generic concepts in wide areas of the family, and a good proportion of genera are now represented by monophyletic units. This is not to say that no problems remain. Indeed, a substantial number of major difficulties still confront the ant taxonomist at genus level. These fall into the following main categories.

(1) Unrecognized synonymy. Some genera currently treated as valid are probably junior or senior synonyms. For instance the genus *Acanthomyops* is probably synonymous with *Lasius;* the widely distributed, large genus *Smithistruma* may be a junior synonym of the small, obscure *Pentastruma,* and other small smithistrumiform dacetonines may also fall into this synonymy; *Lordomyrma* appears to be the senior synonym of *Ancyridris;* and the two Australian myrmicines *Machomyrma* and *Anisopheidole* may represent a single genus.

(2) Compound genera. In some cases a single genus name may conceal more than one real genus. As an example, the genus *Leptothorax* currently has several junior synonyms, but some of these names may represent valid taxa. The genera *Acropyga, Anoplolepis,* and *Dorylus* contain disparate subgenera, some of which should be accorded genus rank. *Camponotus* currently has about 50 subgenera, some of which may be worthy of elevation to genus rank, others of which are certainly synonyms.

(3) Insufficiently characterized genera. A number of genera, currently known from many species and universally but informally recognized as uniform groups, lack convincing all-embracing diagnostic characters. The main difficulty here is usually that accretion of new species, both described and undescribed, has extended the limits of the genus well beyond those conceived by its original author. Such genera need full taxonomic surveys to establish their limits, and these surveys must include all species-rank taxa. Genera falling into this category include *Vollenhovia, Oligomyrmex,* and all the components of tribe Camponotini.

(4) Zoogeographical confusion. A number of genera are currently recognized more by zoogeography than by unique characters, and this is a very unsatisfactory state of affairs. For instance, the relation-

ship of the Neotropical genera *Antichthonidris* and *Nothidris* to the Australasian groups of *Monomorium* awaits analysis. *Antichthonidris*-like species exist in Australia but are currently classified with *Monomorium,* while many Australian *Monomorium* are morphologically very similar to the Neotropical *Nothidris.*

These examples are by no means an exhaustive list of all the genus-level problems remaining in the Formicidae; there are many others of the same magnitude. Detailed analysis of these problems remains to be undertaken but, as I hope the examples illustrate, their investigation must be undertaken on a worldwide basis if accurate conclusions are to be drawn. The solution of all such problems will eventually produce an accurate classification, but the reader should be aware that the keys and syntheses presented here represent nothing more than a state-of-the-art synopsis based on our current level of knowledge. They are not set in granite but rather will be subject to modification and improvement as our understanding increases.

Therefore when using the keys (explained at the end of this section), keep in mind that if you come across a specimen that does not convincingly fit either lug of a terminal couplet, it does not necessarily mean that you have a new genus. There is always the possibility that better interpretation of diagnostic characters and improved diagnoses of the available names will establish the identity in an already-named taxon. A golden rule, often overlooked or ignored by taxonomists, and sometimes apparently not even realized, is this: *Any name proposed in the genus group (genus or subgenus name) must demonstrate and maintain its validity on a worldwide basis.* If it does not, it is worthless.

The synoptic classifications presented here include some new modifications and, it is hoped, present an advance on previous synopses. Readers interested in the history of synoptic classifications are referred to the following publications: Ashmead (1905), Brown (1973), Dlussky and Fedoseeva (1988), Emery (1877, 1895, 1896, 1901, 1910b, 1911, 1912, 1914, 1915, 1921, 1922a,b, 1925a). Forel (1878, 1893, 1917), Hölldobler and Wilson (1990), Snelling (1981), G. C. Wheeler and J. Wheeler (1985), W. M. Wheeler (1922).

Outline of Zoogeography

As far as the ant fauna is concerned the world can be split into eight zoogeographical regions. Some of these are better defined than others in terms of faunal distinctness, but each has a reasonable number of endemics, taxa found there and nowhere else. Many of the regional names are widely used by zoogeographers and do not need redefining here—for instance, Palaearctic, Nearctic, and Neotropical—but the limits of others deserve a few comments as they show some divergence from classical usage.

The Afrotropical (= Ethiopian) region includes all subsaharan Africa and the southern half of the Saudi Arabian Peninsula but excludes Madagascar and its nearby islands, which constitute the Malagasy region. The classical Oriental region is here divided into two, Oriental and Indo-Australian (= Malesian). The latter includes the Malay Peninsula, Philippines, East Malaysia, Indonesia up to and including the island of New Guinea, and the island systems of the Pacific Ocean. The Oriental region consists of Pakistan, Sri Lanka, the whole Indian subcontinent to the Himalayas, southern China and Taiwan, and the countries of Burma (Myanmar), Thailand, Cambodia, Laos, and Vietnam. The Australasian region contains the Australian continent, New Caledonia, and New Zealand.

Table 1 indicates the numbers of extant genera, arranged by subfamily, that occupy each of these eight zoogeographical regions. Genera of which a single species has been transferred artificially into a region other than that of its origin, for instance by human commercial activity, are only recorded from their region of origin.

Abbreviations of the names of the regions are PAL = Palaearctic, AFR = Afrotropical, MAL = Malagasy, ORI = Oriental, INA = Indo-Australian, AUS = Australasian, NEA = Nearctic, NEO = Neotropical.

Many genera naturally occur in more than one zoogeographical region; some occur in all the regions. If these are subtracted from the totals in Table 1, the result (Table 2) gives the numbers of the *endemic* genera of the regions. As can be seen, the absolute numbers of genera in a region and the numbers of endemic genera do not rank the regions in the same way. In terms of total numbers of genera the order is INA (126), NEO (118), ORI (101), AUS (94), AFR (89), PAL (70), NEA (62), MAL (46). But in terms of endemicity, expressed here as a percentage of the total number of genera, the order is NEO (51%), AFR (33%), AUS (22%), INA (18%), PAL (17%), MAL (7%), NEA (5%), ORI (5%); a radically different order in which only NEA occupies the same relative position. It is interesting, though perhaps not surprising, to note that the regions that formed part of the ancient Gondwanaland (NEO, AFR, AUS) today show the highest levels of endemicity.

Preparation of Specimens

If any key is to be used with some hope of success, specimens must be prepared and mounted in such a way that the characters necessary for identification are easily visible. This may seem self-evident but is frequently overlooked or considered of minimal importance. Good mounting technique is, on the contrary, of prime importance, as badly mounted specimens actively hinder identification.

Ants bear identification characters on all parts of the body, but the area around the clypeus, mandibles, and mouthparts is particularly important at genus and species rank. The areas of least taxonomic value are the midventral alitrunk and coxal apices. Therefore the optimum mounting technique for ants is one that uses this area alone and leaves the remainder of the body free for easy examination: this is the card-point technique. It requires some practice to accomplish well but is worth the effort in terms of superbly presented, easily compared, and examined specimens.

Basically the technique involves glueing the ant to the upper apex of a small triangle of stiff card or bristol board. The best gum to use is a water-soluble glue such as *Seccotine,* diluted to the appropriate consistency (the smaller the ant the more dilute the glue). A minimum amount of glue is applied to the upper apex of the card and the ant is placed transversely on it, venter down, so that only the coxal apices are in contact with the glue and so that the head, waist, and gaster project freely and their ventral surfaces are visible. Always orient the specimen so that the head is to the right when the apex

Table 1 Numbers of Genera by Zoogeographical Region

Subfamily	PAL	AFR	MAL	ORI	INA	AUS	NEA	NEO	Total
Aenictinae	1	1	—	1	1	1	—	—	1
Aenictogitoninae	—	1	—	—	—	—	—	—	1
Aneuretinae	—	—	—	1	—	—	—	—	1
Apomyrminae	—	1	—	—	—	—	—	—	1
Cerapachyinae	1	3	3	2	3	2	1	4	5
Dolichoderinae	5	4	3	8	12	13	5	8	22
Dorylinae	1	1	—	1	1	—	—	—	1
Ecitoninae	—	—	—	—	—	—	3	5	5
Formicinae	16	15	7	16	21	18	10	9	49
Leptanillinae	3	1	—	4	6	1	—	—	7
Leptanilloidinae	—	—	—	—	—	—	—	1	1
Myrmeciinae	—	—	—	—	—	1	—	—	1
Myrmicinae	31	38	22	46	58	35	31	66	155
Nothomyrmeciinae	—	—	—	—	—	1	—	—	1
Ponerinae	11	23	10	21	23	21	11	24	42
Pseudomyrmecinae	1	1	1	1	1	1	1	2	3
Totals by region	70	89	46	101	126	94	62	118	
Total number of ant genera of the world									296

Table 2 Numbers of Endemic Genera by Zoogeographical Region

Subfamily	PAL	AFR	MAL	ORI	INA	AUS	NEA	NEO
Aenictinae	—	—	—	—	—	—	—	—
Aenictogitoninae	—	1	—	—	—	—	—	—
Aneuretinae	—	—	—	1	—	—	—	—
Apomyrminae	—	1	—	—	—	—	—	—
Cerapachyinae	—	—	—	—	—	—	—	2
Dolichoderinae	—	2	—	—	1	2	—	3
Dorylinae	—	—	—	—	—	—	—	—
Ecitoninae	—	—	—	—	—	—	—	2
Formicinae	3	5	—	—	6	8	2	4
Leptanillinae	—	—	—	—	3	—	—	—
Leptanilloidinae	—	—	—	—	—	—	—	1
Myrmeciinae	—	—	—	—	—	1	—	—
Myrmicinae	8	12	3	4	12	7	1	39
Nothomyrmeciinae	—	—	—	—	—	1	—	—
Ponerinae	1	8	—	—	—	1	—	9
Pseudomyrmecinae	—	—	—	—	—	—	—	1
Totals by region	12	29	3	5	22	20	3	60

of the card triangle is directed away from the mounter. Push down the legs so that they do not obscure the head, alitrunk, or waist profile. A pin, preferably stout (continental size no. 5), is run through the center base of the card triangle, and the card is pushed up the shaft of the pin with forceps. One or two more specimens (which can be different castes from the same nest) may be staged on triangles below the first, but space should always be left for data labels. Always ensure that specimens to be mounted are free of alcohol and reasonably clean. In many genera the palp formula, mandibular dentition, or both are of importance. One or more specimens in a series should have the mandibles opened and the mouthparts everted prior to mounting.

One final caveat. Ants should never, ever, be mounted flat down in a pool of glue on a card rectangle. Not only does this obscure all ventral characters but also the glue frequently creeps and covers parts of the mandibles, the sides of the alitrunk, and other important features.

Organization and Use of the Book

As noted above, the primary aim of this volume is to provide an accurate identification guide to subfamily and genus rank. Besides this it seeks to provide the reader with synoptic classifications of genera within the various subfamilies and lists of taxonomic works that provide determinations to species rank, where available, to facilitate further identification and study.

First, the family Formicidae is described, and keys are given to identify the extant subfamilies. Then each extant subfamily is treated separately, in alphabetical order. Each subfamily section commences with a newly formulated diagnosis of the group, followed by the keys to genera, synoptic classification, and list of taxonomic references. A few notes on broad distribution are also provided. A short summary of the extinct subfamilies completes this main part of the work.

A bibliography of faunistic studies (that is, works which provide keys for the whole fauna of a country or geographical region, rather than keys for particular taxa) is given separately, in the back of the book, for quick reference. Finally, an extensive glossary of morphological terms used in ant taxonomy is provided, as beginners or students from other disciplines may not be familiar with some of the terms encountered in the keys.

Notes on the Keys

The dichotomous keys to ant subfamilies and genera are presented in a way which, it is hoped, will render their use as simple and quick as possible. All the keys apply to the worker caste, for the following reasons: (1) The worker is numerically overwhelmingly the commonest caste, and therefore by far the most commonly encountered and collected. Because of this it is the form that accounts for almost all of the current taxonomy of the group. (2) Other castes and sexes, queens and males, remain very sparsely known or even unknown in a large number of genera. At present it would be impossible to produce any meaningful key based on these forms.

There is no doubt that people in general, but especially first-time users, are disturbed and somewhat intimidated by very long identification keys. The larger subfamilies (Myrmicinae, Formicinae, Ponerinae, Dolichoderinae) have therefore been presented in separate keys, by zoogeographical region. Some of these treat only a single region per subfamily, but in some instances it was expedient to treat two adjacent regions in a single key. With smaller subfamilies a single key covering the whole world fauna is presented.

Of the zoogeographically oriented keys to large subfamilies, the Palaearctic keys include those genera normally found in the region and those elements of the Afrotropical and Oriental regional faunas that may penetrate the southern portions of the Palaearctic. Other members of these extralimital faunas may occur in the southern Palaearctic but have not yet been detected. For samples collected in these zones the key may be supplemented by reference to the Afrotropical and Malagasy key and the Oriental and Indo-Australian key for any given subfamily.

The Afrotropical (= Ethiopian) and the Malagasy zoogeographical regions are keyed together as most genera are common to both. Only a few Malagasy genera are restricted to that region, relatively more are peculiar to the Afrotropical and do not occur in the Malagasy. The Oriental region and the Indo-Australian region are keyed together but the Australasian keys deal only with Australia, New Zealand, and New Caledonia.

The Nearctic and Neotropical regions are keyed separately for the large subfamilies, but Nearctic genera that penetrate the northern Neotropical region, and Neotropical genera that penetrate the southern Nearctic, are included in the keys to both regions.

Note that two keys to subfamilies are presented. The first of these is based on externally observable characters and should be consulted first. The second is more experimental and requires some dissection or careful disarticulation of specimens. It may be used for confirmation of the first if the need arises.

In many of the keys the name of a single subfamily or genus may occur more than once. This is because characters are chosen for ease of recognition, and some characters may occur in more than one state in a single taxon. For instance, the number of antennal segments in worker ants varies from 4 to 12. The number of segments is usually easily counted and therefore forms a very useful character for subdividing a key. But not all genera have an antennal segment count that is uniform throughout all their species. Therefore if a key is split at some point on the number of antennal segments, say 11 *versus* 12, then the name of any genus containing species with both counts will occur twice. In taxa where this has occurred their names in the keys are followed by the word *(part)*. This is not a disadvantage, as the function of a key is to make identification possible, and any technique that enhances its success rate is valid. It may, however, become a possible source of error in some rare circumstances. Imagine, for example, that you have just discovered a species with 11 antennal segments in a genus where all the previously known species had 12. Obviously that specimen would not run out to the correct identity in the keys; there is a good chance that it would fail to run to any terminal lug of a couplet. In such circumstances it is advisable to go back to the original "11 *versus* 12" couplet and run the specimen through the other half of the key. Do not automatically assume that you have a new genus.

The key to subfamilies and a number of the regional keys to genera given below appeared in prototype form in Hölldobler and Wilson

(1990). These have been revised and updated. The remaining keys are all newly formulated and presented here for the first time.

Students and others who have little previous experience, either with ants or with the use of keys in general, are advised first to run each specimen or sample through the key to subfamilies. Having ascertained the subfamily, the reader can then check the identification against the subfamily diagnosis given at the head of each section. If the subfamily is a small one the specimen can then be run directly through the key to world genera that follows the subfamily diagnosis. If the subfamily is large, however, the reader must turn to the key covering the zoogeographical region of origin of the specimen.

All the keys are dichotomous, which means that at each position the reader is faced with a choice of only two alternatives. A set of two alternatives is called a *couplet,* and each half-couplet is called a *lug.* Compare the specimen to be identified with the morphological characters given in each lug of couplet 1, ascertain which lug matches the specimen, then turn to the couplet number given at the end of that lug. Repeat the process until a lug or couplet is encountered that runs out to the name of a genus.

Users who are more familiar with ants may of course turn directly to the appropriate subfamily or regional key.

The electron microscope photographs used to illustrate the keys are representative of genera. That is, each is chosen to represent the general habitus of a particular genus and cannot show all the morphological variation inherent in a genus.

The Family FORMICIDAE

Diagnosis of the Family FORMICIDAE

Eusocial vespoid aculeates with a wingless worker caste, forming perennial colonies. Head prognathous in female castes (workers and queens). Antenna with 4–12 segments in female castes, with 9–13 segments in males. Antenna geniculate between the long basal segment (scape) and the remaining funicular segments. Second abdominal segment reduced, forming a node or scale (the petiole), isolated from the alitrunk in front and the remaining abdominal segments behind. Frequently the third abdominal segment also reduced and isolated (postpetiole). Wings of alate queens deciduous, shed after mating. Metapleural gland generally present on alitrunk, opening above the metacoxa.

The Extant Subfamilies

The number of extant subfamilies in the Formicidae has risen from the 5 recognized in the earliest classifications to the 16 recognized and discussed here. The foundations of the modern subfamily-level classification of Formicidae were laid by Forel (1878), who organized the ants into 5 great subfamilies. These were, according to the correct spelling convention: Camponotinae, Dolichoderinae, Dorylinae, Ponerinae, and Myrmicinae. These remained unchanged for many years, as for example in the catalogue of Dalla Torre (1893). Camponotinae later had its name properly changed to Formicinae by W. M. Wheeler (1920), because the latter name had temporal priority over the former. The other four names have survived to the present day. Most of the remaining 11 subfamilies were established by splitting disparate groups from these early names, but the species that today constitute a number of subfamilies were unknown to the early classifiers.

Key references in the history of subfamily classification, apart from those already mentioned, include Ashmead (1905), Baroni Urbani, Bolton, and Ward (1992), Bolton (1990a,b,c), Borgmeier (1955), Brown (1954a, 1973, 1975), Clark (1951), Dlussky and Fedoseeva (1988), Emery (1877, 1899, 1901, 1910b, 1912), Forel (1885, 1893, 1917), Shattuck (1992b), M. R. Smith (1952, 1958), Taylor (1978), Ward (1990), W. M. Wheeler (1902).

The accepted subfamily-level classification just prior to the revisionary cladistic studies of Baroni Urbani, Bolton, Shattuck, and Ward is summarized in Hölldobler and Wilson (1990).

How stable, in a taxonomic sense, are the subfamilies recognized here? The recent cladistic analyses, mentioned above, show quite convincingly that all the subfamily groupings used here represent monophyletic taxa. And the definitions given at the start of each subfamily section show their consistent, and often quite striking, morphological differences. But this does not ensure that all represent the same taxonomic grade. For instance, an argument can be made for unifying all subfamilies in the doryline section, as they exhibit a high number of synapomorphies which together could constitute a single subfamily diagnosis. This would mean that Aenictinae, Aenictogitoninae, Cerapachyinae, Dorylinae, Ecitoninae, and Leptanilloidinae could be combined into a single subfamily, which would have to take the name Dorylinae as that is the first available name for the group as a whole. Similarly the Apomyrminae and Leptanillinae, distinct as defined here, could be combined into a single subfamily based on uniquely shared apomorphies.

There are no simple answers to such problems; no definitive rights or wrongs. At this level there are only differences of opinion about grade, and these may vary with time or depending on how the characters chosen are weighted both within and between groups. Of course, the opinion must be based on comprehensive analysis and the characters utilized must be carefully defined and, most importantly, be universal. Questions of grade among the higher taxa are very much left to the conscience and consensus opinion of professional taxonomists. The grades utilized must be based on overall knowledge of the world fauna and the relationships between the various groups. They should express what is currently considered to be the best possible method of subdividing the family and indicate not only the diversity of the group as a whole, but also the relationships of its component parts. In my opinion the subfamilies recognized here reflect such a system, although the possibility of future modification is not ruled out.

Two complementary keys to subfamilies are presented below. The first, general, key is a modified and improved version of that first published in Hölldobler and Wilson (1990) and relies solely on characters of externally observable morphology. The second contains characters visible only after special treatment or dissection of specimens. An updated version of the "new format" key first published by Bolton (1990c), it requires a more detailed acquaintance with ant morphology than does the general key.

Key to Subfamilies Based on External Morphology

Characters used in this key are chosen solely for clarity and ease of visibility. As a result, some subfamily names run out at more than one place in the key. No dissection or preparation of the specimen is necessary other than correct routine mounting as described in the Introduction.

Note that the subfamily Aenictogitoninae, restricted to the Afrotropical region, is omitted as it is known only from males.

1 Body with a single reduced or isolated segment (the petiole) between alitrunk and gaster; first gastral segment either entirely confluent with the second (Figs. 4, 6, 29, 37, 99, 133, 175, 461, 482) or separated from it only by a narrow girdling constriction (Figs. 10, 14, 425, 451, 473, 490, 500), and if the latter then the first gastral segment is not markedly reduced in size **2**

— Body with 2 reduced or isolated segments (the petiole and postpetiole) between alitrunk and gaster; either both segments are much reduced (Figs. 2, 71, 167, 171, 183, 215, 267, 305, 341, 389, 417, 520), or the second is somewhat larger than the first (Figs. 12, 173), and if the latter then the postpetiole is distinctly smaller than the first gastral segment and separated from it by an extensive, deep girdling constriction **12**

2 Apex of gaster with a semicircular to circular acidopore formed from the hypopygium, this structure often projecting as a nozzle and fringed with setae (Fig. 160); sometimes the acidopore is concealed by a projection of the pygidium, but if so the antennal insertions are located well behind the posterior clypeal margin (Figs. 84, 96, 98). Sting absent, replaced by an acid-projecting system of which the acidopore is the orifice. (Worldwide) . . . . **FORMICINAE**

— Apex of gaster with hypopygium lacking an acidopore. Sting present or absent; when present it is usually visible (Figs. 10, 14, 447, 455, 468, 476) but when reduced or vestigial (or present but completely retracted) the hypopygium forms a smooth posterior margin (not modified as above), and the antennal sockets abut the posterior margin of the clypeus (Figs. 3, 20, 28, 36, 52, 59, 70) **3**

3 Either pygidium or hypopygium armed with peg-like teeth or short spines. If the pygidium is armed it is usually, but not always, transversely flattened to impressed and has either a single pair of short, posterolaterally situated spines (Figs. 61–63) or a marginal row of short spines or peg-like teeth (Figs.

Morphological Key to Subfamilies *(continued)*

17–19). If the hypopygium is armed then its margin on each side has a row of teeth or spines, generally in the apical half, which project dorsally outside the pygidium (Figs. 507, 508) **4**

— Pygidium and hypopygium both unarmed. Pygidium transversely convex and rounded, lacking either a posterolateral pair of short spines or a marginal row of short spines or peg-like teeth. Hypopygium with its lateral margins smooth and without spines **6**

4 Gastral spiracles 3–5 (= abdominal spiracles 5–7) exposed, not overlapped nor concealed by the tergites of the preceding segments (Figs. 16–19, 60–62). Metapleural gland orifice overhung and concealed from above by a cuticular lip or flange, the latter extending obliquely upwards and forwards on the metapleuron as a rim or ridge (Figs. 12, 14, 16, 60). Helcium sternite convex and bulging ventrally, visible in profile **5**

— Gastral spiracles 3–5 (= abdominal spiracles 5–7) concealed, overlapped and hidden by the tergites of the preceding segments (e.g., Figs. 429, 450, 508). Metapleural gland orifice not overhung nor concealed from above by a cuticular lip or flange, and without a rim or ridge extending obliquely upwards and forwards from the gland orifice (e.g., Figs. 450, 472, 477, 478). Helcium sternite reduced and retracted, not visible in profile. (World tropics) . . . . **PONERINAE** (part)

5 Propodeal spiracle high on side and situated far forward on the sclerite, the spiracular orifice subtended by a longitudinal impression (Fig. 60). Propodeal lobes absent. Armament of pygidium consisting solely of a single pair of posteriorly directed short spines that are situated posterolaterally (Figs. 61–63). Promesonotal suture always distinct (Fig. 60). (Old World tropics and subtropics except Madagascar and Australia) . . . . **DORYLINAE**

— Propodeal spiracle low on side and usually behind the midlength of the sclerite, the spiracular orifice not subtended by a longitudinal impression (Figs. 8, 10, 12, 14, 16). Propodeal lobes present. Armament of pygidium consisting of an apical row or marginal rows of short peg-like teeth or spines (Figs. 17–19). Promesonotal suture usually completely absent, only extremely rarely visible (Figs. 8, 10, 12, 14, 16). (World tropics and subtropics) . . . . **CERAPACHYINAE**

6 Sting vestigial or absent, in any case not visible without dissection. Tergite of helcium with an extensive U- or V-shaped emargination dorsally in its anterior margin (visible if gaster slightly depressed) (Fig. 58). (Worldwide) . . . . **DOLICHODERINAE**

— Sting present and functional, often projecting in dead specimens; in many species the sting shaft is visible through the cuticle of the gastral apex ventrally even when fully retracted. Tergite of helcium entire, its anterior margin dorsally without a U- or V-shaped emargination **7**

7 Pretarsal claws with a tooth, or sometimes several teeth, on the inner curvature behind the apical point (Figs. 515, 516). . **8**

— Pretarsal claws simple, without teeth on the inner curvature behind the apical point (Fig. 514) **10**

8 Stridulatory system present ventrally on gaster, on first and second sternites. Palp formula 6,4. First gastral segment confluent with second, without an impression between them. Mandibles elongate-triangular, blade-like, and multidentate, the masticatory margins meeting along their entire length (Figs. 174, 175). (Southern Australia) . . . . **NOTHOMYRMECIINAE**

— Stridulatory system absent, or present dorsally on gaster on first and second tergites. Palp formula variable but usually lower than 6,4; if palp formula 6,4 then the gaster has an impression between the first and second segments, or the mandibles are not constructed as above, or both **9**

9 Frontal lobes absent; vertical, low carinae may be present between the antennal sockets, but the latter are completely exposed (Fig. 64). Antennal sockets very close to anterior margin of head, the clypeus extremely narrow in front of them. Spiracles of gastral segments 4 and 5 visible. Helcium sternite convex and bulging, visible in profile. (New World tropics and temperate regions) . . . . . . **ECITONINAE** (part)

— Frontal lobes present and horizontally aligned, partially to completely concealing the antennal sockets (Figs. 436, 449, 453, 463, 491). Antennal sockets a notable distance behind the anterior margin of the head, the clypeus well developed in front of them. Spiracles of gastral segments 4 and 5 concealed. Helcium sternite reduced and retracted, not visible in profile. (World tropics) . . . . **PONERINAE** (part)

10 Major tibial spur of hind leg simple or with a few minute barbules. Metathoracic spiracles on dorsal surface when alitrunk viewed from the side (Fig. 4). Palp formula 3,4. Petiole with a long, narrow anterior peduncle and propodeum armed with a pair of spines (Fig. 4). (Sri Lanka) . . . . **ANEURETINAE**

— Major tibial spur of hind leg broadly and distinctly pectinate (Figs. 511–513), or the metathoracic spiracles not on the dorsal surface of the alitrunk when viewed from the side (either not visible or on the lateral surface), or both. Palp formula variable, generally not 3,4 but if so then either the petiole lacks a long anterior peduncle or the propodeum lacks spines, or both **11**

11 Labrum with peg-like teeth (Fig. 5). Propodeal lobes absent. Frontal lobes absent (Fig. 5). Eyes absent. Promesonotal suture present, deeply impressed and flexible (Fig. 6). (West Africa) . . . . **APOMYRMINAE**

— Labrum without peg-like teeth. Propodeal lobes present. Frontal lobes and eyes usually present, but if either is absent then the promesonotal suture is also absent. (Worldwide) . . . . **PONERINAE** (part)

12 Pygidium transversely flattened or impressed and armed laterally, posteriorly, or both, with a row of short spines or peg-like teeth that usually project vertically (Figs. 17–19). (World tropics and subtropics) . . . . **CERAPACHYINAE** (part)

— Pygidium transversely rounded, may be very small, not armed laterally or posteriorly with a row of short spines or peg-like teeth **13**

13 Frontal lobes either absent or very reduced and vertical; in either case the antennal sockets are completely exposed in full-face view and are not at all concealed or covered by the frontal lobes (Figs. 1, 66, 68, 70, 72, 166, 168, 170, 300, 519) **14**

— Frontal lobes present, horizontal to somewhat elevated; the antennal sockets are always partially or completely covered by the frontal lobes in full-face view and are never completely exposed (Figs. 172, 182, 184, 218, 234, 250, 278, 308, 318, 352, 521) **19**

14 Eyes present and conspicuous, with many distinct ommatidia (Figs. 300, 301, 519) **15**

— Eyes absent or at most represented by a single ommatidium or small, featureless blister (Figs. 1, 66, 68, 70, 72, 166, 168, 170) **16**

15 Promesonotal suture present, freely flexible. Hind tibia with a conspicuous pectinate apical spur. Posterior margin of median portion of clypeus not projecting back between antennal sockets (Fig. 519). (World tropics and subtropics) . . . . **PSEUDOMYRMECINAE**

— Promesonotal suture vestigial (fused and inflexible) to absent. Hind tibia without a pectinate apical spur; spur either simple or (usually) absent. Posterior margin of median portion of clypeus projecting back between antennal sockets (Fig. 300). (Worldwide) . . . . **MYRMICINAE** (part)

16 Promesonotal suture present and very conspicuous in dorsal view; usually deeply impressed and always freely flexible in fresh specimens **17**

— Promesonotal suture vestigial to absent, usually the latter but very rarely a faint transverse line is visible in dorsal view; suture never impressed nor flexible **18**

17 Pygidium large and conspicuous (Figs. 167, 169). Spiracles of gastral segments 3 and 4 concealed by preceding tergites. Sternite of helcium concealed, not visible in profile. Gaster without deep girdling constrictions between the segments (Figs. 167, 169). (Old World tropical and temperate regions) . . . . **LEPTANILLINAE**

— Pygidium reduced, small, and inconspicuous (Fig. 171). Spiracles of gastral segments 3 and 4 exposed, not concealed by preceding tergites. Sternite of helcium convex, bulging ventrally, and visible in profile. Gaster with a deep girdling constriction between the first and second segments, and a similar constriction between the third and fourth segments (Fig. 171). (New World tropics) . . . . . . **LEPTANILLOIDINAE**

18 Antenna with 8–10 segments. Spiracle of postpetiole behind midlength of tergite (Fig. 2). Gastral spiracles circular. First gastral segment with a narrow neck-like constriction behind the articulation with the postpetiole. (Old World tropics and subtropics except Madagascar) .... **AENICTINAE**

— Antenna with 12 segments. Spiracle of postpetiole in front of midlength of tergite (Figs. 67, 69, 71, 73). Gastral spiracles oval to slit-shaped. First gastral segment without a narrow neck-like constriction behind the articulation with the postpetiole. (New World tropics and temperate regions) .... **ECITONINAE** (part)

19 Promesonotal suture usually completely absent; rarely with a vestigial remnant of the suture in the form of a feeble, transversely arched impression in dorsal view, but pronotum and mesonotum always fused and immobile with respect to each other. (Worldwide) .... **MYRMICINAE** (part)

— Promesonotal suture present and very conspicuous in dorsal view, the pronotum and mesonotum not fused, mobile with respect to each other **20**

20 Mandibles long, narrow, and blade-like (Fig. 172), serially multidentate on their inner margins, and crossing over apically at full closure; not closing tightly against the clypeus. Stridulitrum absent from pretergite of gastral segment 1. (Australia and New Caledonia) .... **MYRMECIINAE**

— Mandibles short, stout, and compact (Figs. 517, 519, 521), the apical margin dentate; mandibles overlapping at full closure but not crossing over, closing tightly against the clypeus. Stridulitrum present on pretergite of gastral segment 1. (World tropics and subtropics) .... **PSEUDOMYRMECINAE** (part)

New Format Key to Subfamilies

Before this key can be used some special preparation of specimens is necessary as a number of the characters are apparent only following dissection or disarticulation. Specimens to be examined should be macerated in sodium hydroxide until all soft tissues have been removed, then thoroughly washed and returned to alcohol for study.

In this key frequent mention is made of the abdominal, rather than the gastral, segment number. This is to obviate possible confusion in terminology because of the variable number of separated waist segments (see the Glossary).

Note that the subfamily Aenictogitoninae, restricted to the Afrotropical region, is omitted as it is known only from males.

This key is to some extent provisional. It has been tested against workers of most, but not all, genus-rank taxa. It is therefore presented here somewhat experimentally, for testing and for comment.

1 Abdominal segment 4 with tergosternal fusion, the 2 sclerites rigidly fused and immobile with respect to each other. (Worldwide) .... **PONERINAE**

— Abdominal segment 4 with tergite and sternite not fused, the 2 sclerites free and mobile with respect to each other **2**

2 Abdominal segment 3 with tergosternal fusion posterior to helcium. Tergite and sternite of helcium also fused **3**

— Abdominal segment 3 with tergite and sternite not fused posterior to helcium. Tergite and sternite of helcium may be fused or unfused **9**

3 Abdominal spiracles 5–7 on posttergites, visible without distension or dissection of abdomen. Sternite of helcium large, convex, and bulging ventrally, visible in profile. Pygidium specialized by armament, depression of posttergite posteromedially, extreme reduction, or a combination of these **4**

— Abdominal spiracles 5–7 on pretergites, not visible without distension or dissection of abdomen. Sternite of helcium small, concealed, not bulging ventrally, not visible in profile. Pygidium unspecialized, large, evenly biconvex, unarmed .. **8**

4 Propodeal spiracle situated low down on side and behind the midlength of the sclerite in profile **5**

— Propodeal spiracle situated high up on side and in front of the midlength of the sclerite in profile **6**

5 Pygidium large, not overhung by tergite of abdominal segment 6. Dorsum of pygidium flattened, margins of flattened area laterally, posteriorly, or both, armed with denticles or short spinules. Promesonotal suture usually absent; if present (1 species) the suture is fused and inflexible. (World tropics and subtropics.) .... **CERAPACHYINAE**

— Pygidium much reduced, small and overhung by tergite of abdominal segment 6. Dorsum of pygidium not flat, without denticles or spinules. Promesonotal suture fully developed and flexible. (New World tropics) . **LEPTANILLOIDINAE**

6 Promesonotal suture absent. Pygidium a reduced, U-shaped sclerite, small and often somewhat overhung by the tergite of abdominal segment 6 .... **7**

— Promesonotal suture present and conspicuous, but fused and inflexible. Pygidium large, the posttergite indented or depressed posteromedially and bidentate laterally. (Old World tropics and subtropics except Madagascar and Australia) .... **DORYLINAE**

7 Antenna with 8–10 segments. Posttergite of abdominal segment 4 with a neck-like anterior constriction. Spiracle of abdominal segment 3 (postpetiole) behind midlength of segment. (Old World tropics and subtropics except Madagascar) .... **AENICTINAE**

— Antenna with 12 segments. Posttergite of abdominal segment 4 without a neck-like anterior constriction. Spiracle of abdominal segment 3 (postpetiole) in front of, or very rarely at, the midlength of the segment. (New World tropics to temperate regions) .... **ECITONINAE**

8 Waist of a single, separated segment (petiole). Abdominal segment 4 without differentiated presclerites. Abdominal sternite 3 immediately behind the helcium with a transverse suture. (West Africa) .... **APOMYRMINAE**

— Waist of 2 segments (petiole and postpetiole). Abdominal segment 4 with differentiated presclerites. Abdominal sternite 3 immediately behind the helcium without a transverse suture. (Old World tropics to temperate regions) . . . . **LEPTANILLINAE**

9 Metacoxal cavities open; the insertion cavity of each metacoxa in the ventral alitrunk not completely surrounded by cuticle, so that the cavity medially is confluent with the cavity in which the petiole articulates **10**

— Metacoxal cavities closed; the insertion cavity of each metacoxa in the ventral alitrunk completely surrounded by cuticle, so that the cavity medially is separated from the cavity in which the petiole articulates **12**

10 Abdomen constricted between segments 3 and 4; abdominal segment 4 with differentiated presclerites. (Australia and New Caledonia) . . . . **MYRMECIINAE**

— Abdomen not constricted between segments 3 and 4; abdominal segment 4 without differentiated presclerites **11**

11 Pretarsal claws with a tooth behind the apical point. Stridulitrum present ventrally on abdominal segment 4. Mandibles elongate. (Southern Australia) . **NOTHOMYRMECIINAE**

— Pretarsal claws unarmed behind the apical point. Stridulitrum absent from abdomen. Mandibles short. (Sri Lanka) . . . . **ANEURETINAE**

New Format Key to Subfamilies *(continued)*

12 Abdomen not constricted between segments 3 and 4; abdominal segment 4 without differentiated presclerites. Anterodorsal margin of helcium usually emarginate to deeply indented medially, only extremely rarely without this character . . **13**

— Abdomen constricted between segments 3 and 4; abdominal segment 4 with differentiated presclerites. Anterodorsal margin of helcium entire, never emarginate nor indented . . . **14**

13 Acidopore present. (Worldwide) . . . . **FORMICINAE**

— Acidopore absent. (Worldwide) . . . . . **DOLICHODERINAE**

14 Tergite and sternite of abdominal segment 2 fused. Pronotum and mesonotum fused, without a flexible suture. In frontal view the sternite of the helcium meeting the inverted-U-shaped tergite at the apices of the tergal arms on each side; the sternite evenly transversely convex ventrally. (Worldwide) . . . . **MYRMICINAE**

— Tergite and sternite of abdominal segment 2 not fused. Pronotum and mesonotum not fused, with a flexible suture. In frontal view the sternite of the helcium meeting the inverted-U-shaped tergite some distance up the inner surface from the apices of the tergal arms on each side; the sternite sinuate, not evenly transversely convex ventrally. (World tropics and subtropics) . . . . **PSEUDOMYRMECINAE**

Subfamily AENICTINAE

Diagnosis of Worker (Figs. 1, 2)

Ants with the following combination of characters together.

1 Clypeus reduced, narrow from front to back especially in front of the antennal insertions, bringing the antennal sockets very close to the anterior margin of the head.
2 Antennal sockets horizontal, in the plane of the transverse axis of the head, exposed in full-face view.
3 Frontal lobes absent; narrow vertical carinae may be present between the antennal sockets.
4 Narrow neck joining condylar bulb of antennal scape to shaft of scape proper straight, not sharply angled or bent downwards in frontal or full-face view.
5 Eyes absent; antenna with 8–10 segments; gena between antennal socket and lateral margin of head usually with a short longitudinal carina.
6 Promesonotal suture absent, the pronotum and mesonotum fused together.
7 Metapleural gland orifice in lower posterior corner of metapleuron, opening laterally, the orifice concealed behind a ventrally directed cuticular flange or flap.
8 Propodeal lobes present.
9 Propodeal spiracle situated high on the side of the sclerite and far forward, not subtended by an endophragmal pit or a longitudinal impression.
10 Metatibial glands present.
11 Metacoxal cavities closed; cuticular annulus around each cavity broad and complete, the annulus not broken medioventrally nor with a flexible suture traversing the annulus from the coxal cavity to the cavity in which the petiole articulates.
12 Waist of two separated segments, the petiole and postpetiole (= abdominal segments 2 and 3).
13 Spiracle on side of postpetiole situated at or usually behind the midlength of the segment.
14 Petiole sessile to subsessile, the tergite and sternite not fused; sternite of petiole with a simple posterior margin and simple articulation to the postpetiole.
15 Abdominal stridulatory system absent.
16 Abdominal spiracles 5–7 (= gastral spiracles 2–4) shifted backwards, not concealed by the posterior margins of the preceding segments and visible without distension or dissection of the abdomen.
17 Helcium sternite large and convex, bulging ventrally, visible in profile in normally mounted specimens.
18 Abdominal segment 3 (= postpetiole) with tergosternal fusion; tergites and sternites of following abdominal segments (4–7 = gastral segments 1–4) not fused.
19 Abdominal segment 4 (= gastral segment 1) with presclerites sharply defined and differentiated from the postsclerites, the former fitting tightly within the narrow posterior end of the third segment. Abdominal segment 4 immediately behind the presclerites constricted into a narrowed neck.
20 Pygidium (tergite of abdominal segment 7 = gastral segment 4) very small, reduced to a narrow U-shaped sclerite.
21 Sting developed and functional.

Synoptic Classification

Subfamily **AENICTINAE.**
Tribe **Aenictini.** Genus: ***Aenictus*** (Figs. 1, 2) (= *Enictus* (misspelling), = *Paraenictus,* = *Typhlatta*).

Distribution

The single genus included in this subfamily occurs throughout the Afrotropical, Oriental, and Indo-Australian regions. A few species also occur in the southern Palaearctic and the northeastern part of the Australasian region. It is absent from the Malagasy, Nearctic, and Neotropical regions.

Taxonomic References

Identification of extant species
Aenictus: Wilson (1964) [Oriental, Indo-Australian, Australasian]; Terayama and Yamane (1989) [Sumatra].

Other taxonomic references
Aenictinae: Gotwald (1982); Hölldobler and Wilson (1990); Bolton (1990c); Baroni Urbani, Bolton, and Ward (1992).
See also References to Faunistic Studies.

Subfamily AENICTOGITONINAE

This small subfamily contains only seven species in a single genus *(Aenictogiton),* is known only from its morphologically peculiar males, and is restricted to Central Africa. As this survey is based on the worker caste, no formal diagnosis of the subfamily is offered here. What little is currently known of the aenictogitonines can be obtained from Santschi (1924), which includes a key to species, Brown (1975), and Baroni Urbani, Bolton, and Ward (1992).

One of the major outstanding tasks in systematic myrmecology is the discovery of the elusive workers and queens of this enigmatic subfamily.

Subfamily ANEURETINAE

Diagnosis of Worker (Figs. 3, 4)

Ants with the following combination of characters together.

1 Clypeus broad from front to back so that antennal sockets are well behind anterior margin of head. Median portion of clypeus extended backwards between the antennal sockets.

2 Antennal sockets inclined, the portion of the socket margin and torulus closest to the dorsal midline of the head on a higher level than the portion of the margin most distant from the midline.

3 Frontal carinae present, not expanded into frontal lobes anteriorly; antennal sockets exposed and torular sclerites visible.

4 Narrow neck joining condylar bulb of antennal scape to shaft of scape proper straight.

5 Eyes present; ocelli absent; antenna with 12 segments.

6 Promesonotal suture present and flexible.

7 Metapleural gland orifice present, situated in lower posterior corner of metapleuron, opening posterolaterally.

8 Metanotum and its spiracles present on dorsal alitrunk.

9 Metacoxal cavities open; without a thin, continuous strip of cuticle separating the metacoxal cavity from the cavity in which the petiole articulates (this character awaits confirmation).

10 Propodeal lobes absent.

11 Waist of 1 segment, the petiole (= abdominal segment 2); petiole with a long anterior peduncle.

12 Helcium tergite dorsally without a U-shaped emargination of its leading edge. Helcium sternite small, retracted, concealed by the tergite.

13 Abdominal stridulatory system absent.

14 Abdominal segment 4 (= gastral segment 2) without differentiated presclerites.

15 Abdominal spiracles 5–7 (= gastral spiracles 3–5) apparently concealed by posterior margins of preceding tergites.

16 Abdominal segment 2 (petiole) with tergosternal fusion; abdominal segments 3–7 (= gastral segments 1–5) without tergosternal fusion.

17 Pygidium (tergite of abdominal segment 7 = gastral segment 5) large, simple.

18 Hypopygium (sternite of abdominal segment 7 = gastral segment 5) not modified into an acidopore apically.

19 Sting present and distinct, functional.

Synoptic Classification

A name prefixed by * indicates an extinct taxon.

Subfamily **ANEURETINAE.**

Tribe **Aneuretini.** Genera: ***Aneuretellus, Aneuretus*** (Figs. 3, 4), ****Mianeuretus, *Paraneuretus, *Protaneuretus.***

[Material of the unavailable name Prodolichoderinae is in part referable to Aneuretinae.]

Distribution

The single extant genus and species of this subfamily is restricted to the island of Sri Lanka. The extinct genera listed above come from the Baltic Amber, Russia, and the U.S.A., indicating a much wider distribution of the group in Tertiary times.

Taxonomic References

Aneuretus: Wilson, Eisner, G. C. Wheeler and J. Wheeler (1956); Hölldobler and Wilson (1990); Shattuck (1992b); Baroni Urbani, Bolton, and Ward (1992).

Subfamily APOMYRMINAE

Diagnosis of Worker (Figs. 5, 6)

Ants with the following combination of characters together.

1 Labrum with numerous peg-like teeth present.
2 Clypeus narrow from front to back, bringing the antennal sockets close to the anterior margin of the head.
3 Eyes absent; antenna with 12 segments.
4 Antennal sockets horizontal, in the plane of the transverse axis of the head, exposed in full-face view.
5 Frontal lobes absent.
6 Narrow neck joining condylar bulb of antennal scape to shaft of scape proper straight, not sharply angled or bent downwards in frontal or full-face view.
7 Promesonotal suture present and flexible, deeply impressed; the pronotum capable of movement relative to the mesonotum.
8 Metapleural gland orifice in lower posterior corner of metapleuron, opening ventrally.
9 Propodeal lobes absent.
10 Propodeal spiracle far back on sclerite, low down on the side.
11 Metacoxal cavities closed; cuticular annulus around each cavity broad and complete, the annulus not broken medioventrally nor with a flexible suture traversing the annulus from the coxal cavity to the cavity in which the petiole articulates.
12 Waist of 1 separated segment, the petiole (= abdominal segment 2), which is pedunculate anteriorly.
13 Petiole tergite constituting most of the segment, the sternite reduced to a minute, medioventral sclerite posteriorly.
14 Abdominal stridulatory system absent.
15 Abdominal spiracles 5–7 (= gastral spiracles 3–5) concealed by the posterior margins of the preceding segments and not visible without distension or dissection of the abdomen.
16 Helcium sternite small and retracted, concealed by tergite and not visible in profile.
17 Abdominal sternite 3 (= gastral sternite 1) close behind the helcium with a transverse sulcus running the width of the sclerite.
18 Abdominal segment 3 (= gastral segment 1) with tergosternal fusion; tergites and sternites of following abdominal segments (4–7 = gastral segments 2–5) not fused.
19 Abdominal segment 4 (= gastral segment 2) without differentiated presclerites.
20 Pygidium (tergite of abdominal segment 7 = gastral segment 5) large and simple, unarmed, convex across and downcurved posteriorly.
21 Sting large, well developed, and functional.

Synoptic Classification

Subfamily **APOMYRMINAE.**
Tribe **Apomyrmini.** Genus: ***Apomyrma*** (Figs. 5, 6).

Distribution

The single genus of this subfamily is represented by two currently known West African species. They have been recorded from Ivory Coast, Ghana, Benin, Nigeria, and Cameroon.

Taxonomic References

Apomyrma: Brown, Gotwald, and Levieux (1971); Dlussky and Fedoseeva (1988); Hölldobler and Wilson (1990); Bolton (1990b); Baroni Urbani, Bolton, and Ward (1992).

Figures 1–6 Worker heads in full-face view and bodies in profile:

1–2, AENICTINAE, *Aenictus*

3–4, ANEURETINAE, *Aneuretus*

5–6, APOMYRMINAE, *Apomyrma.*

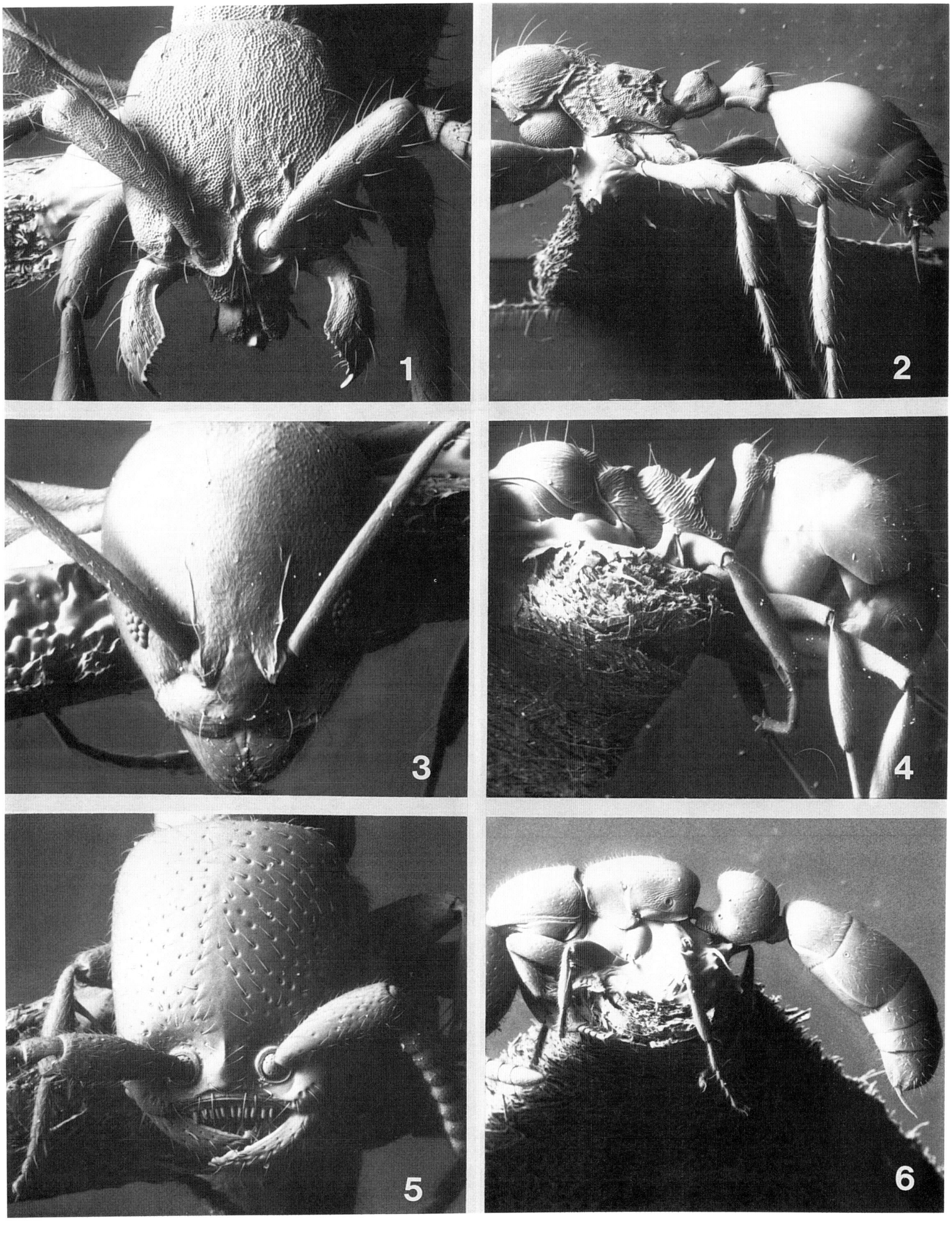
1
2
3
4
5
6

Subfamily CERAPACHYINAE

Diagnosis of Worker (Figs. 7–19)

Ants with the following combination of characters together.

1 Clypeus, from front to back, moderately broad to reduced; in the latter bringing the antennal sockets close to the anterior margin of the head.

2 Antennal sockets horizontal, in the plane of the transverse axis of the head, partially or wholly exposed in full-face view.

3 Frontal lobes rarely weakly present, predominantly vestigial to absent; usually narrow vertical carinae are all that are present between the antennal sockets.

4 Narrow neck joining condylar bulb of antennal scape to shaft of scape proper straight, not sharply angled or bent downwards in frontal or full-face view.

5 Eyes usually present but frequently reduced or absent in some groups; antenna with 9–12 segments.

6 Promesonotal suture usually absent (present in just one species, where it is fused and inflexible), the alitrunk fusiform and box-like.

7 Metapleural gland orifice in lower posterior corner of metapleuron, opening laterally, the orifice concealed behind a ventrally directed cuticular flange or flap.

8 Propodeal spiracle situated low down on the side of the sclerite, at or behind the midlength.

9 Metatibial glands present.

10 Metacoxal cavities closed; cuticular annulus around each cavity broad and complete, the annulus not broken medioventrally nor with a flexible suture traversing the annulus from the coxal cavity to the cavity in which the petiole articulates.

11 Propodeal lobes present.

12 Waist of 1 or 2 segments. Generally of a single, separated segment, the petiole (= abdominal segment 2); usually also with a deep constriction between abdominal segments 3 and 4. Abdominal segment 3 with a graded morphoclinal reduction from a full-sized segment to a small postpetiole, thus making the waist 2-segmented.

13 Petiole sessile to subsessile, the tergite and sternite not fused; sternite of petiole with a simple posterior margin and simple articulation to the third abdominal segment.

14 Abdominal stridulatory system absent.

15 Abdominal spiracles 5–7 shifted backwards, not concealed by the posterior margins of the preceding segments and visible without distension or dissection of the abdomen.

16 Helcium sternite large and convex, bulging ventrally, visible in profile in normally mounted specimens; tergite of helcium lacking a notch or impression in its dorsal margin anteriorly.

17 Abdominal segment 3 [*Note:* petiole is abdominal segment 2] with tergosternal fusion; tergites and sternites of following abdominal segments (4–7) not fused.

18 Abdominal segment 4 with presclerites sharply defined and differentiated from the postsclerites, the former fitting tightly within the posterior end of the third segment.

19 Pygidium (tergite of abdominal segment 7, the last visible tergite) large, its dorsum flattened or impressed; lateral and/or posterior margins of flattened area armed with a series or row of denticles, small teeth, or peg-like spines, with more than 2 pairs present (Figs. 17–19).

20 Sting large and strongly developed.

Key to World CERAPACHYINAE (Workers)

1 With head in full-face view the gena between the antennal socket and the lateral margin of the head with a curved to approximately straight longitudinal carina, which ends at the posterior margin of the lateral portion of the clypeus (Figs. 11, 15) **2**

— With head in full-face view the gena between the antennal socket and the lateral margin of the head plain, without a longitudinal carina ending at the posterior margin of the lateral portion of the clypeus (Figs. 7, 9) **4**

2 Gaster with strong girdling constrictions between segments 2 and 3 and between segments 3 and 4, as well as between segments 1 and 2 (Fig. 16). (World tropics) ***Sphinctomyrmex***

— Gaster only with a strong girdling constriction between segments 1 and 2; segments 2, 3, and 4 without girdling constrictions separating them (Figs. 12, 14) **3**

3 Tibial spurs absent from middle legs. Pretarsal claws usually with a single preapical tooth. Hind basitarsus with a longitudinal impression or groove, which may be glandular, on its ventral (inner) surface close to the tibia. (Afrotropical, Malagasy, Indo-Australian) . . . . ***Simopone***

— Tibial spurs present on middle legs. Pretarsal claws always simple. Hind basitarsus without a longitudinal impression or glandular groove on its ventral (inner) surface close to the tibia. (Worldwide in tropics and subtropics) . . . ***Cerapachys***

4 Antennal scrobes present (Fig. 9). Side of head without a longitudinal groove running posteriorly from the mandibular articulation. Middle and hind tibiae each with 2 pectinate spurs, one large, the other smaller and less conspicuous. Helcium approximately at midheight of first gastral segment (Fig. 10). Eyes distinct. (Neotropical) . . . . ***Cylindromyrmex***

— Antennal scrobes absent (Fig. 7). Side of head with a longitudinal groove running posteriorly from the mandibular articulation. Middle and hind tibiae each with a single spur, which is large and pectinate. Helcium above midheight of first gastral segment (Fig. 8). Eyes minute to vestigial. (Neotropical, southern Nearctic) . . . . ***Acanthostichus***

Synoptic Classification

Subfamily **CERAPACHYINAE.**

Tribe **Acanthostichini.** Genus: ***Acanthostichus*** (Figs. 7, 8, 17) (= *Ctenopyga*).

Tribe **Cylindromyrmecini.** Genus: ***Cylindromyrmex*** (Figs. 9, 10) (= *Holcoponera* (homonym), = *Hypocylindromyrmex,* = *Metacylindromyrmex*).

Tribe **Cerapachyini** (= Eusphinctinae, = Lioponerini). Genera: ***Cerapachys*** (Figs. 11, 12) (= *Ceratopachys,* = *Chrysapace,* = *Cysias,* = *Lioponera,* = *Neophyracaces,* = *Ooceraea,* = *Parasyscia,* = *Phyracaces,* = **Procerapachys,* = *Syscia*), ***Simopone*** (Figs. 13, 14, 19), ***Sphinctomyrmex*** (Figs. 15, 16, 18) (= *Aethiopopone,* = *Eusphinctus,* = *Nothosphinctus,* = *Zasphinctus*).

[Material of the unavailable name Prodorylinae is referable to Cerapachyinae.]

Distribution

Two of the five cerapachyine genera, *Acanthostichus* and *Cylindromyrmex,* are mostly restricted to the Neotropical region, though the former extends its range into the extreme southern Nearctic; the other three are more widely distributed. *Simopone* is found in the Afrotropical, Malagasy, and Indo-Australian regions, with the greatest number of species in the first of these. *Sphinctomyrmex* is predominantly Australasian but a few species of this genus also occur in the Afrotropical, Oriental, Indo-Australian, and Neotropical regions. *Cerapachys* is the largest and most widely distributed genus of the subfamily, with representatives in all zoogeographical regions. The greatest number of species occur in the Indo-Australian region, but only very few are found in the Palaearctic, Nearctic, and Neotropical regions.

Taxonomic References

Identification of extant species

Acanthostichus: Kusnezov (1962); Kempf (1964a); MacKay (in preparation). ***Cerapachys***: Brown (1975) [world]; Ogata (1983) [Japan]. ***Cylindromyrmex:*** Brown (1975). ***Simopone:*** Brown (1975) [Afrotropical]. ***Sphinctomyrmex:*** Brown (1975) [Oriental, Indo-Australian, Australasian].

Other taxonomic references

Cerapachyinae: Brown (1954a, 1975); Dlussky and Fedoseeva (1988); Hölldobler and Wilson (1990); Bolton (1990a); Baroni Urbani, Bolton and Ward (1992).

See also References to Faunistic Studies.

Figures 7–19 CERAPACHYINAE workers. Figs. 7–16, heads in full-face view and bodies in profile:

7–8, **Acanthostichini,** *Acanthostichus*

9–10, **Cylindromyrmecini,** *Cylindromyrmex*

11–16, **Cerapachyini:** 11–12, *Cerapachys;* 13–14, *Simopone;* 15–16, *Sphinctomyrmex*

17–19, apex of gaster to show armament of pygidium: 17, *Acanthostichus;* 18, *Sphinctomyrmex;* 19, *Simopone.*

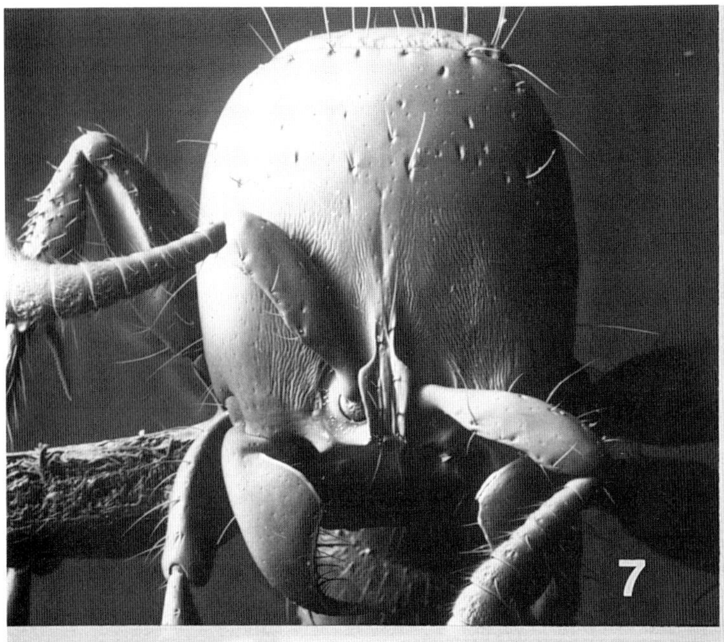
7

8
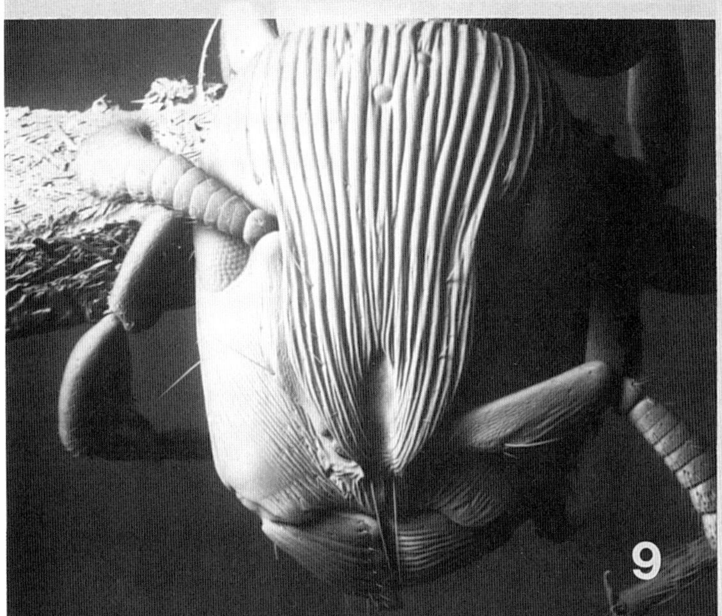
9
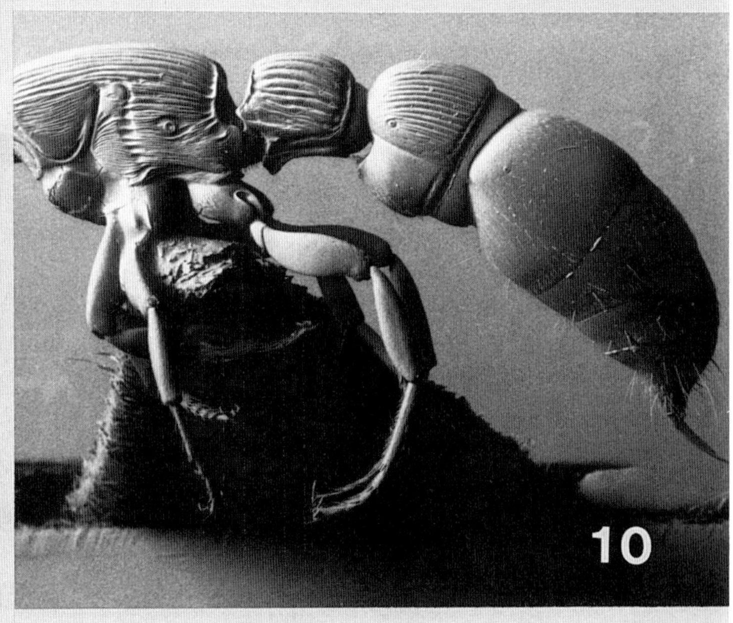
10
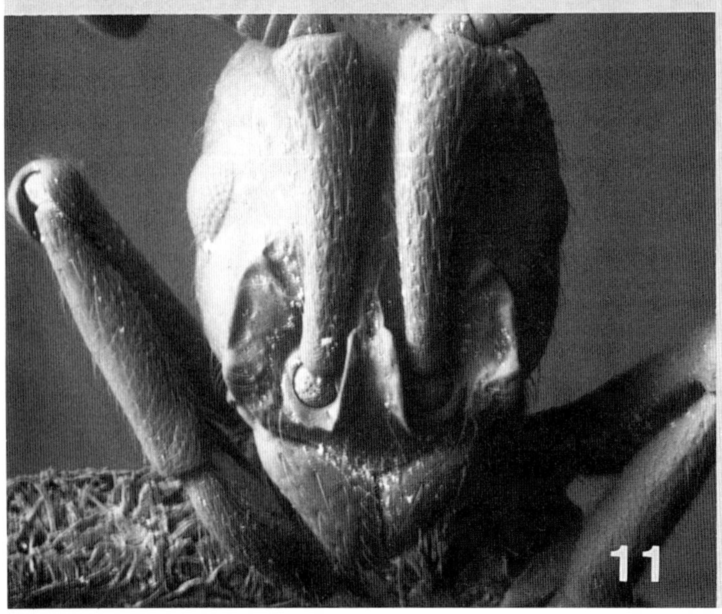
11

12

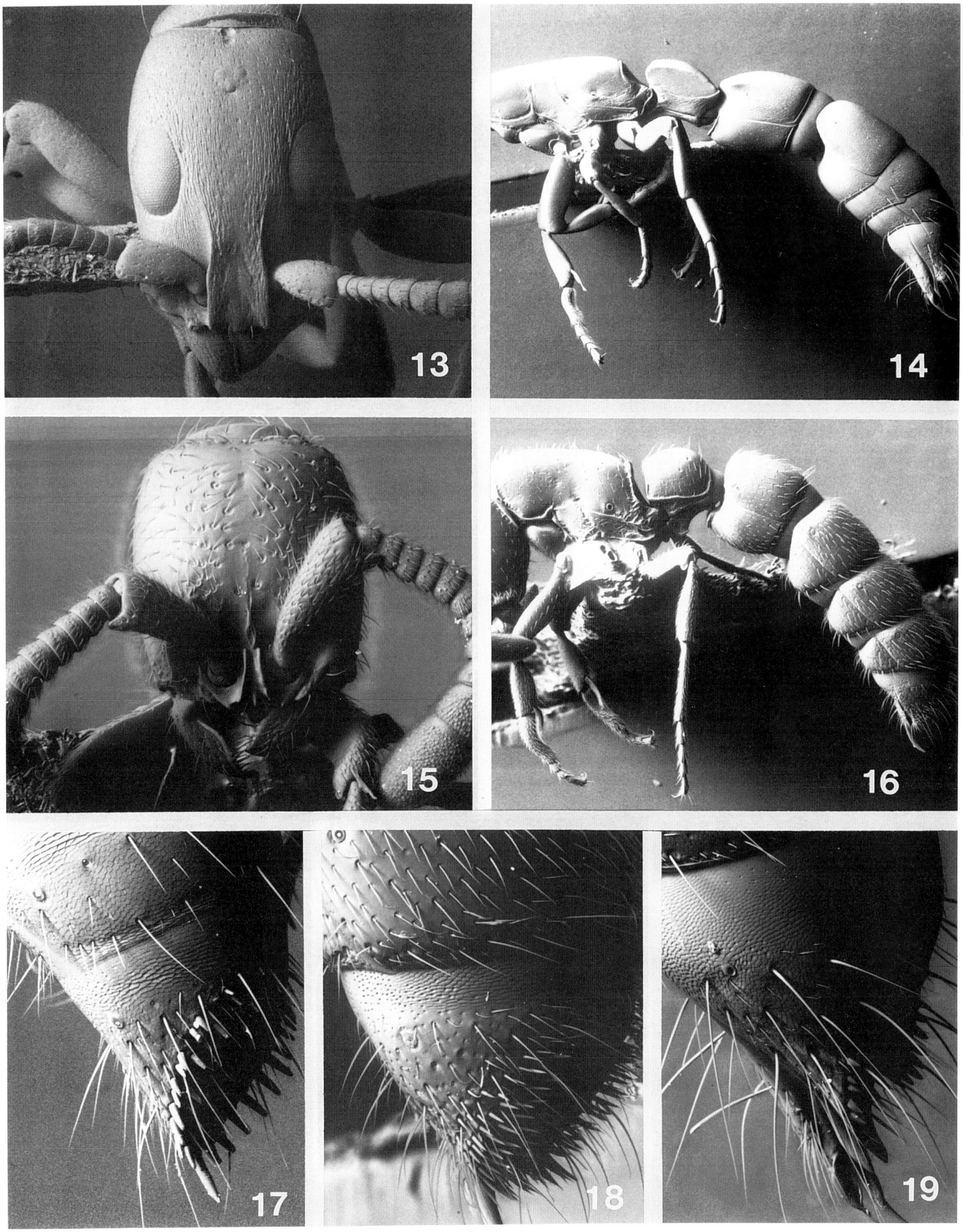
13
14
15
16
17
18
19

Subfamily DOLICHODERINAE

Diagnosis of Worker (Figs. 20–58)

Ants with the following combination of characters together.

1 Median portion of clypeus broad from front to back so that antennal sockets are well behind anterior margin of head. Median portion of clypeus usually extended backwards between the antennal sockets, sometimes not. Often a postclypeal frontal triangle present but in some the frontoclypeal suture obliterated.

2 Antennal sockets inclined, the portion of the socket margin and torulus closest to the dorsal midline of the head on a higher level than the portion of the margin most distant from the midline; torular sclerites usually exposed.

3 Frontal carinae usually present, rarely absent; when present varying from a pair of simple carinae or the margins of a raised plateau to narrow flanges; antennal sockets partly or wholly exposed. Only very rarely the carinae expanded into narrow frontal lobes.

4 Narrow neck joining condylar bulb of antennal scape to shaft of scape proper straight.

5 Eyes usually present, only rarely vestigial or absent; ocelli rarely present; antenna usually with 12 segments but very rare counts of 8 and 11 are known.

6 Promesonotal suture present and usually flexible.

7 Metapleural gland orifice present, situated in lower posterior corner of metapleuron, opening laterally or posterolaterally; the orifice commonly with guard setae crossing its aperture.

8 Metanotum and its spiracles frequently present on dorsal alitrunk.

9 Metacoxal cavities closed; a thin, continuous strip of cuticle separates the metacoxal cavity from the cavity in which the petiole articulates.

10 Propodeal lobes absent.

11 Waist of 1 segment, the petiole (= abdominal segment 2).

12 Helcium tergite dorsally with an extensive U-shaped emargination of its leading edge (Fig. 58), the emargination often reaching back almost the whole length of the sclerite. Helcium sternite small, retracted, concealed by the tergite.

13 Abdominal stridulatory system absent.

14 Abdominal segment 4 (= gastral segment 2) without differentiated presclerites.

15 Abdominal spiracles 4–7 (= gastral spiracles 2–5) sometimes concealed by posterior margins of preceding tergites; frequently abdominal spiracles 4 and 5 visible and sometimes also 6 and 7 visible without distension of the abdomen.

16 Abdominal segment 2 (petiole) with tergosternal fusion; abdominal segments 3–7 (= gastral segments 1–5) without tergosternal fusion.

17 Pygidium (tergite of abdominal segment 7 = gastral segment 5) small or very small, simple; pygidium often reflexed so that it is on the ventral surface of the gaster, often overhung and partially to almost entirely concealed by the tergite of abdominal segment 6 (= gastral segment 4).

18 Hypopygium (sternite of abdominal segment 7 = gastral segment 5) not modified into an acidopore apically.

19 Sting vestigial to absent, nonfunctional, and not detectable without dissection.

Key to Palaearctic DOLICHODERINAE (Workers)

1 Petiole in profile usually a simple, transversely flattened strip, sometimes slightly swollen anterodorsally but never equipped with a standing scale (Figs. 53, 55). Petiole overhung by first gastral segment and not visible in dorsal view when alitrunk and gaster are in the same plane **2**

— Petiole in profile surmounted by a scale that may be high and erect or lower and somewhat inclined forward, but scale always present and conspicuous (Figs. 27, 29, 41, 43, 47). Petiole scale not or only weakly overhung by first gastral segment, usually visible in dorsal view when alitrunk and gaster are in the same plane **3**

2 In dorsal view only 4 gastral segments visible (Fig. 53). Fifth tergite reflexed below the fourth, visible in ventral view where it forms a transverse plate abutting the fifth sternite; the anal and associated orifices are thus situated ventrally . . . . ***Tapinoma***

— In dorsal view 5 gastral tergites visible (Fig. 55), the fifth small but continuing the line of the gaster and not reflexed below the fourth; the anal and associated orifices are thus situated apically . . . . ***Technomyrmex***

3 Palp formula 4,3 or 2,2 . . . . ***Bothriomyrmex***

— Palp formula 6,4 . . . . **4**

4 Integument thick, hard, and strongly sculptured; the head with foveolate punctures present. Hypostoma with an anterolateral tooth-like prominence on each side . . . . . ***Dolichoderus***

— Integument thin and flexible, densely but weakly sculptured; the head without foveolate punctures. Hypostoma without an anterolateral tooth-like prominence on each side **5**

5 With alitrunk in profile the metanotal groove impressed; mesonotum and propodeum not forming a continuous surface (Figs. 41, 47). Metathoracic spiracles dorsal **6**

— With alitrunk in profile the metanotal groove not impressed, reduced to a transverse suture so that the mesonotum and propodeum form a continuous surface (Fig. 43). Metathoracic spiracles lateral . . . . ***Liometopum***

6 With the alitrunk in profile the propodeal declivity concave (Fig. 47). Fourth gastral sternite flat across entire posterior border . . . . ***Ochetellus***

— With the alitrunk in profile the propodeal declivity convex (Fig. 41). Fourth gastral sternite keel-shaped posteriorly . . . . ***Linepithema***

Key to Afrotropical and Malagasy DOLICHODERINAE (Workers)

1 Propodeum armed with a pair of teeth, tubercles, or acute angles posterolaterally (Fig. 23), these sometimes linked across by a narrow carina. Between the tubercles, teeth, or angles, or somewhat more dorsally, the propodeum at its midwidth with a prominent tubercle, projecting plate, longitudinal ridge, or tooth . . . . ***Axinidris***

— Propodeum unarmed, without posterolateral teeth, tubercles, or acute angles; without a median prominence of any description on the dorsum (Figs. 41, 55) **2**

2 Scale of petiole well developed, in profile very distinct (Figs. 41, 47). Scale somewhat inclined forward but not reduced and not overhung by an anterior projection of the first gastral segment above it **3**

— Scale of petiole very reduced or vestigial, in profile very small and low or even more or less absent (Figs. 53, 55). Petiole overhung and concealed by an anterior projection of the first gastral segment above it **5**

3 With propodeum in profile the declivity concave (Fig. 47). (Mauritius only) . . . . ***Ochetellus***

— With propodeum in profile the declivity convex (Fig. 41) . . **4**

4 Each mandible with 5–8 teeth plus 5–13 denticles (Fig. 40). Anterior clypeal margin concave in full-face view. Fourth gastral sternite keel-shaped posteriorly . . . . ***Linepithema***

— Each mandible with 7 teeth and no denticles. Anterior clypeal margin convex in full-face view. Fourth gastral sternite not keel-shaped posteriorly . . . . ***Ecphorella***

5 In dorsal view only 4 gastral tergites visible. Fifth tergite reflexed below the fourth, visible in ventral view where it forms a transverse plate abutting the fifth sternite; the anal and associated orifices are thus situated ventrally . . . . ***Tapinoma***

— In dorsal view 5 gastral tergites visible, the fifth small but continuing the line of the gaster and not reflexed below the fourth; the anal and associated orifices are thus situated apically . . . . ***Technomyrmex***

Key to Oriental and Indo-Australian DOLICHODERINAE (Workers)

1 Petiole in profile usually a simple, transversely flattened strip, sometimes slightly swollen anterodorsally but never equipped with a standing or anteriorly inclined scale (Figs. 53, 55). Petiole overhung by first gastral segment and usually not visible in dorsal view when alitrunk and gaster are in the same plane **2**

— Petiole in profile surmounted by a node or scale that may be high and erect or lower and somewhat inclined forward, but scale (or node) always present and conspicuous (Figs. 27, 29, 39, 41, 43, 47, 57). Petiole not or partially overhung by first gastral segment, usually visible in dorsal view when alitrunk and gaster are in the same plane **3**

2 In dorsal view 5 gastral tergites visible, the fifth small but continuing the line of the gaster and not reflexed below the fourth; the anal and associated orifices are thus situated apically . . . . ***Technomyrmex***

— In dorsal view only 4 gastral tergites visible. Fifth tergite reflexed below the fourth, visible in ventral view where it forms a transverse plate abutting the fifth sternite; the anal and associated orifices are thus situated ventrally . . . . ***Tapinoma***

3 Palp formula 4,3 or 2,2 . . . . ***Bothriomyrmex***

— Palp formula 6,4 or 5,3 . . . . **4**

4 Head and alitrunk extremely elongate and slender (Figs. 38, 39); appendages narrow and extremely long. Hypostoma notched medially. Petiole nodiform. Mandibles elongate-triangular and slender, outer margins shallowly concave through the central section of their length . . . ***Leptomyrmex***

— Head and alitrunk usually broad and stocky, not extremely elongate (Figs. 20, 21, 28, 29, 44, 45, 48, 49); appendages not extremely long. Hypostoma not notched medially. Petiole usually a scale of some form, only rarely nodiform. Mandibles triangular, their outer margins generally convex but sometimes straight or slightly concave through the central section of their length **5**

5 Hypostoma with an anterolateral tooth-like prominence at each side. Integument thick, hard, and armor-like, the surface varying from smooth (very rare) to strongly and coarsely sculptured . . . . ***Dolichoderus***

— Hypostoma without an anterolateral tooth-like prominence at each side. Integument thin and flexible, not armor-like, the surface usually very finely and densely sculptured, only extremely rarely more or less smooth **6**

6 Posterodorsal angles of propodeum drawn out into short tubercles or prominences, the propodeal spiracles situated at the apices of the tubercles or prominences (Fig. 57) . . ***Turneria***

— Posterodorsal angles of propodeum not drawn out into short tubercles or prominences. Propodeal spiracles lateral (Figs. 21, 37, 45, 49, 51) **7**

7 With alitrunk in profile metanotal groove impressed and metathoracic spiracles dorsal (Figs. 21, 37, 41, 45, 47, 49, 51) **8**

— With alitrunk in profile metanotal groove not impressed and metathoracic spiracles lateral (Fig. 43) . . . . ***Liometopum***

8 Palp formula 5,3 . . . . ***Papyrius***

— Palp formula 6,4 . . . . **9**

9 With propodeum in profile the declivity concave (Fig. 47) . . . . ***Ochetellus***

— With propodeum in profile the declivity flat to convex (Figs. 21, 37, 45, 51) **10**

10 Fourth gastral sternite flat across its entire posterior margin. Anterolateral clypeal border with a shoulder and posterior to the mediolateral region; central anterior clypeal margin with a projection (Figs. 36, 50) **11**

— Fourth gastral sternite keel-shaped posteromedially. Anterolateral clypeal border even with, or anterior to, the mediolateral region; central anterior clypeal margin without a projection, broadly concave to convex (Figs. 20, 40, 44) **12**

11 With head in full-face view the occipital margin usually convex, only rarely weakly concave (Fig. 36). Dorsum of petiole vertical to moderately inclined anteriorly (Fig. 37). Usually monomorphic species (rarely otherwise) with eyes placed relatively posteriorly on head . . . . ***Iridomyrmex***

— With head in full-face view the occipital margin markedly concave (Fig. 50). Dorsum of petiole strongly inclined anteriorly (Fig. 51). Polymorphic species with eyes placed relatively anteriorly on head . . . . ***Philidris***

12 First gastral tergite projecting anteriorly, almost entirely concealing the petiole in dorsal view. In profile the propodeal declivity much longer than the dorsum, the two surfaces separated by a distinct angular ridge or carina (Fig. 45) . . . . ***Loweriella***

— First gastral tergite not projecting anteriorly, not at all concealing the petiole in dorsal view. In profile the propodeal declivity subequal to or shorter than the dorsum, the 2 surfaces rounding together (Figs. 21, 41) **13**

13 Anterior clypeal margin with a broad median concavity (Fig. 40). Apical (masticatory) margin of mandible with 5–8 teeth and 5–13 denticles. Apical tooth of mandible elongate, much longer than the preapical tooth . . . . ***Linepithema***

— Anterior clypeal margin flat to convex medially (Fig. 20). Apical (masticatory) margin of mandible with 5–10 teeth and 0–4 denticles. Apical tooth of mandible at most only slightly longer than the preapical . . . . ***Anonychomyrma***

Key to Australasian DOLICHODERINAE (Workers)

1 Petiole in profile usually a simple, transversely flattened strip, sometimes slightly swollen anterodorsally but never equipped with a standing scale or node (Figs. 53, 55). Petiole overhung by first gastral segment and usually not visible in dorsal view when alitrunk and gaster are in the same plane **2**

— Petiole in profile surmounted by a node or scale that may be high and erect or lower and somewhat inclined forward, but scale (or node) always present and conspicuous (Figs. 21, 27, 29, 35, 37, 39, 41, 47, 49, 51, 57). Petiole not or only weakly overhung by first gastral segment, usually visible in dorsal view when alitrunk and gaster are in the same plane **3**

2 In dorsal view 5 gastral tergites visible, the fifth small but continuing the line of the gaster and not reflexed below the fourth; the anal and associated orifices are thus situated apically . . . . ***Technomyrmex***

— In dorsal view only 4 gastral tergites visible. Fifth tergite reflexed below the fourth, visible in ventral view where it forms a transverse plate abutting the fifth sternite; the anal and associated orifices are thus situated ventrally . . . . ***Tapinoma***

3 Head and alitrunk extremely elongate and slender (Figs. 38, 39); appendages narrow and extremely long. Hypostoma notched medially . . . . ***Leptomyrmex***

— Head and alitrunk usually broad and stocky, not extremely elongate (Figs. 20, 21, 28, 29, 34–37, 46–51); appendages not extremely long. Hypostoma not notched medially **4**

4 Hypostoma with an anterolateral tooth-like prominence at each side. Integument thick, hard, and armor-like, the surface varying from smooth (rare) to strongly and coarsely sculptured . . . . ***Dolichoderus***

— Hypostoma without an anterolateral tooth-like prominence at each side. Integument usually thin and flexible, only rarely armor-like, the surface generally finely and densely sculptured **5**

5 With alitrunk in profile the propodeal spiracle located dorsally (Figs. 35, 57) **6**

— With alitrunk in profile the propodeal spiracle located laterally (Figs. 27, 37, 47, 51) **7**

6 Propodeum bispinose, the propodeal spiracles located near the bases of the spines (Fig. 35) . . . . ***Froggattella***

— Propodeum with posterodorsal angles drawn out into short tubercles or prominences, the propodeal spiracles located at the apices of the tubercles or prominences (Fig. 57) . . . . ***Turneria***

7 Palp formula 4,3 or 2,2 . . . . ***Bothriomyrmex***

— Palp formula 6,4 or 5,3 . . . . **8**

8 Palp formula 5,3. Anterior clypeal margin with 8–20 very short, straight setae which only slightly project beyond the margin (Fig. 48). Apical (masticatory) margin of mandible with 11–14 teeth and no denticles. Metanotal groove U-shaped (Fig. 49). Monomorphic . . . . ***Papyrius***

— Palp formula usually 6,4 but if 5,3 then the species polymorphic. Anterior clypeal margin with 2–16 short, straight setae which project considerably beyond the margin (Figs. 20, 36, 40, 46). Apical (masticatory) margin of mandible with 5–12 teeth and usually also a few denticles. Metanotal groove a distinct angle, a narrow groove, or a notch (Figs. 21, 37, 41, 47, 51) . . . **9**

9 With propodeum in profile the declivity concave. Dorsal alitrunk more or less flat, the metanotal groove notch-like (Fig. 47) . . . . ***Ochetellus***

— With propodeum in profile the declivity convex, or very rarely almost flat. Dorsal alitrunk with propodeum depressed below the level of the promesonotum, the metanotal groove angular (Figs. 21, 37, 41, 51) **10**

10 Anterolateral clypeal border with a shoulder and posterior to the mediolateral region (Figs. 36, 50). Anterior clypeal margin centrally with a projection, which may be strongly or weakly developed **11**

— Anterolateral clypeal border even with, or anterior to, the mediolateral region (Figs. 20, 40). Anterior clypeal margin centrally without a projection, the margin broadly concave to convex **12**

11 With head in full-face view the occipital margin usually convex, only occasionally weakly concave (Fig. 36). Dorsum of petiole vertical to moderately inclined anteriorly (Fig. 37). Eyes situated relatively posteriorly on head . . . . ***Iridomyrmex***

— With head in full-face view the occipital margin markedly concave (Fig. 50). Dorsum of petiole strongly inclined anteriorly (Fig. 51). Eyes situated relatively anteriorly on head . . . . ***Philidris***

12 Anterior clypeal setae moderately curved ventrally. In profile the dorsal surface of the propodeum shorter than the declivity. Fourth gastral sternite flat posteriorly. Mandible with 4 or 5 teeth and 4 or 5 denticles . . . . ***Doleromyrma***

— Anterior clypeal setae straight. In profile the dorsal surface of the propodeum subequal to or longer than the declivity. Fourth gastral sternite keel-shaped posteriorly. Mandible with 5–10 teeth and 0–13 denticles **13**

13 Anterior clypeal border with a broad median concavity (Fig. 40). Mandible with 5–8 teeth and 5–13 denticles. Apical tooth of mandible elongate, much longer than the preapical tooth . . . . ***Linepithema***

— Anterior clypeal border flat or convex medially (Fig. 20). Mandible with 5–10 teeth and 0–4 denticles. Apical tooth of mandible at most only slightly longer than the preapical tooth . . . . ***Anonychomyrma***

Key to Nearctic DOLICHODERINAE (Workers)

1 Petiole in profile usually a simple, transversely flattened strip, sometimes slightly swollen anterodorsally but never equipped with a standing scale or node (Figs. 53, 55). Petiole overhung by first gastral segment and usually not visible in dorsal view when the alitrunk and gaster are in the same plane **2**

— Petiole in profile surmounted by a node or scale that may be high and erect or lower and somewhat inclined forward, but scale (or node) always present and conspicuous. Petiole not or only weakly overhung by first gastral segment, usually visible in dorsal view when alitrunk and gaster are in the same plane **3**

2 In dorsal view only 4 gastral tergites visible. Fifth tergite reflexed below the fourth, visible in ventral view where it forms a transverse plate abutting the fifth sternite; the anal and associated orifices are thus situated ventrally . . . . ***Tapinoma***

— In dorsal view 5 gastral tergites visible, the fifth small but continuing the line of the gaster and not reflexed below the fourth; the anal and associated orifices are thus situated apically . . . . ***Technomyrmex***

3 Integument thick, hard, and strongly sculptured; the head with foveolate punctures. Hypostoma with an anterolateral tooth-like prominence on each side . . . . ***Dolichoderus***

— Integument thin and flexible, densely but weakly sculptured; the head without foveolate punctures. Hypostoma without an anterolateral tooth-like prominence on each side **4**

4 With propodeum in profile the angle between dorsum and declivity extended into a single raised tooth or spine (Fig. 31). With head in full-face view the clypeal setae projecting forward beyond the apices of the closed mandibles. Psammophore present and apical mandibular tooth enlarged, much larger than the preapical tooth (Fig. 30) . . . . . . ***Dorymyrmex***

— With propodeum in profile the angle between dorsum and declivity rounded to angulate, but never extended into a raised tooth or spine (Figs. 33, 41, 43). With head in full-face view the clypeal setae not projecting forward beyond the apices of the closed mandibles. Psammophore absent and apical mandibular tooth small, at most only slightly larger than the preapical tooth (Figs. 32, 40, 42) **5**

5 With alitrunk in profile the metanotal groove not impressed, reduced to a transverse suture so that the mesonotum and propodeum form a continuous surface (Fig. 43) . . . . ***Liometopum***

— With alitrunk in profile the metanotal groove impressed; mesonotum and propodeum not forming a continuous surface (Figs. 33, 41) **6**

6 Anterior clypeal margin with 2–12 ventrally curved setae, which are approximately the same length as the closed mandibles (Fig. 32). Basal margin of mandible unarmed. First gastral tergite projecting anteriorly over the petiole (Fig. 33) . . . . ***Forelius***

— Anterior clypeal margin with 2–6 short setae, which are much shorter than the closed mandibles (Fig. 40). Basal margin of mandible with denticles close to the basal angle. First gastral tergite vertical, not projecting over the petiole (Fig. 41) . . . . ***Linepithema***

Key to Neotropical DOLICHODERINAE (Workers)

1 Petiole in profile usually a simple, transversely flattened strip, sometimes slightly swollen anterodorsally but never equipped with a standing scale or node (Figs. 53, 55). Petiole overhung by first gastral segment and usually not visible in dorsal view when the alitrunk and gaster are in the same plane **2**

— Petiole in profile surmounted by a node or scale that may be high and erect or lower and somewhat inclined forward, but scale (or node) always present and conspicuous (Figs. 25, 29, 31, 33, 41). Petiole not or only weakly overhung by first gastral segment, usually visible in dorsal view when alitrunk and gaster are in the same plane **3**

2 In dorsal view only 4 gastral tergites visible. Fifth tergite reflexed below the fourth, visible in ventral view where it forms a transverse plate abutting the fifth sternite; the anal and associated orifices are thus situated ventrally . . . . ***Tapinoma***

— In dorsal view 5 gastral tergites visible, the fifth small but continuing the line of the gaster and not reflexed below the fourth; the anal and associated orifices are thus situated apically . . . . ***Technomyrmex***

3 Hypostoma with an anterolateral tooth-like prominence on each side. Integument thick, hard, and often strongly sculptured . . . . ***Dolichoderus***

— Hypostoma without an anterolateral tooth-like prominence on each side. Integument thin and flexible, usually densely but weakly sculptured **4**

4 With propodeum in profile the angle between dorsum and declivity extended into a single raised tooth or spine (Fig. 31). With head in full-face view the clypeal setae projecting forward beyond the apices of the closed mandibles. Psammophore present and apical mandibular tooth enlarged, much larger than the preapical tooth (Fig. 30) . . . . . ***Dorymyrmex***

— With propodeum in profile the angle between dorsum and declivity rounded to angulate, but never extended into a raised tooth or spine (Figs. 25, 33, 41). With head in full-face view the clypeal setae not projecting forward beyond the apices of the closed mandibles. Psammophore absent and apical mandibular tooth small, at most only slightly larger than the preapical tooth (Figs. 24, 32, 40) **5**

5 Eyes absent. Palp formula 2,3 . . . . ***Anillidris***

— Eyes present. Palp formula 6,4 . . . . **6**

6 Anterior clypeal margin with 2–12 ventrally curved setae, which are approximately the same length as the closed mandibles. First gastral tergite projecting anteriorly over the petiole (Figs. 32, 33) . . . . ***Forelius***

— Anterior clypeal margin with 2–6 short setae which are much shorter than the closed mandibles. First gastral tergite vertical, not projecting anteriorly over the petiole (Figs. 24, 25, 40, 41) **7**

7 Anterior clypeal margin with a broad median concavity (Fig. 40). Mandible with 5–8 teeth and 5–13 denticles, the apical mandibular tooth much longer than the preapical . . . . ***Linepithema***

— Anterior clypeal margin flat or convex medially (Fig. 24). Mandible with 5–10 teeth and 0–4 denticles, the apical mandibular tooth at most only slightly longer than the preapical . . . . ***Azteca***

Synoptic Classification

A name prefixed by * indicates an extinct taxon.

Subfamily **DOLICHODERINAE.**

Tribe **Dolichoderini** (= Anonychomyrmini, = Axinidrini, = Leptomyrmicini, = Liometopini, = *Miomyrmicini, = *Pityomyrmecini, = Tapinomini, = *Zherichiniini). Genera: ***Anillidris***, ***Anonychomyrma*** (Figs. 20, 21), *****Asymphylomyrmex***, ***Axinidris*** (Figs. 22, 23), ***Azteca*** (Figs. 24, 25) (= *Aztecum* (misspelling)), ***Bothriomyrmex*** (Figs. 26, 27) (= *Chronoxenus*), ***Doleromyrma***, ***Dolichoderus*** (Figs. 28, 29) (= *Acanthoclinea*, = *Diabolus* (homonym), = *Diceratoclinea*, = *Hypoclinea*, = *Karawajewella*, = *Monacis*, = *Monoceratoclinea*), ***Dorymyrmex*** (Figs. 30, 31) (= *Ammomyrma*, = *Araucomyrmex*, = *Biconomyrma*, = *Conomyrma*, = *Psammomyrma*, = *Spinomyrma*), ***Ecphorella***, *****Elaeomyrmex***, *****Eotapinoma***, ***Forelius*** (Figs. 32, 33) (= *Amyrmex*, = *Neoforelius*), ***Froggattella*** (Figs. 34, 35), ***Iridomyrmex*** (Figs. 36, 37, 58), *****Kotshkorkia***, ***Leptomyrmex*** (Figs. 38, 39), *****Leptomyrmula***, ***Linepithema*** (Figs. 40, 41), ***Liometopum*** (Figs. 42, 43) (= **Ctenobethylus*), ***Loweriella*** (Figs. 44, 45), *****Miomyrmex***, ***Ochetellus*** (Figs. 46, 47), ***Papyrius*** (Figs. 48, 49), *****Petraeomyrmex***, ***Philidris*** (Figs. 50, 51), *****Pityomyrmex***, *****Protazteca***, ***Tapinoma*** (Figs. 52, 53) (= *Micromyrma*, = *Neoclystopsenella*, = *Semonius*, = *Tapinoptera*, = *Zatapinoma*), ***Technomyrmex*** (Figs. 54, 55) (= *Aphantolepis*, = *Engramma*), ***Turneria*** (Figs. 56, 57), *****Zherichinius***.

The unidentifiable name *Hypochira* may refer to a dolichoderine ant. [Material of the unavailable names Eudolichoderinae and Prodolichoderinae (in part) is referable to Dolichoderinae.]

Distribution

The subfamily Dolichoderinae has a worldwide distribution, as indicated in the table given in the Introduction. The numbers of genera shared between the various zoogeographical regions are shown below, excluding endemic genera and those artificially introduced by human activities. In the table PAL = Palaearctic, AFR = Afrotropical, MAL = Malagasy, ORI = Oriental, INA = Indo-Australian, AUS = Australasian, NEA = Nearctic, NEO = Neotropical.

	PAL	AFR	MAL	ORI	INA	AUS	NEA
AFR	2						
MAL	3	2					
ORI	6	2	3				
INA	5	2	3	7			
AUS	5	2	3	7	11		
NEA	3	1	1	3	2	2	
NEO	3	2	2	3	3	3	4

Taxonomic References

Identification of extant species

Some older references have a suffixed comment "[out of date]." These references are included because they contain the only identification keys ever attempted for the taxon in question. They should be used with great caution as, for the most part, they contain numerous infraspecific and infrasubspecific taxa that are no longer recognized. Older references that have been superseded, or those rendered useless by the volume of later descriptions and synonymies, are omitted.

Axinidris: Shattuck (1991). ***Azteca:*** Longino (1989, 1992) [*Cecropia*-inhabiting species]. ***Bothriomyrmex:*** Emery (1925c) [out of date]. ***Dolichoderus:*** Clark (1930) [Australasian, out of date]; Kempf (1959a, 1972b) [former *Monacis* species, Neotropical]; Kempf (1969) [*Dolichoderus, sensu stricto,* Neotropical]; Lattke (1987) [*bispinosus*-group, Neotropical]; MacKay (1993) [Nearctic, Neotropical]; Johnson (1989a) [North America]. ***Dorymyrmex:*** Gallardo (1916) [Argentina, out of date]; Trager (1988), Johnson (1989b) [Nearctic]. ***Leptomyrmex:*** W. M. Wheeler (1934a) [out of date]. ***Liometopum:*** Kupyanskaya (1988) [east Palaearctic]. ***Tapinoma:*** Emery (1925b) [Palaearctic, out of date]. ***Technomyrmex:*** W. M. Wheeler (1922) [former *Engramma* species, out of date]. ***Turneria:*** Shattuck (1990).

Other taxonomic references

Dolichoderinae: Brown (1973); Snelling (1981); Dlussky and Fedoseeva (1988); Hölldobler and Wilson (1990); Shattuck (1992a, 1992b, 1992c [review of subfamily]); Baroni Urbani, Bolton, and Ward (1992).

See also References to Faunistic Studies.

Figures 20–58 DOLICHODERINAE workers. Figs. 20–57, heads in full-face view and bodies in profile:

20–21, *Anonychomyrma;* 22–23, *Axinidris;* 24–25, *Azteca;* 26–27, *Bothriomyrmex;* 28–29, *Dolichoderus;* 30–31, *Dorymyrmex;* 32–33, *Forelius;* 34–35, *Froggattella;* 36–37, *Iridomyrmex;* 38–39, *Leptomyrmex;* 40–41, *Linepithema;* 42–43, *Liometopum;* 44–45, *Loweriella;* 46–47; *Ochetellus;* 48–49, *Papyrius;* 50–51, *Philidris;* 52–53, *Tapinoma;* 54–55, *Technomyrmex;* 56–57, *Turneria*

58, dorsal view of petiole, helcium, and base of first gastral segment to show emargination of helcium in *Iridomyrmex.*

20

21

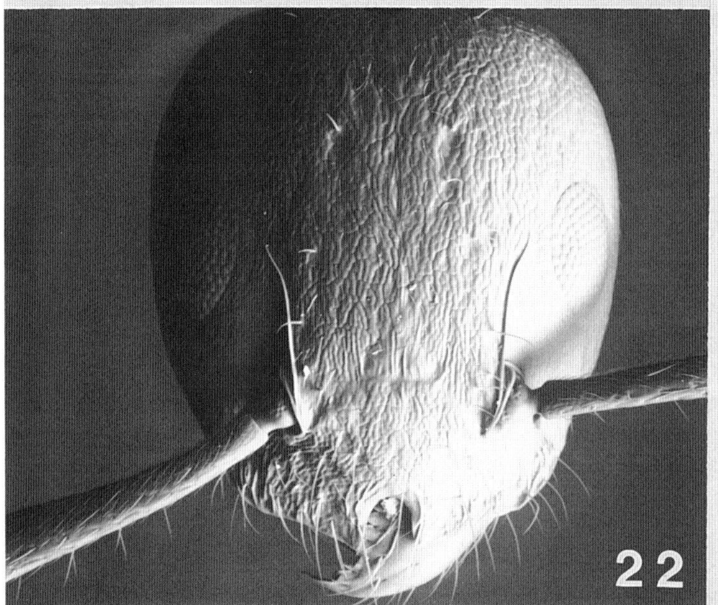

22

23

24

25

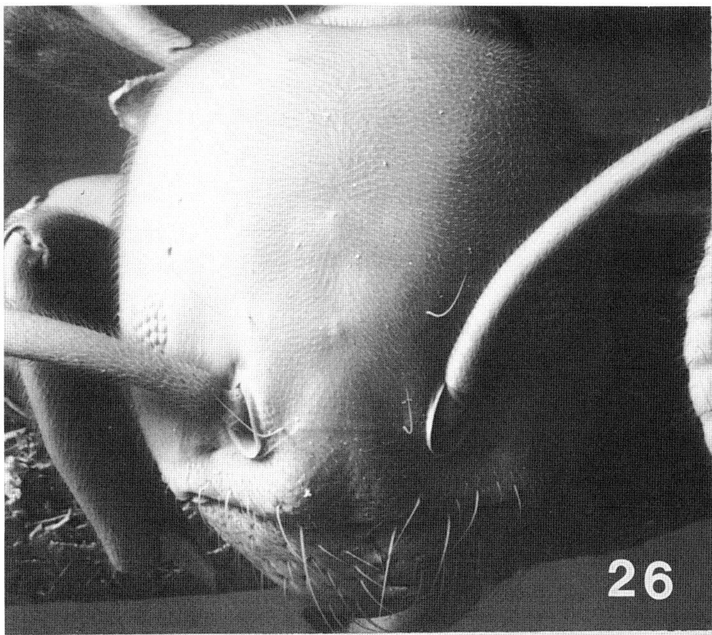
26

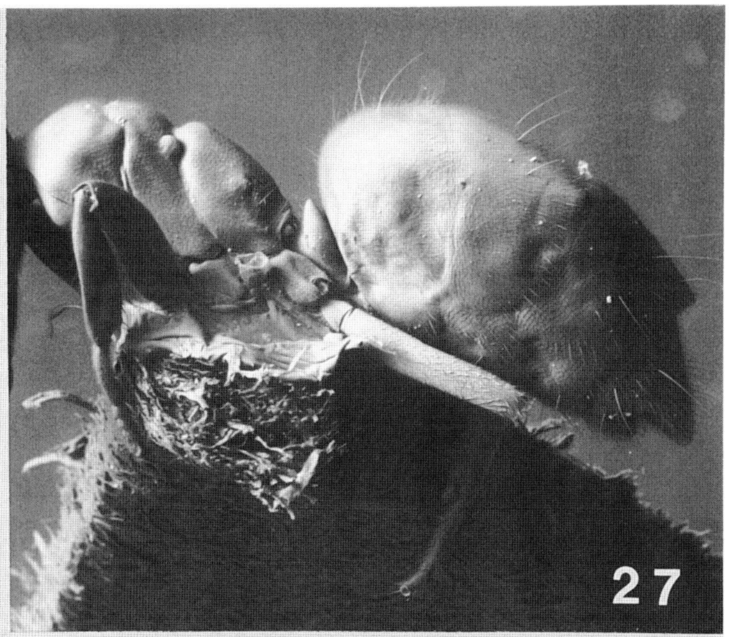
27

28

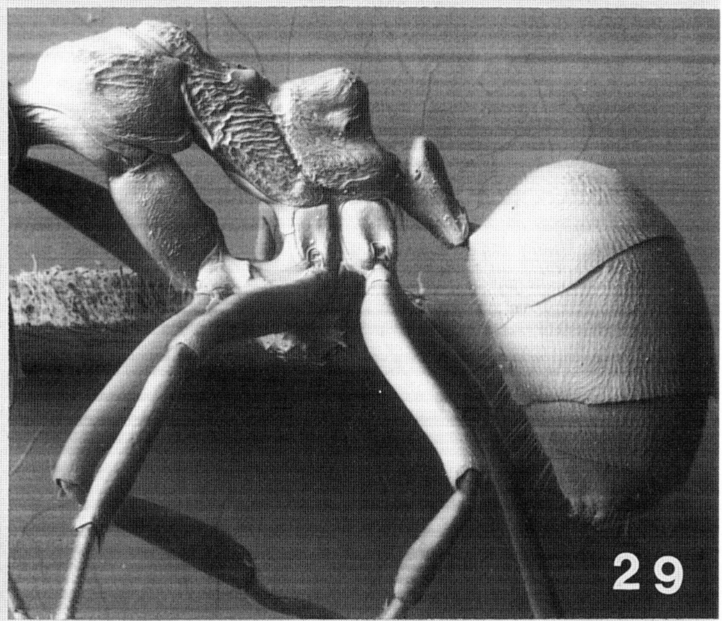
29

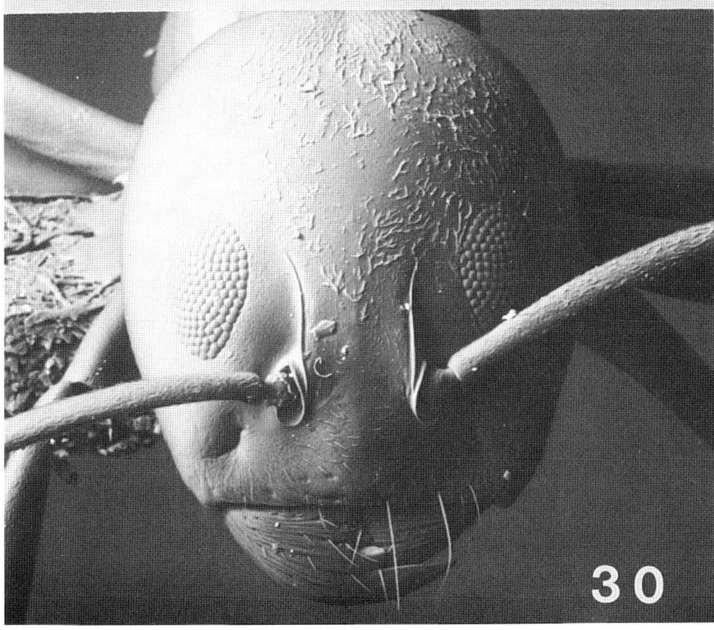
30

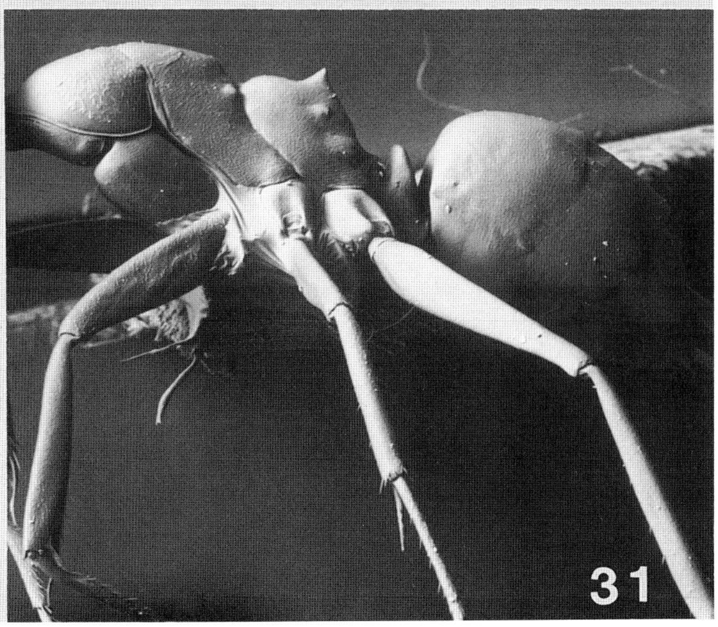
31

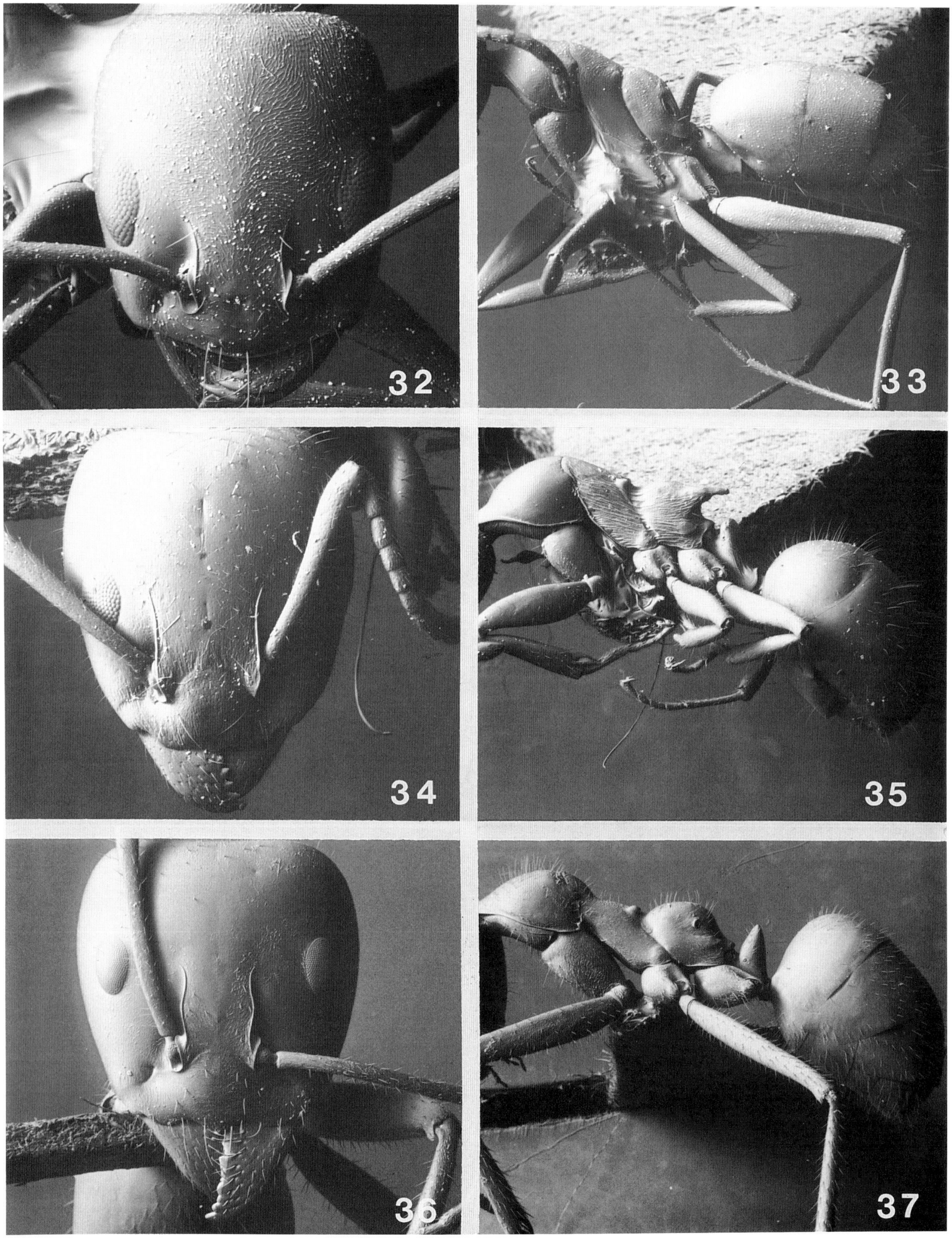
32
33
34
35
36
37

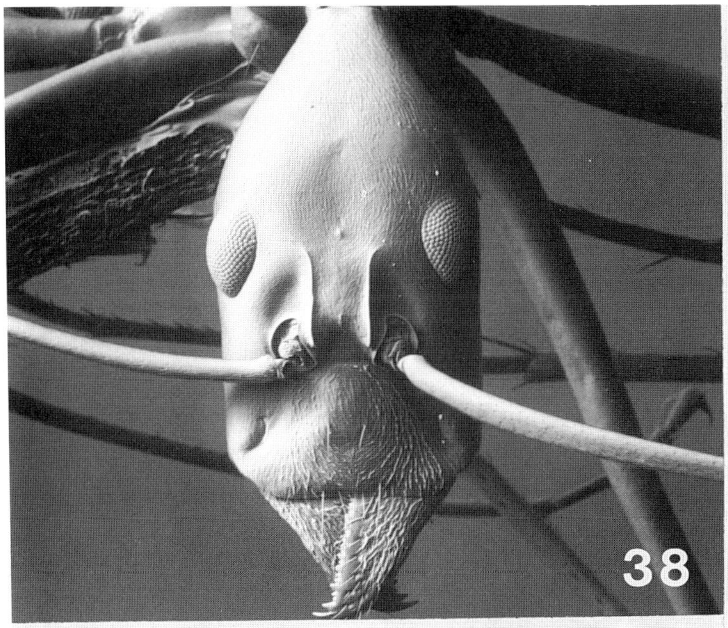
38
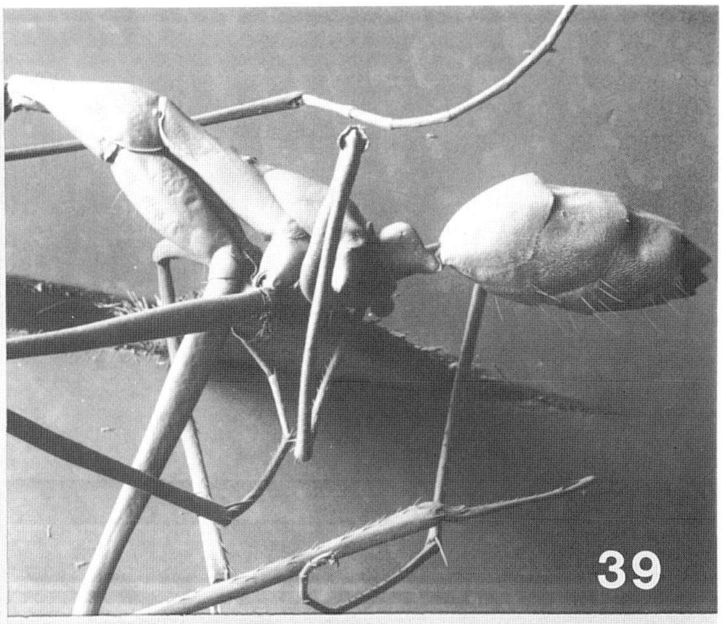
39
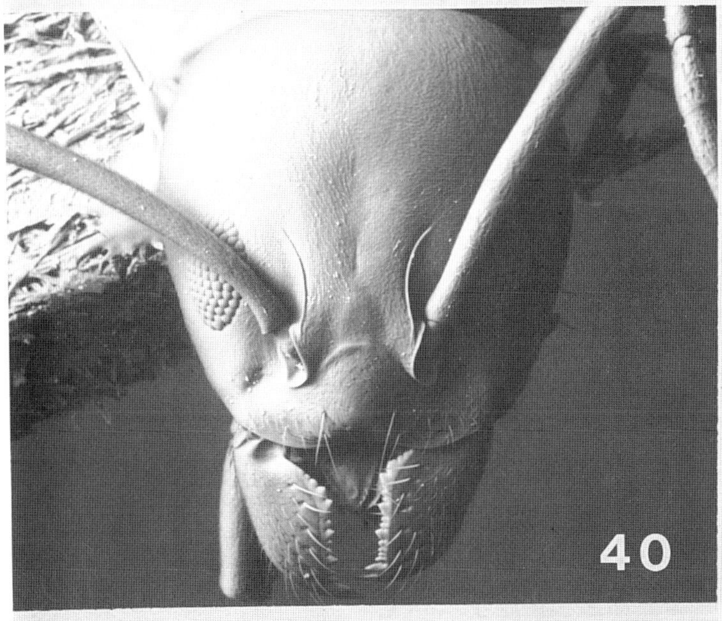
40

41
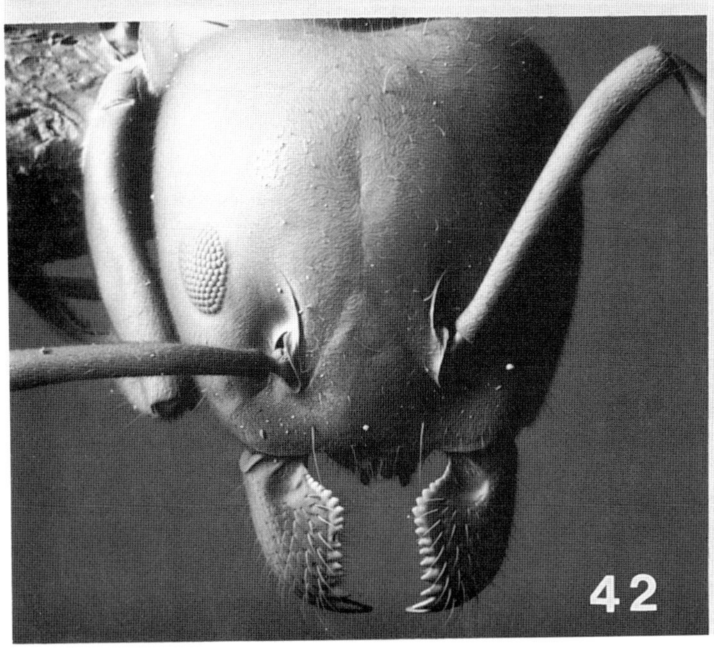
42
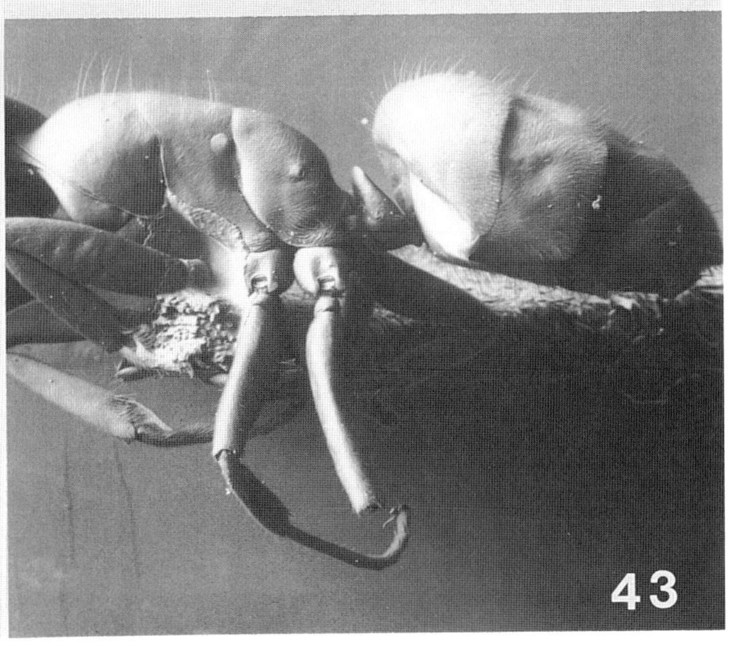
43

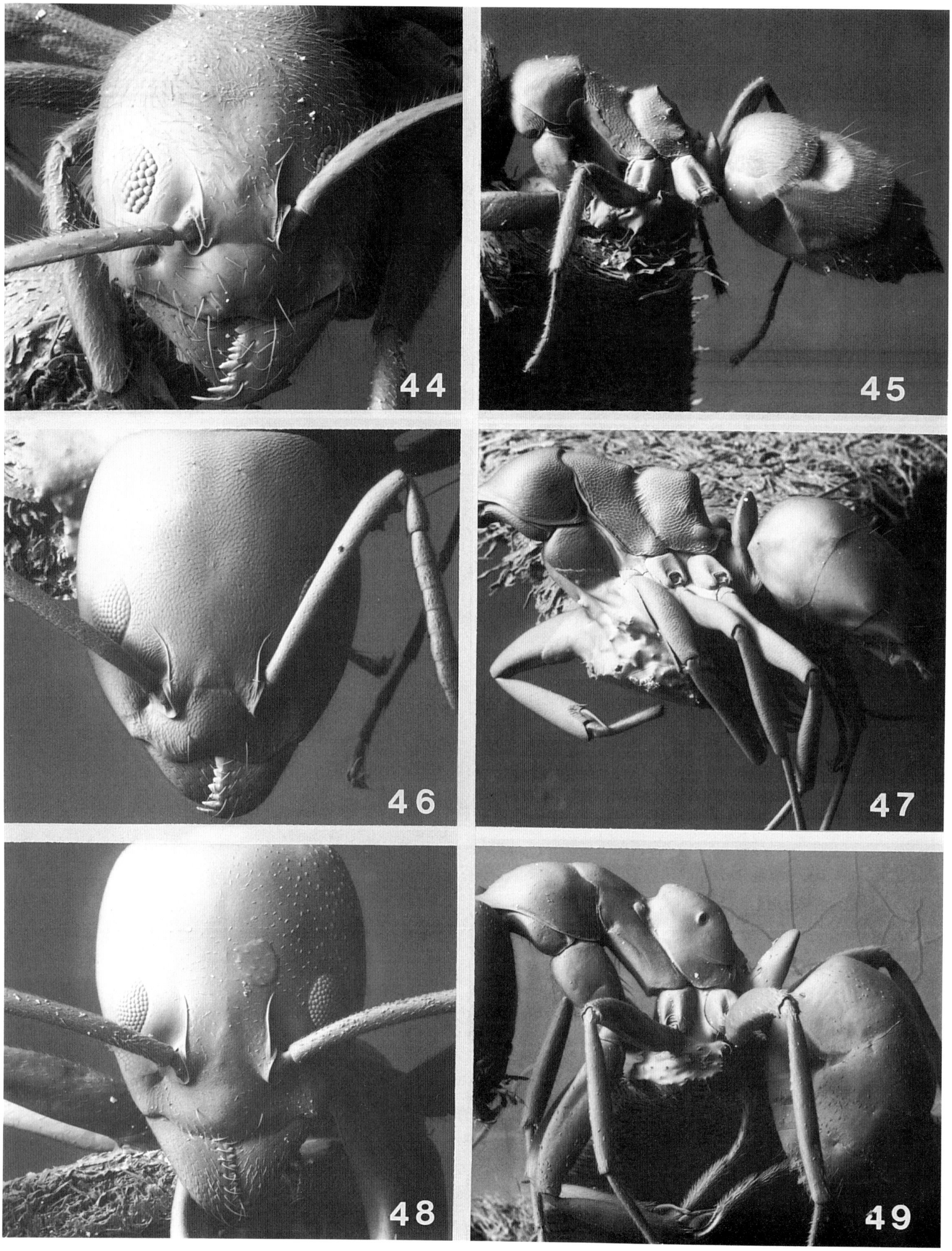
44
45
46
47
48
49

50

51

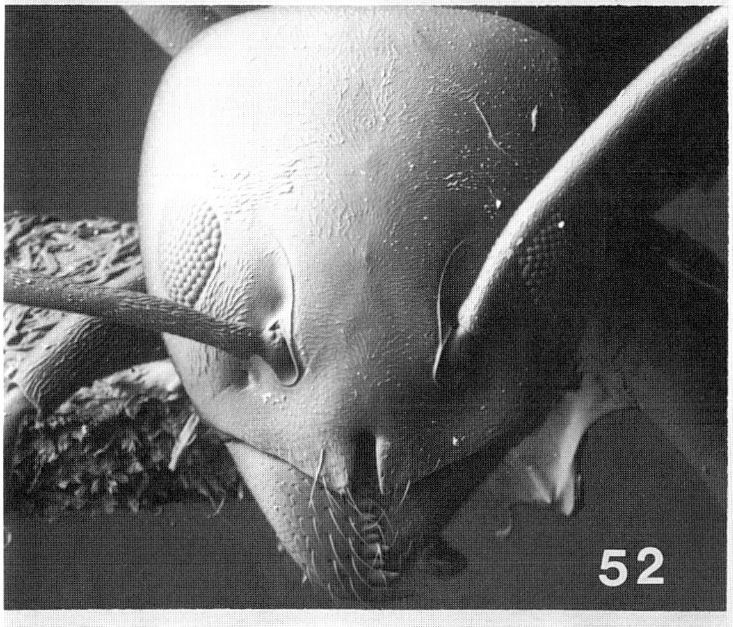

52

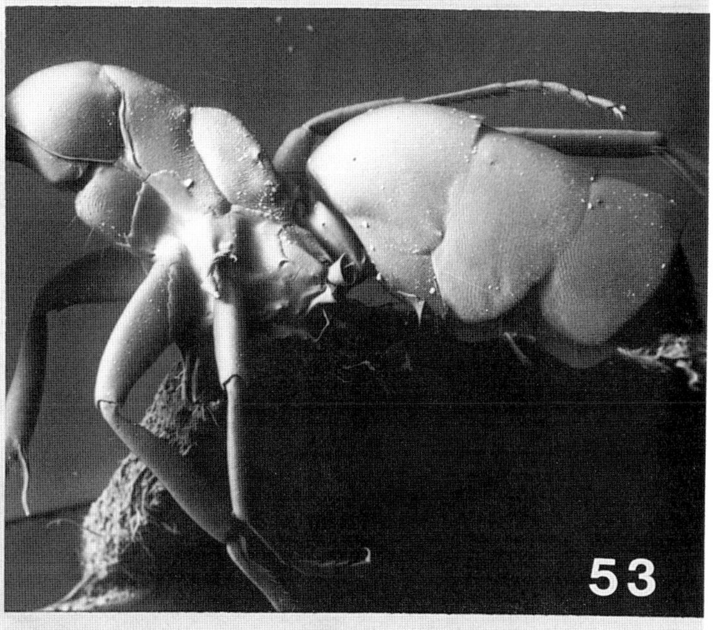

53

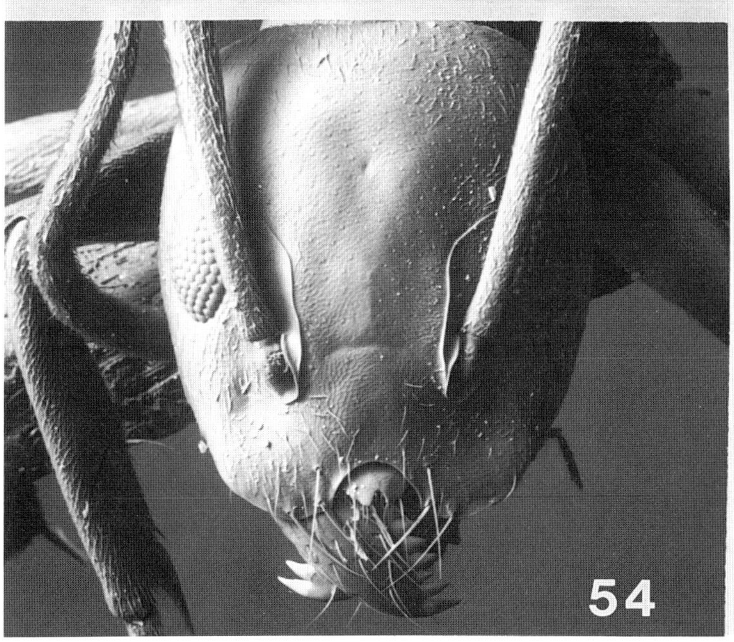

54

55

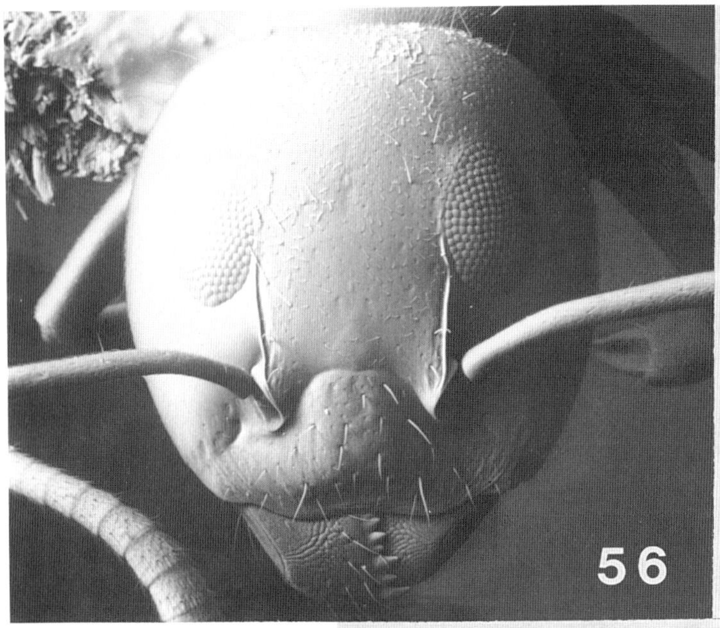
56

57

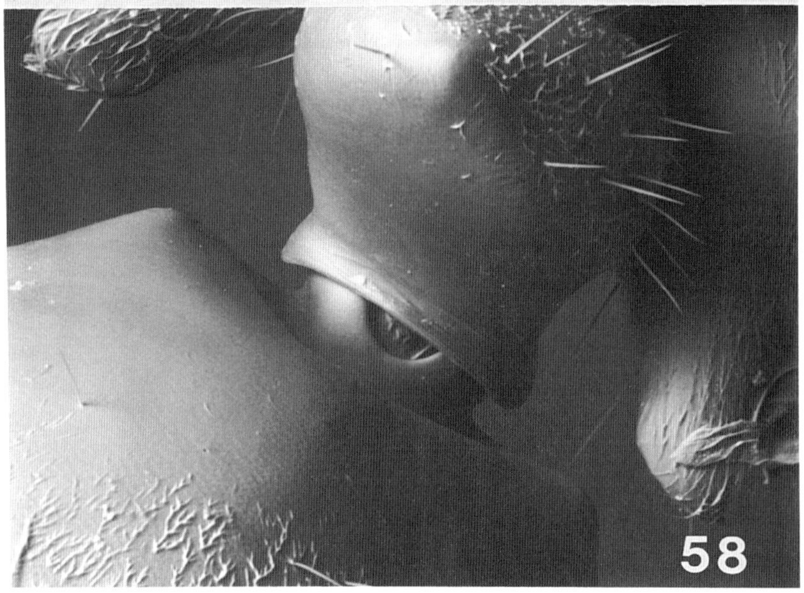
58

Subfamily DORYLINAE

Diagnosis of Worker (Figs. 59–63)

Ants with the following combination of characters together.

1 Clypeus reduced, narrow from front to back especially in front of the antennal insertions, bringing the antennal sockets very close to the anterior margin of the head.

2 Antennal sockets horizontal, in the plane of the transverse axis of the head, exposed in full-face view.

3 Frontal lobes vestigial to absent; usually narrow vertical carinae between the antennal sockets are all that are present.

4 Narrow neck joining condylar bulb of antennal scape to shaft of scape proper straight, not sharply angled or bent downwards in frontal or full-face view.

5 Eyes absent; antenna with 7–12 segments.

6 Promesonotal suture present and conspicuous but the suture fused so that pronotum and mesonotum are immobile with respect to one another.

7 Metapleural gland orifice in lower posterior corner of metapleuron, opening laterally, the orifice concealed behind a ventrally directed cuticular flange or flap.

8 Propodeal lobes absent.

9 Propodeal spiracle situated high on side and far forward; the spiracle subtended by a longitudinal impression and an endophragmal pit.

10 Metatibial glands present.

11 Metacoxal cavities closed; cuticular annulus around each cavity broad and complete, the annulus not broken medioventrally nor with a flexible suture traversing the annulus from the coxal cavity to the cavity in which the petiole articulates.

12 Waist of a single, separated segment, the petiole (= abdominal segment 2).

13 Petiole sessile to subsessile, the tergite and sternite not fused; sternite of petiole with a simple posterior margin and simple articulation to the third abdominal segment.

14 Abdominal stridulatory system absent.

15 Abdominal spiracles 5–7 (= gastral spiracles 3–5) shifted backwards, not concealed by the posterior margins of the preceding segments and visible without distension or dissection of the abdomen.

16 Helcium sternite large and convex, bulging ventrally, visible in profile in normally mounted specimens; tergite of helcium lacking a notch or impression in its dorsal margin anteriorly.

17 Abdominal segment 3 (= gastral segment 1) with tergosternal fusion; tergites and sternites of following abdominal segments (4–7 = gastral segments 2–5) not fused.

18 Abdominal segment 4 (= gastral segment 2) with presclerites sharply defined and differentiated from the postsclerites, the former fitting tightly within the posterior end of the third segment.

19 Strongly differentiated presclerites present on abdominal segments 5–7 (= gastral segments 3–5).

20 Pygidium (tergite of abdominal segment 7 = gastral segment 5) large, flattened, or impressed dorsally; the lateral margin of the impressed area armed with a single pair of teeth or short spines posteriorly (Figs. 61–63).

21 Sting reduced and nonfunctional as a weapon.

Synoptic Classification

Subfamily **DORYLINAE.**

Tribe **Dorylini.** Genus: ***Dorylus*** (Figs. 59–63) [subgenera: nominal plus *Alaopone* (= *Shuckardia*), *Anomma* (= *Sphegomyrmex*, = *Sphecomyrmex* (misspelling)), *Dichthadia, Rhogmus, Typhlopone* (= *Cosmaecetes*, = *Cosmaegetes* (misspelling))].

[Material of the unavailable name Eudorylinae is referable to Dorylinae.]

Distribution

The single genus of this subfamily is found in the southern Palaearctic, Afrotropical, Oriental, and Indo-Australian regions, with the vast majority of species in the Afrotropical. It is absent from the Malagasy and Australasian regions and does not occur anywhere in the New World.

Taxonomic References

Identification of extant species
Dorylus: Wilson (1964) [Oriental, Indo-Australian].

Other taxonomic references
Dorylinae: Brown (1973); Snelling (1981); Gotwald (1982); Hölldobler and Wilson (1990); Bolton (1990c); Baroni Urbani, Bolton, and Ward (1992).

Figures 59–63 DORYLINAE workers. Figs. 59–60, head in full-face view and body in profile of *Dorylus.* Figs. 61–63, gastral apex in *Dorylus* to show pygidium (61–62, in lateral views; 63, in dorsal view).

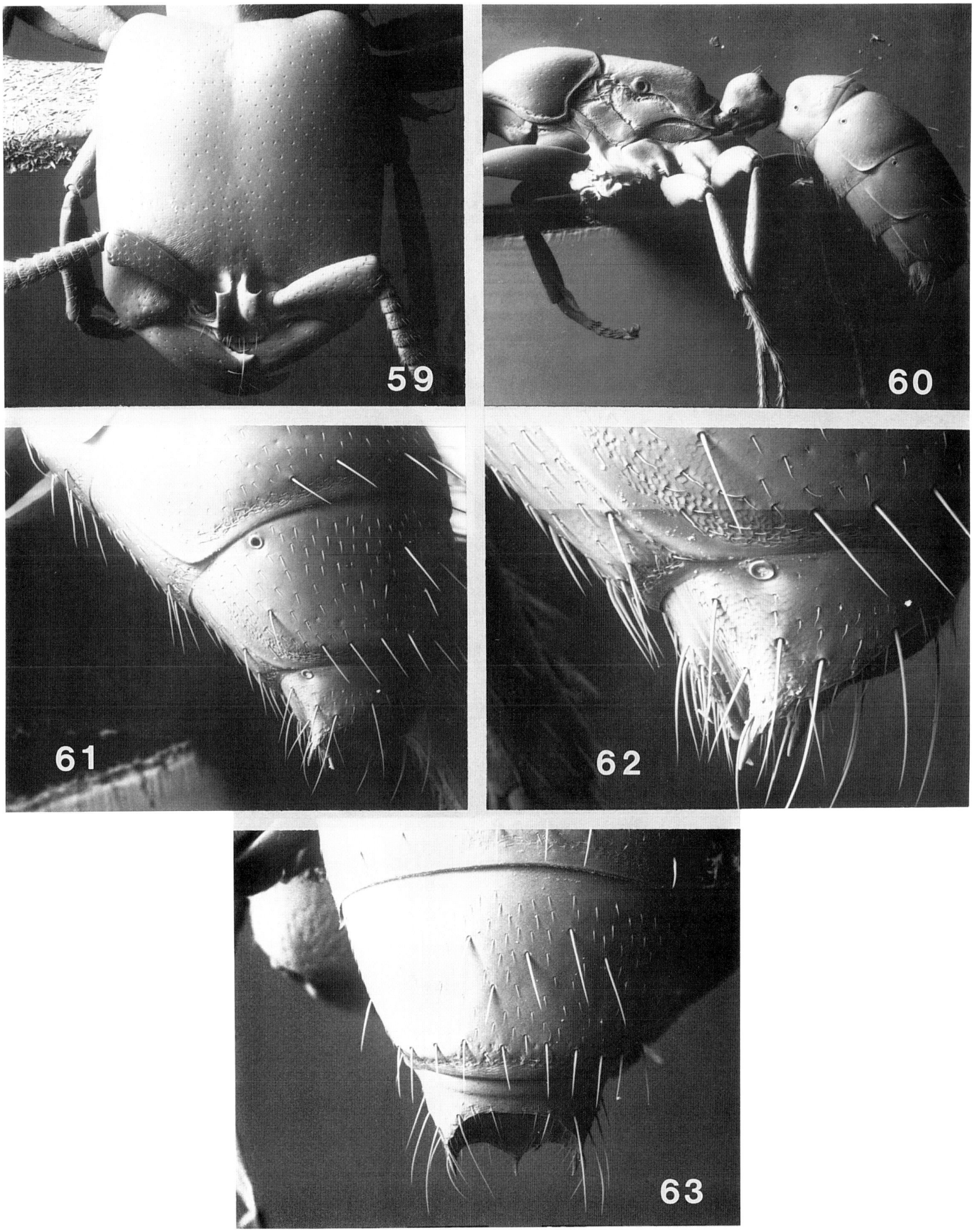
59
60
61
62
63

Subfamily ECITONINAE

Diagnosis of Worker (Figs. 64–73)

Ants with the following combination of characters together.

1 Clypeus reduced, narrow from front to back especially in front of the antennal insertions, bringing the antennal sockets close to the anterior margin of the head.

2 Antennal sockets horizontal, in the plane of the transverse axis of the head, exposed in full-face view.

3 Frontal lobes absent; usually narrow vertical carinae between the antennal sockets are all that are present.

4 Narrow neck joining condylar bulb of antennal scape to shaft of scape proper straight, not sharply angled or bent downwards in frontal or full-face view.

5 Eyes represented by a small transparent blister, by a single ommatidium, or vestigial, or absent; antenna with 12 segments.

6 Promesonotal suture absent, the pronotum and mesonotum fused together.

7 Metapleural gland orifice in lower posterior corner of metapleuron, opening laterally, the orifice concealed behind a ventrally directed cuticular flange or flap.

8 Propodeal lobes present.

9 Propodeal spiracle situated high on the side and far forward.

10 Metatibial glands present.

11 Metacoxal cavities closed; cuticular annulus around each cavity broad and complete, the annulus not broken medioventrally nor with a flexible suture traversing the annulus from the coxal cavity to the cavity in which the petiole articulates.

12 Waist of 1 or 2 separated segments; petiole, or petiole plus postpetiole (= abdominal segment 2, or 2 plus 3).

13 Petiole sessile to subsessile, the tergite and sternite not fused; sternite of petiole with a simple posterior margin and simple articulation to the third abdominal segment.

14 Abdominal stridulatory system absent.

15 Abdominal spiracles 5–7 shifted backwards, not concealed by the posterior margins of the preceding segments, and visible without distension or dissection of the abdomen.

16 Orifices of spiracles on abdominal segments 5–7 oval to slit-shaped and directed posteriorly.

17 Helcium sternite large and convex, bulging ventrally, visible in profile in normally mounted specimens; tergite of helcium lacking a notch or impression in its dorsal margin anteriorly.

18 Abdominal segment 3 (the segment following the petiole) with tergosternal fusion; tergites and sternites of following abdominal segments (4–7) not fused.

19 Abdominal segment 4 with presclerites sharply defined and differentiated from the postsclerites, the former fitting tightly within the posterior end of the third segment.

20 Pygidium (tergite of abdominal segment 7, the last visible tergite) small or very small, usually reduced to a narrow, U-shaped sclerite.

21 Sting strongly developed and functional.

Key to World ECITONINAE (Workers)

1 One separated, reduced segment (the petiole) between alitrunk and gaster. Abdominal segment 3 (= first gastral segment) not constricted behind; no separate postpetiole delimited (Fig. 65). (Neotropical) . . . . ***Cheliomyrmex***

— Two separated, reduced segments between alitrunk and gaster. Abdominal segment 3 deeply constricted behind, delimiting a postpetiole, so that abdominal segment 4 is the first gastral segment (Figs. 67, 69, 71, 73) **2**

2 Pretarsal claws of middle and hind legs lacking a preapical tooth on their inner curvatures (as Fig. 514). (Neotropical, Nearctic) . . . . ***Neivamyrmex***

— Pretarsal claws of middle and hind legs with a distinct preapical tooth on their inner curvatures, the tooth usually near the midlength of each claw (as Fig. 515) **3**

3 Limits of metatibial gland visible, the gland appearing as an elongate strip on the ventral surface of the hind tibia immediately behind the spur socket; surface of gland area with distinctly different sculpture, color, or texture from the surrounding cuticle **4**

— Limits of metatibial gland not discernible; no elongate strip of glandular tissue on ventral surface of hind tibia immediately behind spur socket; cuticle of ventral metatibia everywhere uniform in sculpture and texture. (Neotropical, extreme southern Nearctic) . . . . ***Nomamyrmex***

4 Propodeum in profile equipped with a pair of teeth, spines, or projecting lamellae where the dorsal surface meets the declivity (Fig. 67). (Neotropical) . . . . ***Eciton***

— Propodeum in profile with the dorsal surface meeting the declivity in a rounded angle; without teeth, spines, or projecting lamellae (Fig. 69). (Neotropical, southern Nearctic) . . . . ***Labidus***

Synoptic Classification

Subfamily **ECITONINAE.**

Tribe **Cheliomyrmecini.** Genus: ***Cheliomyrmex*** (Figs. 64, 65).

Tribe **Ecitonini.** Genera: ***Eciton*** (Figs. 66, 67) (= *Ancylognathus*, = *Camptognatha*, = *Holopone*, = *Mayromyrmex*), ***Labidus*** (Figs. 68, 69) (= *Nycteresia*, = *Pseudodichthadia*), ***Neivamyrmex*** (Figs. 70, 71) (= *Acamatus* (homonym), = *Woitkowskia*), ***Nomamyrmex*** (Figs. 72, 73).

[Material of the unavailable name Metadorylinae is in part referable to Ecitonini.]

Distribution

The five genera of this subfamily are distributed entirely in the New World, with a vast preponderance of species in the Neotropical region. The genera *Cheliomyrmex* and *Eciton* are restricted to this region but a few species of *Neivamyrmex*, *Nomamyrmex*, and *Labidus* occur in the Nearctic region. Only the first of these three can claim a fairly wide distribution within the Nearctic, the last two being restricted to the extreme south of the region.

Taxonomic References

Identification of extant species

Ecitoninae: Borgmeier (1955) [world]; Watkins (1976) [world]; Watkins (1982) [Mexico]. ***Neivamyrmex:*** Watkins (1972) [U.S.A.]; ***Nomamyrmex:*** Watkins (1977).

Other taxonomic references

Ecitoninae: Brown (1973); Snelling (1981); Gotwald (1982); Hölldobler and Wilson (1990); Bolton (1990c); Baroni Urbani, Bolton, and Ward (1992).

Figures 64–73 ECITONINAE workers, heads in full-face view and bodies in profile:

64–65, **Cheliomyrmecini,** *Cheliomyrmex*

66–73, **Ecitonini:** 66–67, *Eciton;* 68–69, *Labidus;* 70–71, *Neivamyrmex;* 72–73, *Nomamyrmex.*

64

65

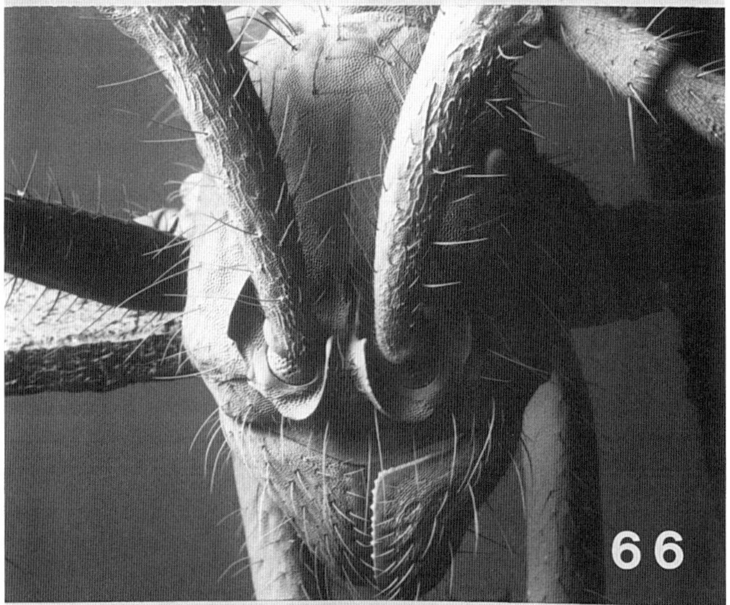
66

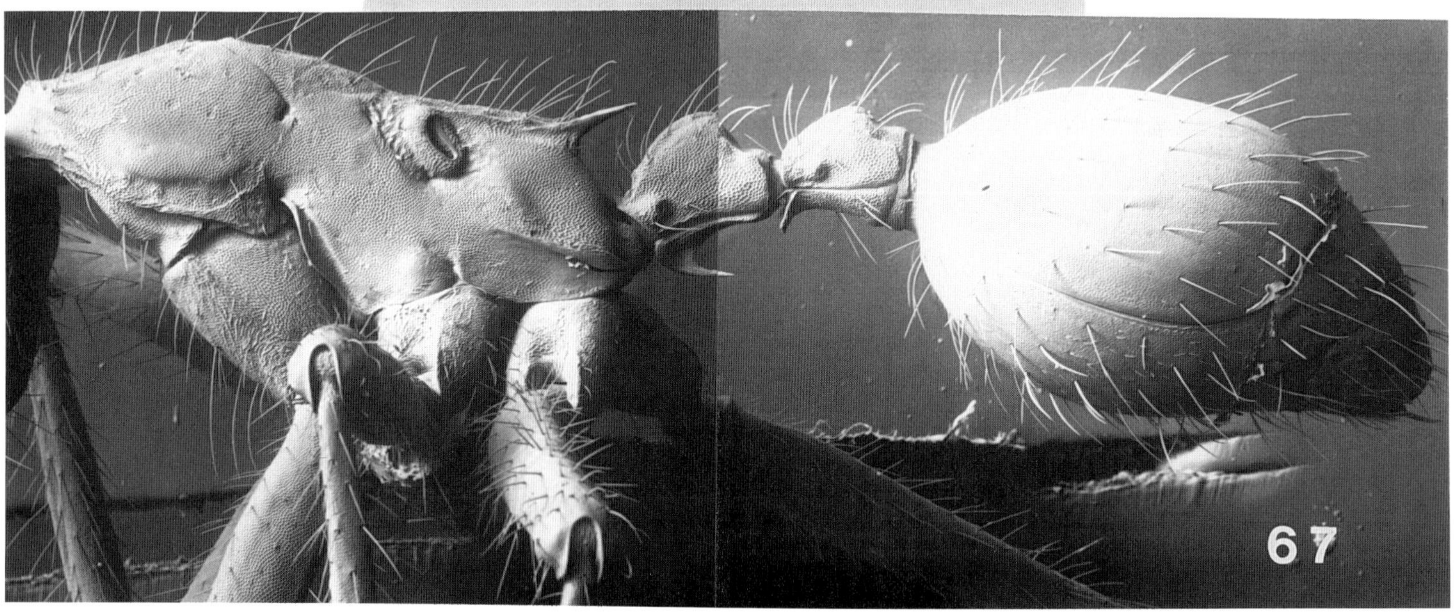
67

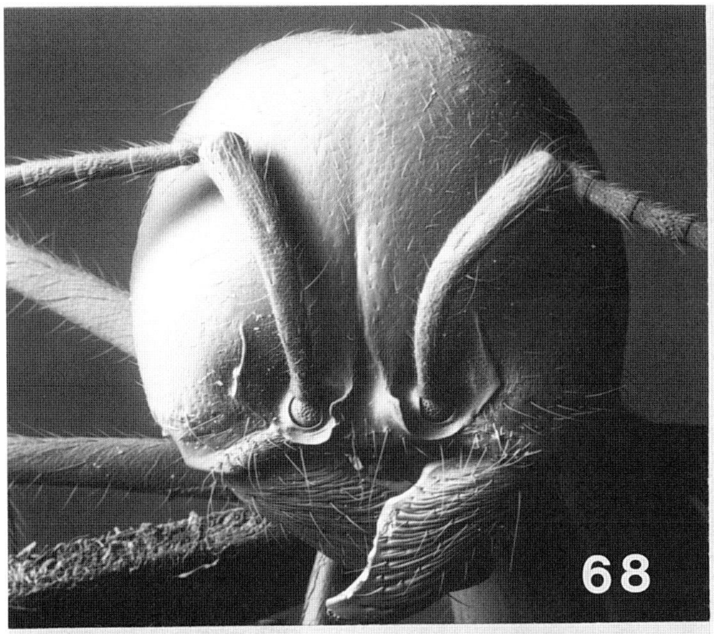
68

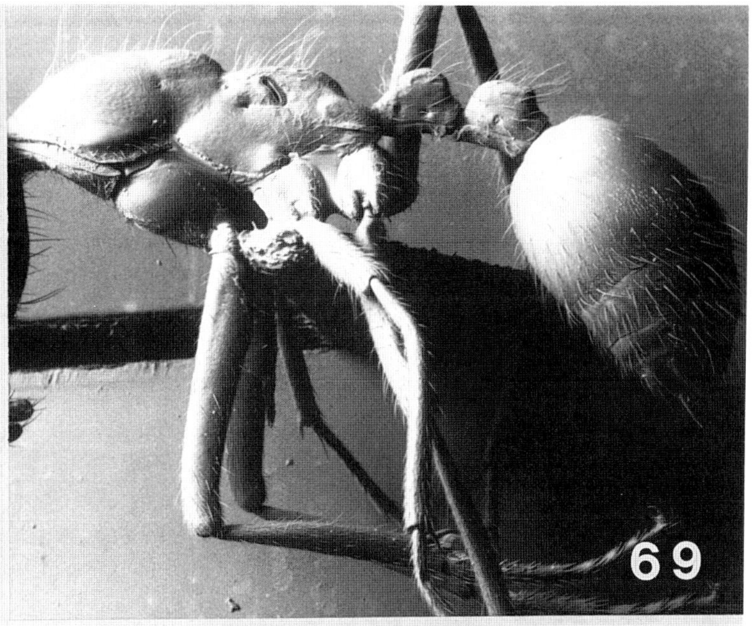
69

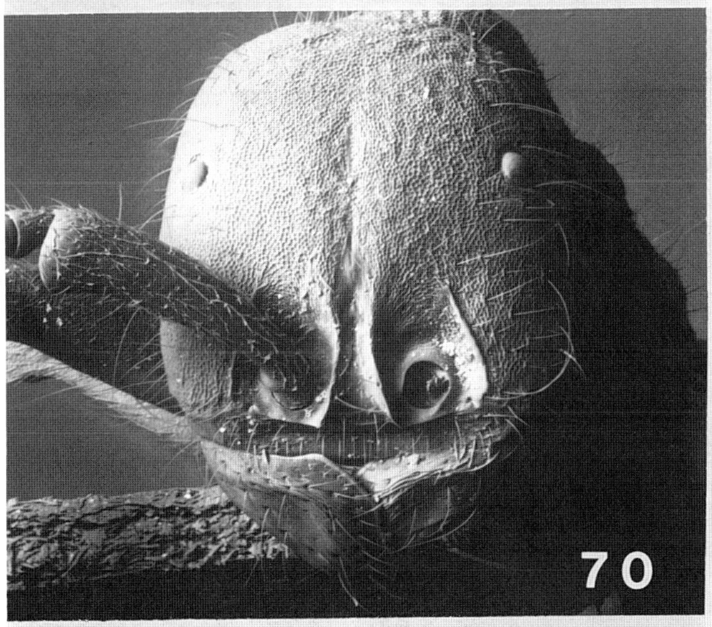
70

71

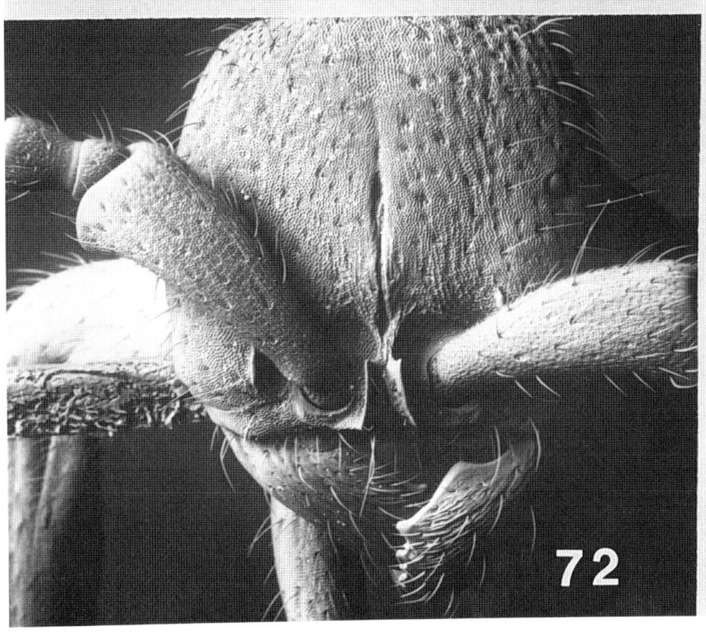
72

73

Subfamily FORMICINAE

Diagnosis of Worker (Figs. 74–165)

Ants with the following combination of characters together.

1 Clypeus broad from front to back so that antennal sockets are well behind anterior margin of head. Median portion of clypeus usually not extended backwards between the frontal carinae, rarely otherwise. Usually a postclypeal frontal triangle present that may project back between the frontal carinae or antennal sockets.

2 Antennal sockets inclined, the portion of the socket margin and torulus closest to the dorsal midline of the head on a higher level than the portion of the margin most distant from the midline.

3 Frontal carinae usually present, rarely absent; when present varying from a pair of simple carinae or the margins of a raised plateau to elevated broad flanges, which however only rarely completely conceal the antennal sockets; the latter are usually partly or wholly exposed. Only very rarely the carinae expanded into frontal lobes that conceal the antennal sockets.

4 Narrow neck joining condylar bulb of antennal scape to shaft of scape proper straight.

5 Eyes usually present, only rarely vestigial or absent; ocelli sometimes present; antenna with 8–12 segments.

6 Promesonotal suture usually present and flexible but sometimes fused; more rarely the suture vestigial or absent.

7 Metapleural gland orifice present or absent; when present situated in lower posterior corner of metapleuron, opening laterally or posterolaterally; the orifice commonly with numerous guard setae crossing its aperture.

8 Metanotum and its spiracles frequently present on dorsal alitrunk.

9 Metacoxal cavities closed; a thin, continuous strip of cuticle separates the metacoxal cavity from the cavity in which the petiole articulates.

10 Propodeal lobes absent.

11 Waist of 1 segment, the petiole (= abdominal segment 2).

12 Helcium tergite dorsally usually with an extensive U-shaped emargination of its leading edge, the emargination often reaching back almost the whole length of the sclerite; this emargination only very rarely absent. Helcium sternite small, retracted, concealed by the tergite.

13 Abdominal stridulatory system absent.

14 Abdominal segment 4 (= gastral segment 2) without differentiated presclerites.

15 Abdominal spiracles 4–7 (= gastral 2–5) sometimes concealed by posterior margins of preceding tergites; frequently abdominal spiracles 4–5 visible and sometimes also 6–7 visible without distension of the abdomen.

16 Abdominal segment 2 (petiole) with tergosternal fusion; abdominal segments 3–7 (= gastral segments 1–5) without tergosternal fusion.

17 Pygidium (tergite of abdominal segment 7 = gastral segment 5) usually large, simple.

18 Hypopygium (sternite of abdominal segment 7 = gastral segment 5) with a U-shaped to almost circular acidopore present apically. Acidopore may be at apex of a nozzle-like extension of the hypopygium and equipped apically with a circlet of hairs (Fig. 160), or may be merely an emargination of the hypopygial apical margin. In the latter case the acidopore may be concealed by the pygidial apex when not in use, and consequently difficult to see without opening out the 2 sclerites.

19 Sting absent, replaced by formic-acid-projecting system of which the acidopore is the orifice.

Key to Palaearctic FORMICINAE (Workers)

1 Antenna with 9 segments . . . . ***Brachymyrmex***
— Antenna with 11 or 12 segments **2**

2 Antenna with 11 segments **3**
— Antenna with 12 segments **6**

3 Propodeum armed with a pair of spines, teeth, or tubercles (Fig. 155). Dorsal edge of petiole usually armed with a pair of teeth or spines but sometimes only emarginate . . . . ***Lepisiota***
— Propodeum and petiole unarmed, without trace of spines, teeth, or tubercles (Figs. 151, 153) **4**

4 Palp formula 6,4 .... **5**
— Palp formula 5,3 or less .... ***Acropyga***

5 With alitrunk in dorsal view the mesonotum separated from the metanotum by a conspicuous transverse groove or impression, so that the metanotum forms a distinctly isolated sclerite (also visible in profile, Fig. 157) .... ***Plagiolepis***
— With alitrunk in dorsal view the mesonotum fused with the metanotum, the two not separated by a transverse impression or groove, the metanotum not forming an isolated sclerite (also visible in profile, Fig. 153) .... ***Anoplolepis***

6 Metapleuron with a distinct, wide orifice for the metapleural gland, situated above the hind coxa and below the level of the propodeal spiracle (Fig. 164). Orifice of metapleural gland protected by a line or tuft of guard setae, which are usually very conspicuous. Antennal sockets situated very close to posterior margin of clypeus (Figs. 104, 108, 120, 124) . . . **7**
— Metapleural gland orifice absent, the surface of the metapleuron uninterrupted by a gland orifice above the hind coxa and below the level of the propodeal spiracle (Fig. 165). Antennal sockets situated well behind the posterior margin of the clypeus (Figs. 84, 98) **17**

7 With gaster in ventral view the first sternite with a transverse sulcus immediately behind the helcium (Figs. 161, 162). Hind coxae close together in ventral view, their inner margins touching or almost touching when the coxae are directed outwards. Orifice of propodeal spiracle oval, elliptical, or an elongate slit, which is near-vertical or inclined from the vertical (Figs. 164, 103, 105, 107, 109, 111). With alitrunk in absolute profile the propodeal spiracle well in front of the point where the propodeal side rounds into the declivity . **8**
— With gaster in ventral view the first sternite entire, without a transverse sulcus behind the helcium (Fig. 163). Hind coxae widely separated in ventral view, their inner margins far apart when the coxae are directed outwards. Orifice of propodeal spiracle circular to subcircular (Figs. 121, 125, 127). With alitrunk in absolute profile the propodeal spiracle bordering or actually on the curvature where the propodeal side rounds into the declivity **15**

8 Mandible a narrow, sickle-like to sabre-like curved blade that tapers to a sharp apical point (Fig. 106). Inner margin of each blade with a single tooth or with a minutely denticulate or weakly jagged margin; without conspicuously differentiated serial teeth **9**
— Mandible triangular to elongate-triangular and armed with 5 or more conspicuously differentiated serial teeth (Figs. 100, 102, 104, 108, 110) **10**

9 Palp formula 6,4. Maxillary palpi extremely long and very distinct .... ***Cataglyphis*** (part)
— Palp formula 4,2. Maxillary palpi short and inconspicuous .... ***Polyergus***

10 Apical (masticatory) margin of mandible usually with 8 teeth but sometimes with more. Third tooth of mandible, counting from the apex, always distinctly smaller and shorter than the fourth; the fourth tooth larger than all the remaining teeth to the basal angle .... ***Formica***
— Apical (masticatory) margin of mandible usually with 5–7 teeth, only very rarely with more. If more than 5 teeth present then the third tooth, counting from the apex, larger and longer than the fourth; teeth after the fourth either irregular or evenly decreasing to the basal angle **11**

11 Propodeal spiracle an elongate, narrow slit (Fig. 103). Basal segment of maxillary palp flattened. Setae at base of stipes longer than half the length of the stipes . ***Cataglyphis*** (part)
— Propodeal spiracle oval to elliptical (Figs. 101, 109, 111, and as Fig. 164), not an elongate, narrow slit. Basal segment of maxillary palp round. Setae at base of stipes shorter than half the length of the stipes **12**

12 Petiole squamiform, with lateral and dorsal crests (Fig. 109) **13**
— Petiole blocky to high-nodiform, lacking lateral and dorsal crests (Figs. 101, 111) **14**

13 Mandible with 5 teeth which evenly decrease in size from apex to base so that the third tooth is larger than the fourth (Fig. 108). Workers dimorphic .... ***Proformica***
— Mandible with 5 teeth which do not evenly decrease in size from apex to base; instead the third tooth is distinctly smaller than the fourth. Workers monomorphic .... ***Bajcaridris*** [*Note:* this name is unavailable here; it will be formally described by D. Agosti in a forthcoming publication.]

14 Mandible with 7 or 8 teeth. With head in full-face view the occipital margin concave and the occipital corners projecting as a pair of blunt, posteriorly directed processes (Fig. 110). Body surfaces smooth and shining. Frontal carinae reduced to a pair of minute ridges .... ***Rossomyrmex***
— Mandible with 5 teeth. With head in full-face view the occipital margin strongly convex, without differentiated, projecting occipital corners (Fig. 100). Body surfaces sculptured and dull. Frontal carinae conspicuous .... ***Alloformica***

15 With the head in full-face view the eyes at or in front of the midlength of the sides of the head (Fig. 124). Head and alitrunk with stout, coarse setae arranged in distinct pairs (Fig. 125) .... ***Paratrechina***
— With the head in full-face view the eyes distinctly behind the midlength of the side of the head (Figs. 120, 126). Setae on head and alitrunk not distinctly paired (Figs. 121, 127) . **16**

16 Mandible with 6 teeth, very rarely with 7. Anterior face of first gastral segment broadly transversely concave throughout its height (Fig. 127). Antennal scapes relatively very long; when laid straight back from their insertions at least half of their length projects beyond the occipital margin. Mesothorax constricted immediately behind pronotum (Fig. 127) .... ***Prenolepis***
— Mandible with at least 7 teeth, usually with more than 7. An-

terior face of first gastral segment with a small concave area immediately above the helcium, but the face not broadly transversely concave throughout its height (Fig. 121). Antennal scapes much shorter; when laid straight back from their insertions with much less than half their length projecting beyond the occipital margin. Mesothorax not constricted immediately behind pronotum (Fig. 121) . . . . ***Lasius***

17 First gastral tergite short, accounting for less than half the length of the gaster in dorsal view or in profile; the first gastral tergite at most only slightly longer than the second (Fig. 85). Petiole a node or scale, never armed with spines or teeth. Propodeum and humeral angles of pronotum unarmed. Polymorphic species . . . . ***Camponotus***

— First gastral tergite large, accounting for at least half the length of the gaster in dorsal view or in profile; the first gastral tergite distinctly much longer than the second (Fig. 99). Petiole armed with 1 or 2 pairs of teeth or spines. Propodeum and humeral angles of pronotum usually armed with teeth or spines. Monomorphic species . . . . ***Polyrhachis***

Key to Afrotropical and Malagasy FORMICINAE (Workers)

1 Antenna with 9 segments **2**

— Antenna with 10–12 segments **4**

2 Palp formula 6,4. Basal margin of mandible edentate . . . . ***Brachymyrmex***

— Palp formula less than 6,4 (either 5,3 or 3,3). Basal margin of mandible with a small tooth **3**

3 Palp formula 5,3. Polymorphic species with roughly rectangular head capsule in full-face view (Fig. 74). In this view the eyes situated a considerable distance in from the lateral margins of the head . . . . ***Aphomomyrmex***

— Palp formula 3,3. Monomorphic species with heart-shaped head capsule in full-face view (Fig. 80). In this view the eyes situated very close to the lateral margins of the head . . . . ***Petalomyrmex***

4 Antenna with 10 segments . . . . ***Agraulomyrmex***

— Antenna with 11 or 12 segments **5**

5 Antenna with 11 segments **6**

— Antenna with 12 segments **9**

6 Propodeum armed with a pair of spines, teeth, or tubercles (Fig. 155). Dorsal edge of petiole usually armed with a pair of teeth or spines but sometimes only emarginate . . . . ***Lepisiota***

— Propodeum and petiole unarmed, without trace of spines, teeth, or tubercles (Figs. 151, 153) **7**

7 Palp formula 6,4 . . . . **8**

— Palp formula 5,3 or less . . . . ***Acropyga***

8 With alitrunk in dorsal view the mesonotum separated from the metanotum by a conspicuous transverse groove or impression, so that the metanotum forms a distinctly isolated sclerite (also visible in profile, Fig. 157) . . . . ***Plagiolepis***

— With alitrunk in dorsal view the mesonotum fused with the metanotum, the two not separated by a transverse groove or impression, the metanotum not forming an isolated sclerite (also visible in profile, Fig. 153) . . . . ***Anoplolepis***

9 Eyes enormous, occupying almost all the side of the head. Ventrolateral margin of head with a tooth at each side. . . . . ***Santschiella***

— Eyes smaller, occupying less than one-half the side of the head. Ventrolateral margin of head unarmed, without a tooth at each side **10**

10 Metapleuron with a distinct, wide orifice for the metapleural gland, situated above the hind coxa and below the level of the propodeal spiracle (Fig. 164). Orifice of metapleural gland protected by a line or tuft of guard setae, which are usually very conspicuous. Antennal sockets situated close to posterior margin of clypeus (Figs. 102, 124, 128) **11**

— Metapleural gland orifice absent, the surface of the metapleuron uninterrupted by a gland orifice above the hind coxa and below the level of the propodeal spiracle (Fig. 165). Antennal sockets situated a good distance behind the posterior clypeal margin (Figs. 84, 96, 98, 148) **13**

11 Propodeal spiracle an elongate vertical or near-vertical ellipse or slit (Fig. 103). Ocelli present (Fig. 102). First gastral sternite in ventral view with a transverse sulcus immediately behind the helcium (Fig. 162) . . . . ***Cataglyphis***

— Propodeal spiracle circular to subcircular, usually small (Figs. 125, 129). Ocelli absent (Figs. 124, 128). First gastral sternite immediately behind the helcium entire, without a transverse sulcus **12**

12 Palp formula 6,4. Eyes large and very conspicuous. Dorsal surfaces of head and body with distinctly paired coarse setae (Fig. 125) . . . . ***Paratrechina***

— Palp formula 3,4; 3,3; or 3,2. Eyes small to absent [in Afrotropical species]. Dorsal surfaces of head and body without distinctly paired coarse setae (Fig. 129) . . . . ***Pseudolasius***

13 Palp formula 5,4. Mandible with 10 or more teeth or denticles altogether. Apical tooth disproportionately large and the fourth tooth from the apex much larger than the third and the fifth tooth (Fig. 148). Petiole reduced to an elongate low node, which allows the gaster to be reflexed over the alitrunk (Fig. 149) . . . . ***Oecophylla***

— Palp formula 6,4. Mandible usually with 7 teeth at most, sometimes fewer and only very rarely with more. Whatever the number of teeth they decrease in size from apex to base; the fourth tooth is not enlarged as above (Figs. 84, 96, 98). Petiole an erect node or scale, the gaster not capable of reflexion over the alitrunk (Figs. 85, 97, 99) **14**

14 Polymorphic species. Petiole a node or scale, never armed with teeth or spines. Propodeum unarmed. Humeral angles of pronotum unarmed (Fig. 85) . . . . ***Camponotus***

— Monomorphic species. Petiole armed with teeth, spines, or prominent angles. Propodeum usually bidentate or bispinose but may be unarmed. Humeral angles of pronotum marginate or armed with teeth or spines (Figs. 97, 99) **15**

15 Acidopore formed from the invaginated membranous area of the hypopygium, not fringed with setae, concealed by the pygidium when not in use. Tergite of first gastral segment large, accounting for at least half the length of the gaster in dorsal view or in profile; the first gastral tergite distinctly much larger than the second (Fig. 99) .... ***Polyrhachis***

— Acidopore formed at apex of sclerotized portion of hypopygium, always open and usually fringed with setae; not concealed by pygidium when not in use. Tergite of first gastral segment short, accounting for distinctly less than half the length of the gaster in dorsal view or in profile; the first gastral tergite at most only slightly longer than the second (Fig. 97) .... ***Phasmomyrmex***

Key to Oriental and Indo-Australian FORMICINAE (Workers)

1 Antenna with 8 segments **2**

— Antenna with 9–12 segments **3**

2 Antennal scape, when laid back in its natural resting position, passing below the eye. Apical (masticatory) margin of mandible with more than 4 teeth (Figs. 112–115) . ***Gesomyrmex***

— Antennal scape, when laid back, passing above the eye. Apical (masticatory) margin of mandible with 4 teeth (Figs. 78, 79) .... ***Cladomyrma***

3 Antenna with 9–11 segments **4**

— Antenna with 12 segments **8**

4 Palp formula 5,3 or less .... ***Acropyga***

— Palp formula 6,4 .... **5**

5 Antenna with 9 segments .... ***Brachymyrmex***

— Antenna with 11 segments **6**

6 Propodeum armed with a pair of spines, teeth, or tubercles. Dorsal edge of petiole usually armed with a pair of teeth or spines but sometimes only emarginate (Fig. 155) . ***Lepisiota***

— Propodeum and petiole unarmed, without spines, teeth or tubercles **7**

7 With alitrunk in dorsal view the mesonotum separated from the metanotum by a conspicuous transverse groove or impression, so that the metanotum forms a distinctly isolated sclerite (also visible in profile, Fig. 157) .... ***Plagiolepis***

— With alitrunk in dorsal view the mesonotum fused with the metanotum, the two not separated by a transverse groove or impression, the metanotum not forming an isolated sclerite (also visible in profile, Fig. 153) .... ***Anoplolepis***

8 Mandibles extended into extremely long, slender blades that project far in front of the anterior clypeal margin; the mandibles at least 0.85 times the head length and often exceeding the head length (Fig. 146) .... ***Myrmoteras***

— Mandibles subtriangular to elongate-triangular, not extended into long, slender blades and usually very obviously less than 0.85 times the head length **9**

9 Antennal sockets situated close to the posterior clypeal margin (Figs. 104, 118, 124, 128), and metapleuron with a distinct metapleural gland orifice, the orifice situated above the hind coxa and below the level of the propodeal spiracle (Fig. 164) **10**

— Either antennal sockets situated far behind the posterior clypeal margin (Figs. 82, 94, 98, 148), or the metapleuron lacking a metapleural gland orifice in the position described above (Fig. 165), or sometimes both **16**

10 Maxillary palp with 2–4 segments **11**

— Maxillary palp with 6 segments **12**

11 With alitrunk in profile the mesonotum and anepisternum together forming a roughly triangular, oblique wedge between pronotum and remainder of alitrunk. Lateroventral angle of pronotum posteriorly very nearly touching the katepisternal anterior margin (Fig. 129). Anterior clypeal margin convex or indented medially, not broadly evenly concave. Lateral margin of mandible shallowly curved in apical half; at full closure the apical tooth directed laterally or anterolaterally (Fig. 128) .... ***Pseudolasius***

— With alitrunk in profile the mesonotum and anepisternum together not forming a roughly triangular, oblique wedge between pronotum and remainder of alitrunk. Instead this region narrow and elongated, with a distinct horizontal border ventrally between the lateroventral angle of the pronotum posteriorly and the katepisternal anterior margin (Fig. 119). Anterior clypeal margin broadly and evenly concave across its entire width. External margin of mandible strongly curved in apical half; at full closure the apical tooth directed posterolaterally or posteriorly (Fig. 118) .... ***Euprenolepis***

12 With gaster in ventral view the first sternite with a transverse sulcus immediately behind the helcium (Figs. 161, 162). Orifice of propodeal spiracle oval, elliptical, or an elongate slit, which is near-vertical or inclined from the vertical (Figs. 102, 164). With alitrunk in absolute profile the propodeal spiracle well in front of the point where the propodeal side rounds into the declivity **13**

— With gaster in ventral view the first sternite without a transverse sulcus immediately behind the helcium (Fig. 163). Orifice of propodeal spiracle circular to subcircular (Figs. 121, 125, 127). With alitrunk in absolute profile the propodeal spiracle bordering or actually on the curvature where the propodeal side rounds into the declivity **14**

13 Apical (masticatory) margin of mandible usually with 8 teeth but sometimes with more. Third tooth of mandible, counting from the apex, always distinctly smaller and shorter than the fourth; the fourth tooth larger than all the remaining teeth to the basal angle .... ***Formica***

— Apical (masticatory) margin of mandible with 5–7 teeth. If more than 5 teeth present then the third tooth, counting from the apex, larger and longer than the fourth; teeth after the fourth decreasing in size to the basal angle . . . . ***Cataglyphis***

14 With the head in full-face view the eyes at or in front of the midlength of the sides (Fig. 124). Head and alitrunk with stout setae arranged in distinct pairs (Fig. 125) . . . . ***Paratrechina***

— With the head in full-face view the eyes distinctly behind the midlength of the sides (Figs. 120, 126). Setae on head and alitrunk not distinctly paired and usually not stout (Figs. 121, 127) **15**

15 Mandible with 6 teeth, very rarely with 7. Anterior face of first gastral segment broadly transversely concave throughout its height (Fig. 127). Antennal scapes relatively very long; when laid straight back from their insertions at least half their length projects beyond the occipital margin. Mesothorax constricted immediately behind pronotum (Fig. 127) . . . . ***Prenolepis***

— Mandible with at least 7 teeth, usually with more than 7. Anterior face of first gastral segment with a small concave area immediately above the helcium, but the face not broadly transversely concave throughout its height (Fig. 121). Antennal scapes much shorter; when laid straight back from their insertions with much less than half their length projecting beyond the occipital margin. Mesothorax not constricted immediately behind pronotum (Fig. 121) . . . . ***Lasius***

16 Mandible with 10 or more teeth or denticles in total. Apical tooth disproportionately large and the fourth tooth, counting from the apical, larger than the third and the fifth tooth (Fig. 148). Petiole reduced to an elongate, low node that allows the gaster to be reflexed over the alitrunk (Fig. 149). Palp formula 5,4 . . . . ***Oecophylla***

— Mandible usually with 5–7 teeth at most, only very rarely with more. If 7 or more teeth present they decrease in size from apex to base; the fourth tooth is not enlarged as above (Figs. 82, 84, 98). Petiole an erect node or scale, the gaster not capable of reflexion over the alitrunk (Figs. 83, 85, 89, 91, 95, 99). Palp formula usually 6,4, only extremely rarely reduced to 5,4 or 5,3 (two species) **17**

17 Eyes very large and in an extreme posterolateral position on the head (Fig. 94). In full-face view the occipital corner is formed by the curvature of the eye on each side, and posteriorly the eyes form the lateral portion of the occipital margin. The eyes often project slightly farther posteriorly than the true occipital margin that runs between them . . . . ***Opisthopsis***

— Eyes moderate to large and usually situated behind midlength of sides, but not occupying the occipital corners nor constituting a part of the occipital margin (Figs. 82, 84, 88, 90, 98) **18**

18 With alitrunk in profile the metathoracic spiracles forming tuberculiform prominences which project beyond the outline of the dorsum. Propodeum posteriorly with a raised transverse ridge which appears as a tooth in profile (Fig. 91). Pronotum and petiole node unarmed . . . . ***Forelophilus***

— With alitrunk in profile the metathoracic spiracles usually not forming tuberculiform prominences which project beyond the outline of the dorsum. Rarely, when such spiracles are present, the propodeum posteriorly does not have a raised transverse ridge which appears as a tooth in profile, or the pronotum and petiole are armed with teeth or spines, or both (Figs. 83, 85, 89, 99) **19**

19 Antennal funiculus with apical segments gradually but strongly broadening, forming a club. Proventriculus with sepals of calyx short, scarcely longer than the basal bulb. [One species, Singapore, rare] . . . . ***Overbeckia***

— Apical segments of antennal funiculus not forming a club. Proventriculus with sepals of calyx much longer than the basal bulb **20**

20 Metapleural gland orifice present on side of metapleuron above the hind coxa and below the level of the propodeal spiracle (Fig. 164). Orifice usually preceded by a longitudinal impression which is overhung by a projecting rim of cuticle, the orifice itself usually with a conspicuous tuft of downward-directed guard setae **21**

— Metapleural gland orifice absent from side of metapleuron (Fig. 165). An oblique impression separating metapleuron from propodeum frequently present, but lacking a gland orifice as described above **23**

21 Tergite of first gastral segment extremely large, accounting for considerably more than half the length of the gaster in dorsal view or in profile; sometimes the entire gastral dorsum consisting of the first tergite alone (Fig. 89). Dorsum of petiole usually armed with spines, teeth, or tubercles, the dorsolateral angles frequently dentate or spinose and the sides sometimes spinose; petiole only very rarely nodiform . ***Echinopla***

— Tergite of first gastral segment much smaller, accounting for much less than half the length of the gaster in dorsal view or in profile (Figs. 83, 85). Petiole nodiform to thickly scale-like, never armed with spines or teeth **22**

22 Mandible with 5 teeth. Median portion of clypeus usually shorter and less prominent than the lateral portions, not forming a narrow lobe that projects forward over the mandibles. Metanotal groove deeply impressed . . . . ***Calomyrmex***

— Mandible with more than 5 teeth. Median portion of clypeus much longer than lateral portions, forming a narrow lobe that projects forward over the mandibles. Metanotal groove an unimpressed transverse line . . . . ***Camponotus*** (part)

23 Tergite of first gastral segment large, accounting for at least half the length of the gaster in dorsal view or in profile (Fig. 99); the first tergite distinctly much longer than the second. Spines or teeth present on pronotum, propodeum, petiole, or on two or all of these . . . . ***Polyrhachis***

— Tergite of first gastral segment shorter, accounting for distinctly less than half the length of the gaster in dorsal view or in profile (Fig. 85); the first tergite at most only slightly longer than the second. Spines or teeth usually absent from pronotum, propodeum, and petiole; very rarely one of these locations armed . . . . **Camponotus** (part)

Key to Australasian FORMICINAE (Workers)

1 Antenna with 9–11 segments **2**
— Antenna with 12 segments **6**

2 Antenna with 9 segments . . . . **Brachymyrmex**
— Antenna with 10 or 11 segments **3**

3 Palp formula less than 6,4 (commonly 2,3) . . . . **Acropyga**
— Palp formula 6,4 . . . . **4**

4 Propodeum armed with one or more pairs of spines, teeth, tubercles, or cuticular prominences (Fig. 159) . . . **Stigmacros**
— Propodeum unarmed and rounded, lacking spines, teeth, tubercles, or cuticular prominences **5**

5 With alitrunk in dorsal view the mesonotum separated from the metanotum by a conspicuous transverse groove or impression, so that the metanotum forms a distinctly isolated sclerite (also visible in profile, Fig. 157) . . . . **Plagiolepis**
— With alitrunk in dorsal view the mesonotum fused with the metanotum, the two not separated by a transverse groove or impression, the metanotum not forming an isolated sclerite (also visible in profile, Fig. 153) . . . . **Anoplolepis**

6 Metapleural gland orifice absent, the surface of the metapleuron uninterrupted by a gland orifice above the hind coxa and below the level of the propodeal spiracle (Fig. 165) **7**
— Metapleuron with a distinct wide orifice for the metapleural gland, the orifice situated above the hind coxa and below the level of the propodeal spiracle (Fig. 164). The metapleural gland orifice usually protected by a line or tuft of guard setae, which are generally quite conspicuous **9**

7 Mandible with 10 or more teeth or denticles in total. Apical tooth disproportionately large and the fourth tooth, counting from the apical, larger than the third and fifth tooth (Fig. 148). Petiole reduced to an elongate, low node, which allows the gaster to reflex over the alitrunk (Fig. 149). Palp formula 5,4 . . . . **Oecophylla**
— Mandible usually with 5–7 teeth at most, only very rarely with more. If 7 or more teeth or denticles present in total, then they decrease in size from apical to basal and the fourth tooth is smaller than the third, not enlarged as above (Figs. 84, 98). Petiole a node or scale, the gaster not capable of reflexion over the alitrunk (Figs. 85, 99). Palp formula 6,4 . . . . **8**

8 Tergite of first gastral segment large, accounting for at least half the length of the gaster in dorsal view or in profile (Fig. 99); the first tergite distinctly much longer than the second. Spines or teeth present on pronotum, propodeum, petiole, or on two or all three of these. Monomorphic species . . . . **Polyrhachis**
— Tergite of first gastral segment small, accounting for distinctly less than half the length of the gaster in dorsal view or in profile (Fig. 85); the first tergite at most only slightly longer than the second. Spines or teeth usually absent from pronotum, propodeum, and petiole; only very rarely with one of these locations armed. Polymorphic species . . . **Camponotus**

9 Anteriormost point of torulus far behind the clypeal suture (Figs. 82, 88, 92, 94), the gap separating them at least equal to the basal width of the scape shaft and generally much more **10**
— Anteriormost point of torulus indenting, abutting, or immediately behind the clypeal suture (Figs. 124, 128, 134, 140, 142). If the last of these, then the gap between them very narrow, distinctly less than the basal width of the scape shaft **13**

10 Eyes very large and in an extreme posterolateral position on the head (Fig. 94). In full-face view the occipital corner is formed by the curvature of the eye on each side, and posteriorly the eyes form the lateral portions of the occipital margin. The eyes often project slightly farther posteriorly than the true occipital margin that runs between them . . . . . . **Opisthopsis**
— Eyes moderate to large in size and usually situated behind the midlength of the sides, but in full-face view not occupying the occipital corners nor constituting a part of the occipital margin (Figs. 82, 88, 92) **11**

11 Mandible with 10 or more teeth or denticles in total (Fig. 92). Ocelli present. First gastral sternite in ventral view without a transverse sulcus immediately behind the helcium. . . . . **Notostigma**
— Mandible with 5 or 6 teeth (usually 5) (Figs. 82, 88). Ocelli absent. First gastral sternite in ventral view with a conspicuous transverse sulcus immediately behind the helcium (Figs. 161, 162) **12**

12 Tergite of first gastral segment extremely large, accounting for considerably more than half the length of the gaster in dorsal view or in profile (Fig. 89); sometimes the entire gastral dorsum consisting of the first tergite alone. Dorsum or sides of petiole often armed with spines, teeth, or tubercles, the dorsolateral angles frequently dentate or spinose; petiole only very rarely nodiform . . . . **Echinopla**
— Tergite of first gastral segment much smaller, accounting for much less than half the length of the gaster in dorsal view or in profile (Fig. 83). Petiole nodiform to thickly scale-like, never armed with spines or teeth . . . . **Calomyrmex**

13 In ventral view the first gastral sternite without a transverse sulcus immediately behind the helcium. In the same view the line of tergosternal articulation of the first segment on each side of the helcium directed anteriorly, running far forward then passing through a long, narrowly rounded angle before running posteriorly to the apex of the segment (Fig. 163) **14**
— In ventral view the first gastral sternite with a transverse sulcus immediately behind the helcium. In the same view the line of tergosternal articulation of the first segment on each side of the helcium roughly transverse to quite strongly sinuate,

but not running far forward nor passing through a long, narrowly rounded angle before running posteriorly (Figs. 161, 162) **15**

14 Maxillary palp with 6 segments. Dorsal surfaces of head and body with coarse setae that are arranged in distinct pairs (Fig. 125) . . . . ***Paratrechina***

— Maxillary palp with 2–4 segments. Dorsal surfaces of head and body with soft pilosity, the setae not arranged in distinct pairs (Fig. 129) . . . . ***Pseudolasius***

15 Propodeum armed with a pair of spines at the angle between the dorsum and the declivity (Figs. 131, 143) **16**

— Propodeum unarmed, the dorsum rounding into the declivity (Figs. 135, 137, 139, 141) **17**

16 Petiole bispinose dorsally. Propodeum armed with a second, lower, pair of spines close to the base of the declivity (Fig. 143). With alitrunk in dorsal view the pronotum not expanded laterally into a pair of aliform lobes, the mesonotum not constricted . . . . ***Pseudonotoncus***

— Petiole emarginate dorsally, with a peak or point on each side of the emargination but not bispinose. Propodeum unarmed below the spines at the junction of dorsum and declivity (Fig. 131). With alitrunk in dorsal view the pronotum expanded laterally into a pair of aliform lobes, the mesonotum distinctly constricted . . . . ***Teratomyrmex***

17 Propodeal spiracle usually a slit or elongate oval, sometimes the spiracle comma-shaped (Fig. 135). Psammophore present . . . . ***Melophorus***

— Propodeal spiracle circular to subcircular (Figs. 137, 139, 141). Psammophore absent **18**

18 In ventral view the tergosternal articulation of the first gastral segment, on each side of the helcium, strongly sinuate . . . . ***Prolasius***

— In ventral view the tergosternal articulation of the first gastral segment, on each side of the helcium, shallowly concave to more or less straight (Fig. 161) **19**

19 Alitrunk structure complex (Fig. 139); pronotal humeri usually with a pair of prominent tumuli, mesonotum disproportionately enlarged, metanotum forming a separate swelling or a bifurcate or beveled process behind the mesonotum. Mandible with 6 teeth (Fig. 138) . . . . ***Notoncus***

— Alitrunk structure simple (Fig. 137); sometimes the pronotum, mesonotum, and propodeum forming separate convexities but pronotal humeri never with prominent tumuli, mesonotum never disproportionately enlarged, metanotum not forming a separate swelling or process. Mandible usually with 7–13 teeth (Fig. 136), only very rarely with 6 . . . . ***Myrmecorhynchus***

Key to Nearctic FORMICINAE (Workers)

1 Antenna with 8–11 segments **2**

— Antenna with 12 segments **6**

2 Apical and 1 or 2 preapical antennal segments strikingly broader than the preceding segments and forming a swollen and conspicuous club whose segments are strongly differentiated from the remainder of the funiculus . . . . ***Myrmelachista***

— Antenna not terminating in a club; either the antennal funiculus filiform or the segments gradually enlarging apically . . **3**

3 Palp formula less than 6,4. With head in full-face view the eyes very small, vestigial, or absent (Fig. 150); when present the eyes situated far in front of the midlength of the sides of the head . . . . ***Acropyga***

— Palp formula 6,4. With head in full-face view the eyes well developed and situated just before, at, or behind the midlength of the sides of the head (Figs. 76, 152, 156) . . . **4**

4 Antenna with 9 segments . . . . ***Brachymyrmex***

— Antenna with 11 segments **5**

5 With alitrunk in dorsal view the mesonotum separated from the metanotum by a conspicuous transverse groove or impression, so that the metanotum forms a distinctly isolated sclerite (also visible in profile, Fig. 157) . . . . ***Plagiolepis***

— With alitrunk in dorsal view the mesonotum fused with the metanotum, the two not separated by a transverse groove or impression, the metanotum not forming an isolated sclerite (also visible in profile, Fig. 153) . . . . ***Anoplolepis***

6 Metapleuron with a distinct, wide orifice for the metapleural gland, the orifice situated above the hind coxa and below the level of the propodeal spiracle (Fig. 164). Orifice of metapleural gland usually protected by a line or tuft of guard setae, which are generally very conspicuous. Antennal sockets situated close to posterior margin of clypeus (Figs. 104, 120, 122, 124) **7**

— Metapleural gland orifice absent, the surface of the metapleuron uninterrupted by a gland orifice above the hind coxa and below the level of the propodeal spiracle (Fig. 165). Guard setae absent. Antennal sockets situated well behind the posterior margin of the clypeus (Fig. 84) . . . . ***Camponotus***

7 With the gaster in ventral view the first sternite with a conspicuous transverse sulcus behind the helcium (Fig. 162); tergosternal articulation of first gastral segment directed laterally behind helcium base, then broadly curving posteriorly (Figs. 161, 162). Propodeal spiracle elliptical to broadly oval (Figs. 105, 107, 164) **8**

— With the gaster in ventral view the first sternite without a transverse sulcus behind the helcium; tergosternal articulation of first gastral segment directed anteriorly through a narrow curve from the helcium base and running some distance forward before passing through a narrowly rounded curve or angle and thereafter running posteriorly (Fig. 163). Propodeal spiracle circular to subcircular (Figs. 121, 123, 125, 127) **9**

8 Mandible falcate and edentate (Fig. 106). Palp formula 4,2 . . . . ***Polyergus***

— Mandible triangular and armed with 7 or more sharp teeth of varying size (Fig. 104). Palp formula usually 6,4, rarely reduced but never 4,2 . . . . ***Formica***

9 Palp formula 3,3 . . . . ***Acanthomyops***

— Palp formula 6,4 . . . . **10**

10 Dorsal surfaces of head and body with coarse setae that are arranged in distinct pairs (Fig. 125). With head in full-face view the eyes at or in front of the midlength of the sides (Fig. 124) . . . . ***Paratrechina***

— Dorsal surfaces of head and body with pilosity that is abundant to almost absent, but without coarse setae arranged in distinct pairs (Figs. 121, 123, 127). With head in full-face view the eyes behind the midlength of the sides (Figs. 120, 122, 126) **11**

11 Maxillary palp segments 3 and 4 extremely long, the third segment (counting from the base) at least half the head length and usually more (Fig. 123) . . . . ***Myrmecocystus***

— Maxillary palp segments 3 and 4 short, the third segment (counting from the base) much shorter than half the head length **12**

12 Mandible with 6 teeth, very rarely with 7. Anterior face of first gastral segment broadly transversely concave throughout its height. Antennal scapes relatively very long; when laid straight back from their insertions at least half their length projects beyond the occipital margin. Mesothorax constricted immediately behind pronotum (Fig. 127) . . . . . . ***Prenolepis***

— Mandible with at least 7 teeth, usually with more than 7. Anterior face of first gastral segment with a small concave area immediately above the helcium but the face not broadly transversely concave throughout its height. Antennal scapes much shorter; when laid straight back from their insertions with much less than half their length projecting beyond the occipital margin. Mesothorax not constricted immediately behind pronotum (Fig. 121) . . . . ***Lasius***

Key to Neotropical FORMICINAE (Workers)

1 Antenna with 8–11 segments **2**

— Antenna with 12 segments **6**

2 Apical and 1 or 2 preapical antennal segments strikingly broader than the preceding segments and forming a swollen and conspicuous club whose segments are strikingly differentiated from the remainder of the funiculus . . . . . . ***Myrmelachista***

— Antenna not terminating in a club; either the antennal funiculus filiform or the segments gradually enlarging apically . **3**

3 Palp formula less than 6,4. With head in full-face view the eyes very small, vestigial, or absent (Fig. 150); when present the eyes situated far in front of the midlength of the sides of the head . . . . ***Acropyga***

— Palp formula 6,4. With head in full-face view the eyes well developed and situated just before, at, or behind the midlength of the sides of the head (Figs. 76, 152, 156) . . . **4**

4 Antenna with 9 segments . . . . ***Brachymyrmex***

— Antenna with 11 segments **5**

5 With alitrunk in dorsal view the mesonotum separated from the metanotum by a conspicuous transverse groove or impression, so that the metanotum forms a distinctly isolated sclerite (also visible in profile, Fig. 157) . . . . ***Plagiolepis***

— With alitrunk in dorsal view the mesonotum fused with the metanotum, the two not separated by a transverse impression or groove, the metanotum not forming an isolated sclerite (also visible in profile, Fig. 153) . . . . ***Anoplolepis***

6 Metapleuron with a distinct, wide orifice for the metapleural gland, the orifice situated above the hind coxa and below the level of the propodeal spiracle (Fig. 164). Orifice of metapleural gland usually protected by a line or tuft of guard setae, which are usually very obvious. Antennal sockets situated close to the posterior margin of the clypeus (Figs. 104, 116, 122, 124, 132) **7**

— Metapleural gland orifice absent, the surface of the metapleuron uninterrupted by a gland orifice above the hind coxa and below the level of the propodeal spiracle (Fig. 165). Line or tuft of guard setae absent. Antennal sockets situated well behind the posterior margin of the clypeus (Figs. 84, 86) **15**

7 With the gaster in ventral view the first sternite with a conspicuous flexible transverse sulcus immediately behind the helcium (Figs. 161, 162) **8**

— With the gaster in ventral view the first sternite entire, without a transverse sulcus behind the helcium (Fig. 163) **9**

8 Mandible falcate and edentate (Fig. 106). Palp formula 4,2 . . . . ***Polyergus***

— Mandible triangular and armed with 7 or more sharp teeth of varying size (Fig. 104). Palp formula usually 6,4, rarely reduced but never 4,2 . . . . ***Formica***

9 Eyes enormous; with the head in full-face view each eye extending from the occipital margin to the posterior clypeal margin in the vicinity of the anterior tentorial pit (Fig. 116) . . . . ***Gigantiops***

— Eyes present but smaller; with the head in full-face view the eyes never extending from the occipital margin to the posterior clypeal margin in the vicinity of the anterior tentorial pit (Figs. 120, 122, 124, 126, 132) **10**

10 With the gaster in ventral view the line of the tergosternal articulation on each side of the helcium directed laterally from the helcium base, the line weakly sinuate before curving posteriorly through a broadly rounded angle (as Fig. 161) . . . . ***Lasiophanes***

— With the gaster in ventral view the line of the tergosternal articulation on each side of the helcium directed anteriorly through a narrow curve from the helcium base, the line running forward for a notable distance then passing through a narrowly rounded curve or angle and thereafter running posteriorly (Fig. 163) **11**

11 Palp formula 3,3 .... ***Acanthomyops***
— Palp formula 6,4 .... **12**

12 Dorsal surfaces of head and body with coarse setae that are arranged in distinct pairs (Fig. 125). With head in full-face view the eyes usually at or in front of the midlength of the sides (Fig. 124) .... ***Paratrechina***
— Dorsal surfaces of head and body with pilosity that is abundant to almost absent, but without coarse setae arranged in distinct pairs (Figs. 121, 123, 127). With head in full-face view the eyes usually distinctly behind the midlength of the sides (Figs. 120, 122, 126) **13**

13 Maxillary palp segments 3 and 4 extremely long, the third segment (counting from the base) at least half the length of the head and usually more (Fig. 122) .... ***Myrmecocystus***
— Maxillary palp segments 3 and 4 short, the third segment (counting from the base) distinctly much shorter than half the head length **14**

14 Mandible with 6 teeth, very rarely with 7. Anterior face of first gastral segment broadly transversely concave throughout its height. Antennal scapes relatively very long; when laid straight back from their insertions at least half their length projects beyond the occipital margin. Mesothorax constricted immediately behind pronotum (Fig. 127) ...... ***Prenolepis***
— Mandible with at least 7 teeth, usually with more than 7. Anterior face of first gastral segment with a small concave area immediately above the helcium, but the face not broadly transversely concave throughout its height. Antennal scapes much shorter; when laid straight back from their insertions with much less than half their length projecting beyond the occipital margin. Mesothorax not constricted immediately behind pronotum (Fig. 121) .... ***Lasius***

15 Monomorphic species .... ***Dendromyrmex***
— Dimorphic or polymorphic species .... ***Camponotus***

Synoptic Classification

A name prefixed by * indicates an extinct taxon.

Subfamily **FORMICINAE.**

Tribe **Brachymyrmecini.** Genera: ***Aphomomyrmex*** (Figs. 74, 75) (= *Aphomyrmex* (misspelling)), ***Brachymyrmex*** (Figs. 76, 77) [subgenera: nominal plus *Bryscha*], ***Cladomyrma*** (Figs. 78, 79), ***Petalomyrmex*** (Figs. 80, 81), ***Pseudaphomomyrmex*** (known from queens only).

Tribe **Bregmatomyrmini.** Genus: ***Bregmatomyrma*** (= *Bregmatomyrmex* (misspelling)) (known from queens only).

Tribe **Camponotini** (= Polyrhachidini). Genera: ***Calomyrmex*** (Figs. 82, 83), ****Camponotites, Camponotus*** (Figs. 84, 85) [subgenera: nominal plus *Colobopsis* (= *Campylomyrma* (misspelling), = *Condylomyrma*), *Dinomyrmex* (= *Myrmogigas*), *Hypercolobopsis* (= *Neocolobopsis*), *Karavaievia, Manniella, Mayria, Myrmacrhaphe, Myrmamblys, Myrmaphaenus* (= *Neomyrmamblys,* = *Paracolobopsis*), *Myrmentoma, Myrmepinotus, Myrmepomis* (= *Myrmolophus*), *Myrmespera, Myrmeurynota, Myrmisolepis, Myrmobrachys, Myrmocladoecus, Myrmodirachis, Myrmogonia, Myrmomalis, Myrmonesites, Myrmopalpella, Myrmopelta, Myrmophyma* (= *Myrmocamelus*), *Myrmopiromis, Myrmoplatypus, Myrmoplatys, Myrmopsamma, Myrmopytia, Myrmosaga, Myrmosaulus, Myrmosericus, Myrmosphincta, Myrmostenus, Myrmotarsus, Myrmotemnus, Myrmothrix, Myrmotrema, Myrmoxygenys, Orthonotomyrmex* (= *Orthonotus* (homonym)), *Paramyrmamblys, Pseudocolobopsis, Rhinomyrmex, Tanaemyrmex* (= *Myrmoturba*), *Thlipsepinotus*], ****Chimaeromyrma, Dendromyrmex*** (Figs. 86, 87), ****Drymomyrmex*** (= **Dryomyrmex* (misspelling)), ***Echinopla*** (Figs. 88, 89) (= *Mesoxena*), ***Forelophilus*** (Figs. 90, 91), ***Notostigma*** (Figs. 92, 93), ***Opisthopsis*** (Figs. 94, 95) (= *Myrmecopsis* (homonym)), ***Overbeckia, Phasmomyrmex*** (Figs. 96, 97) [subgenera: nominal plus *Myrmacantha, Myrmorhachis*], ***Polyrhachis*** (Figs. 98, 99) (= *Polyrachis* (misspelling)) [subgenera: nominal plus *Anoplomyrma, Aulacomyrma* (= *Johnia*), *Campomyrma, Chariomyrma, Cyrtomyrma, Dolichorhachis, Hagiomyrma, Hedomyrma* (= *Morleyidris*), *Hemioptica, Myrma* (= *Hoplomyrmus,* = *Pseudocyrtomyrma*), *Myrmatopa* (= *Irenea*), *Myrmhopla* (= *Cephalomyrma,* = *Florencea*), *Myrmothrinax* (= *Evelyna*)], ****Pseudocamponotus.***

Tribe **Formicini.** Genera: ***Alloformica*** (Figs. 100, 101), ***Bajcaridris, Cataglyphis*** (Figs. 102, 103) (= *Eomonocombus,* = *Machaeromyrma,* = *Monocombus,* = *Paraformica*), ***Formica*** (Figs. 104, 105) (= *Adformica,* = *Coptoformica,* = *Formicina,* = *Iberoformica,* = *Neoformica,* = *Raptiformica,* = *Serviformica*), ****Glaphyromyrmex,*** ****Leucotaphus, Polyergus*** (Figs. 106, 107), ***Proformica*** (Figs. 108, 109), ****Protoformica, Rossomyrmex*** (Figs. 110, 111).

Tribe **Gesomyrmecini** (= Dimorphomyrmii, = Gesomyrmini). Genera: ***Gesomyrmex*** (Figs. 112–115) (= *Dimorphomyrmex,* = *Gaesomyrmex*), ****Prodimorphomyrmex.***

Tribe **Gigantiopini.** Genus: ***Gigantiops*** (Figs. 116, 117).

Tribe **Lasiini** (= Acanthomyopsini, = Prenolepidini). Genera: ***Acanthomyops, Euprenolepis*** (Figs. 118, 119) (= *Chapmanella*), ***Lasius*** (Figs. 120, 121) (= *Donisthorpea*) [current subgenera: nominal plus *Austrolasius, Cautolasius, Chthonolasius* (= *Chtonolasius* (misspelling)), *Dendrolasius*], ***Myrmecocystus*** (Figs. 122, 123) (= *Endiodioctes,* = *Eremnocystus*), ***Paratrechina*** (Figs. 124, 125) (= *Andragnathus,* = *Nylanderia,* = *Paraparatrechina*), ***Prenolepis*** (Figs. 126, 127), ****Protrechina, Pseudolasius*** (Figs. 128, 129) (= *Nesolasius*), ***Teratomyrmex*** (Figs. 130, 131).

Tribe **Melophorini** (= Myrmecorhynchini). Genera: ***Lasiophanes*** (Figs. 132, 133), ***Melophorus*** (Figs. 134, 135) (= *Erimelophorus,* = *Trichomelophorus*), ***Myrmecorhynchus*** (Figs. 136, 137), ***Notoncus*** (Figs. 138, 139) (= *Diodontolepis*), ***Prolasius*** (Figs. 140, 141), ***Pseudonotoncus*** (Figs. 142, 143).

Tribe **Myrmelachistini.** Genus: ***Myrmelachista*** (Figs. 144, 145) (= *Decamera* (homonym), = *Hincksidris,* = *Neaphomus*).

Tribe **Myrmoteratini.** Genus: ***Myrmoteras*** (Figs. 146, 147) [subgenera: nominal plus *Myagroteras*].

Tribe **Oecophyllini.** Genus: ***Oecophylla*** (Figs. 148, 149).

Tribe **Plagiolepidini.** Genera: ***Acropyga*** (Figs. 150, 151) [subgenera: nominal plus *Atopodon, Malacomyrma, Rhizomyrma*], ***Agraulomyrmex, Anoplolepis*** (Figs. 152, 153) [subgenera: nominal plus *Mesanoplolepis, Tapinolepis, Zealleyella*], ***Lepisiota*** (Figs. 154, 155) (= *Acantholepis* (homonym), = *Achantilepis* (misspelling), = *Baroniurbania,* = *Pseudacantholepis*), ***Plagiolepis*** (Figs. 156, 157) (= *Aporomyrmex,* = *Paraplagiolepis*) [subgenera: nominal plus *Anacantholepis*], ****Rhopalomyrmex, Stigmacros*** (Figs. 158, 159) (= *Acrostigma* (homonym), = *Campostigmacros,* = *Chariostigmacros,* = *Cyrtostigmacros,* = *Hagiostigmacros,* = *Pseudostigmacros*).

Tribe **Santschiellini.** Genus: ***Santschiella.***

Tribe ***Sicilomyrmecini.** Genus: ****Sicilomyrmex.***

Genus unplaced to tribe: ****Imhoffia.***

[Material of the unavailable names Alloformicinae, Eucamponotinae, Euformicinae, Heteroformicinae, and Mesocamponotinae is referable to Formicinae; that of the unavailable name Procamponotinae is referable to Myrmoteratini.]

Distribution

The subfamily Formicinae has a worldwide distribution, as indicated in the table given in the Introduction. The numbers of genera shared between the various zoogeographical regions are shown below, excluding endemic genera and those artificially introduced by human activities. In the table PAL = Palaearctic, AFR = Afrotropical, MAL = Malagasy, ORI = Oriental, INA = Indo-Australian, AUS = Australasian, NEA = Nearctic, NEO = Neotropical.

AFR	8						
MAL	6	6					
ORI	12	10	6				
INA	8	9	6	12			
AUS	5	7	4	7	10		
NEA	7	3	4	6	4	3	
NEO	4	3	4	4	4	3	5
	PAL	AFR	MAL	ORI	INA	AUS	NEA

Taxonomic References

Identification of extant species

Some older references have a suffixed comment "[out of date]." These references are included because they contain the only identification keys ever attempted for the taxon in question. They should be used with great caution as, for the most part, they contain numerous infraspecific and infrasubspecific taxa that are no longer recognized. Older references that have been superseded, or those rendered useless by the volume of later descriptions and synonymies, are omitted.

Acanthomyops: Wing (1968). ***Acropyga:*** Weber (1944) [Neotropical]. ***Agraulomyrmex:*** Prins (1983). ***Anoplolepis:*** Prins (1982) [*custodiens*-group, partial]. ***Aphomomyrmex:*** Snelling (1979). ***Brachymyrmex:*** Santschi (1923) [out of date]. ***Camponotus:*** Kusnezov (1952b) [Argentina]; Wang, Xiao, and Wu (1989a,b) [China]; Yasumatsu and Brown (1951, 1957) [*herculeanus*-complex, Palaearctic]; Hashmi (1973) [subgenus *Myrmothrix*]; Dumpert (1985) [subgenus *Karavaievia*]; Snelling (1988) [subgenus *Myrmentoma,* Nearctic]; Robertson (1990) [*fulvopilosus*-complex]. ***Cataglyphis:*** Santschi (1929a) [out of date]; Agosti (1990a) [species groups]. ***Cladomyrma:*** Agosti (1991). ***Dendromyrmex:*** Mann (1916). ***Formica:*** Dlussky (1964) [*exsecta*-group, former U.S.S.R.]; Dlussky (1965) [Mongolia, Tibet]; Dlussky (1967) [Palaearctic]; Francoeur (1973) [*fusca*-group, Nearctic]; Dlussky and Pisarski (1971) [Poland]; Wilson and Brown (1955), Buren (1968a), Snelling and Buren (1985) [*sanguinea*-group, Nearctic]; Kupyanskaya (1980) [far eastern Russia]; Wu (1990) [China]; Gösswald (1990) [*rufa*-group]. ***Gesomyrmex:*** Cole (1949). ***Lasiophanes:*** Snelling and Hunt (1976). ***Lasius:*** Wilson (1955) [world]; Collingwood (1982) [Himalayan fauna]; Kupyanskaya (1989) [subgenus *Dendrolasius,* east Palaearctic]; Seifert (1988a, 1990) [subgenus *Chthonolasius,* west Palaearctic]; Seifert (1992) [subgenus *Lasius,* Palaearctic]; Yamauchi and Hayashida (1968, 1970), Yamauchi (1979) [Japan]. ***Myrmecocystus:*** Snelling (1976, 1982b). ***Myrmelachista:*** W. M. Wheeler (1934b) [out of date]. ***Myrmoteras:*** Moffett (1985); Agosti (1992) [Indo-Australian]. ***Notoncus:*** Brown (1955). ***Opisthopsis:*** W. M. Wheeler (1919) [out of date]. ***Paratrechina:*** Trager (1984) [Nearctic]. ***Petalomyrmex:*** Snelling (1979). ***Plagiolepis:*** Radchenko (1989b) [former European U.S.S.R.]. ***Polyergus:*** J. Wheeler (1968) [Nearctic]. ***Polyrhachis:*** Hung (1967) [subgenera]; Hung (1970), Kohout (1988) [subgenus *Polyrhachis*]; Bolton (1973) [Afrotropical]; Wang and Wu (1991) [China]; Bolton (1975c), Kohout (1987) [*sexspinosa*-group]; Kohout (1989) [*relucens*-group, Australia]; Kohout (1990) [*viehmeyeri*-group]. ***Proformica:*** Dlussky (1969) [former U.S.S.R. and contiguous countries]. ***Prolasius:*** McAreavey (1947). ***Pseudolasius:*** Menozzi (1924) [Afrotropical, out of date]. ***Stigmacros:*** McAreavey (1957). ***Teratomyrmex:*** McAreavey (1957).

Other taxonomic references

Formicinae: Brown (1973); Snelling (1981); Dlussky and Fedoseeva (1988); Hölldobler and Wilson (1990); Agosti (1990b, 1991); Shattuck (1992b); Baroni Urbani, Bolton and Ward (1992).

See also References to Faunistic Studies.

Figures 74–165 FORMICINAE workers. Figs. 74–159, heads in full-face view and bodies in profile:

74–81, **Brachymyrmecini:** 74–75, *Aphomomyrmex;* 76–77, *Brachymyrmex;* 78–79, *Cladomyrma;* 80–81, *Petalomyrmex*

82–99, **Camponotini:** 82–83, *Calomyrmex;* 84–85, *Camponotus,* major worker; 86–87, *Dendromyrmex;* 88–89, *Echinopla;* 90–91, *Forelophilus;* 92–93, *Notostigma;* 94–95, *Opisthopsis;* 96–97, *Phasmomyrmex;* 98–99, *Polyrhachis*

100–111, **Formicini:** 100–101, *Alloformica;* 102–103, *Cataglyphis;* 104–105, *Formica;* 106–107, *Polyergus;* 108–109, *Proformica;* 110–111, *Rossomyrmex*

112–115, **Gesomyrmecini:** 112–113, *Gesomyrmex,* minor worker; 114–115, *Gesomyrmex,* major worker

116–117, **Gigantiopini,** *Gigantiops*

118–131, **Lasiini:** 118–119, *Euprenolepis;* 120–121, *Lasius;* 122–123, *Myrmecocystus;* 124–125, *Paratrechina;* 126–127, *Prenolepis;* 128–129, *Pseudolasius;* 130–131, *Teratomyrmex*

132–143, **Melophorini:** 132–133, *Lasiophanes;* 134–135, *Melophorus;* 136–137, *Myrmecorhynchus;* 138–139, *Notoncus;* 140–141, *Prolasius;* 142–143, *Pseudonotoncus*

144–145, **Myrmelachistini,** *Myrmelachista*

146–147, **Myrmoteratini,** *Myrmoteras*

148–149, **Oecophyllini,** *Oecophylla*

150–159, **Plagiolepidini:** 150–151, *Acropyga;* 152–153, *Anoplolepis;* 154–155, *Lepisiota;* 156–157, *Plagiolepis;* 158–159, *Stigmacros*

160, gastral apex to show acidopore in *Cataglyphis*

161–163, ventral view of petiole, helcium, and first gastral segment to show sutures: 161, *Notoncus;* 162, *Cataglyphis;* 163, *Lasius*

164–165, profile of metapleuron, propodeum, and petiole to show presence/absense of metapleural gland orifice: 164, *Formica,* with orifice present; 165, *Camponotus,* with orifice absent.

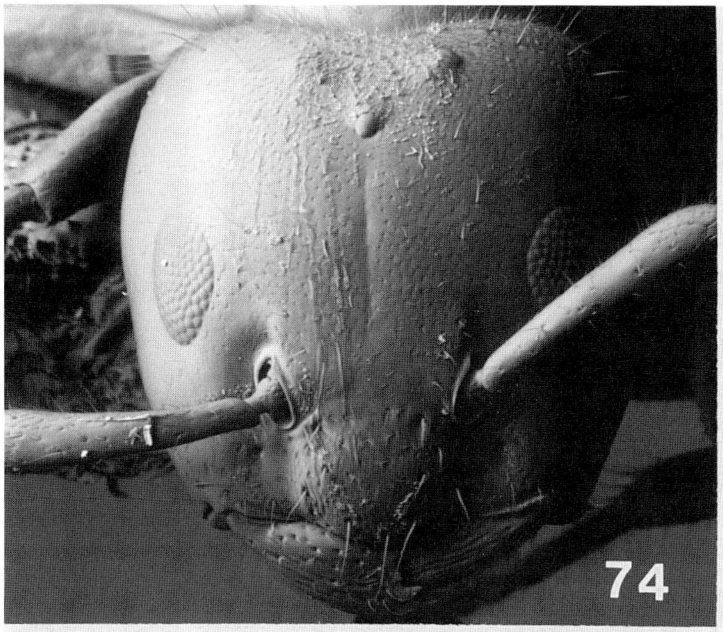
74

75
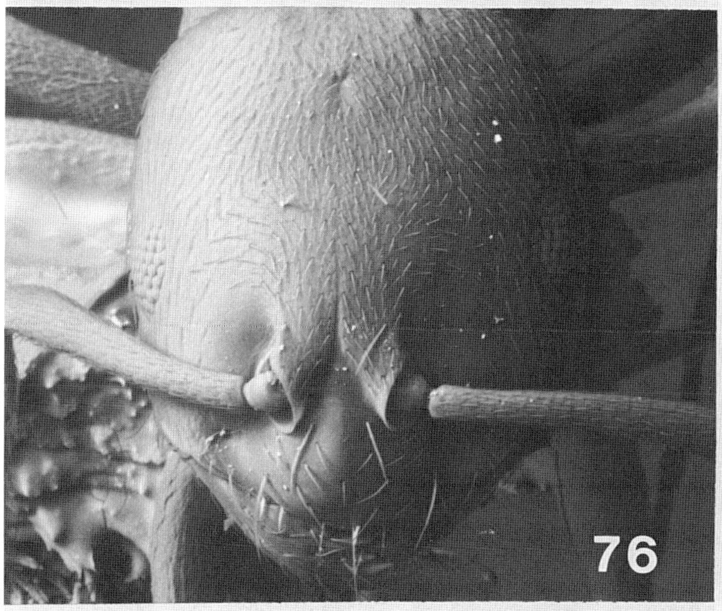
76

77
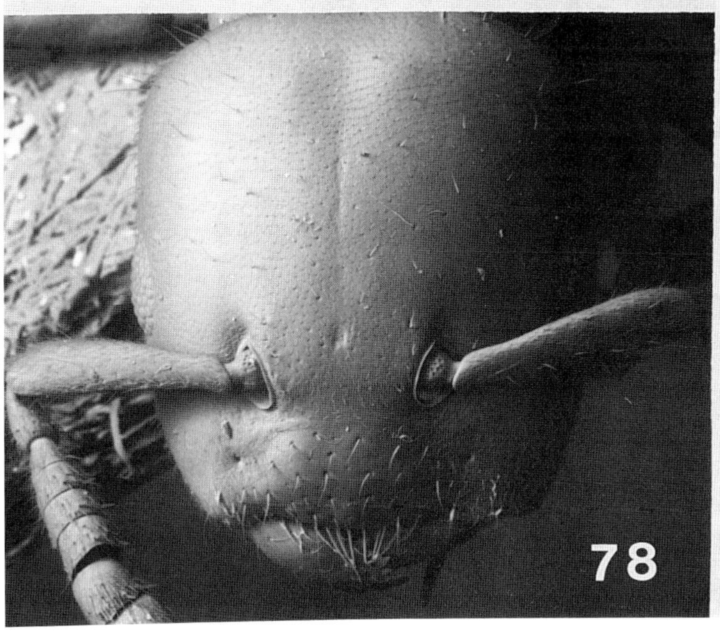
78

79

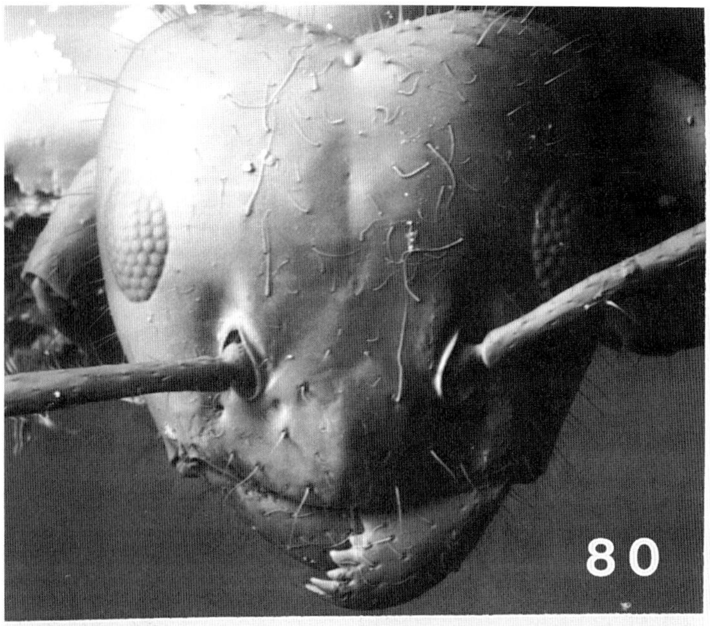
80

81

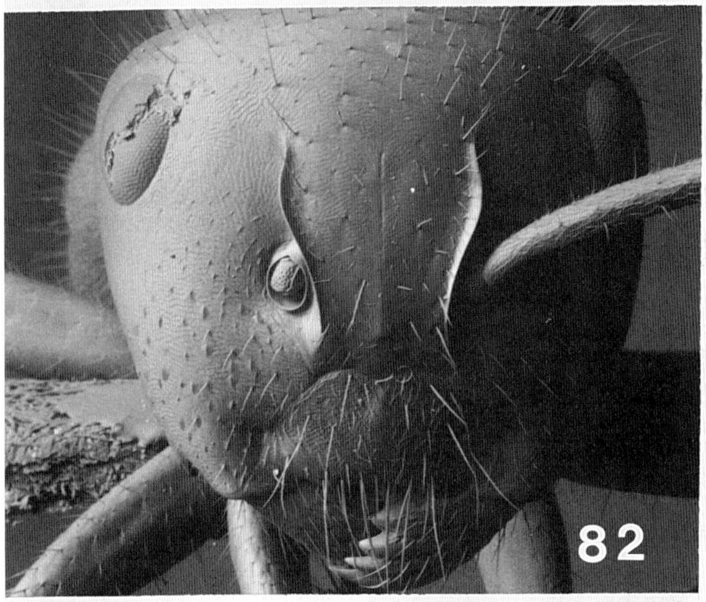
82

83

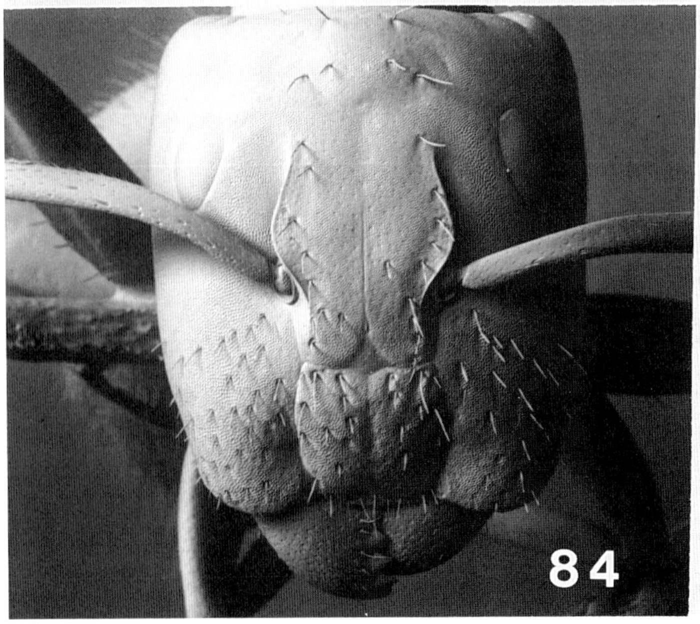
84

85

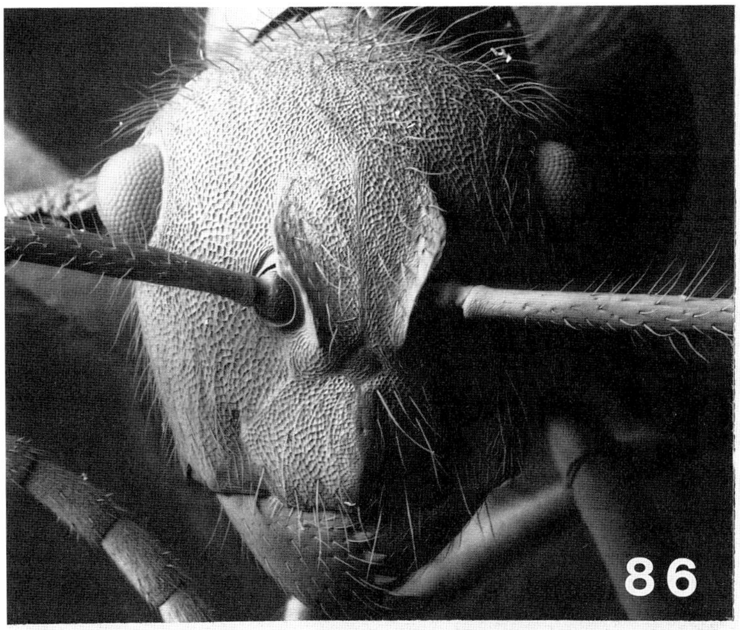
86

87

88

89
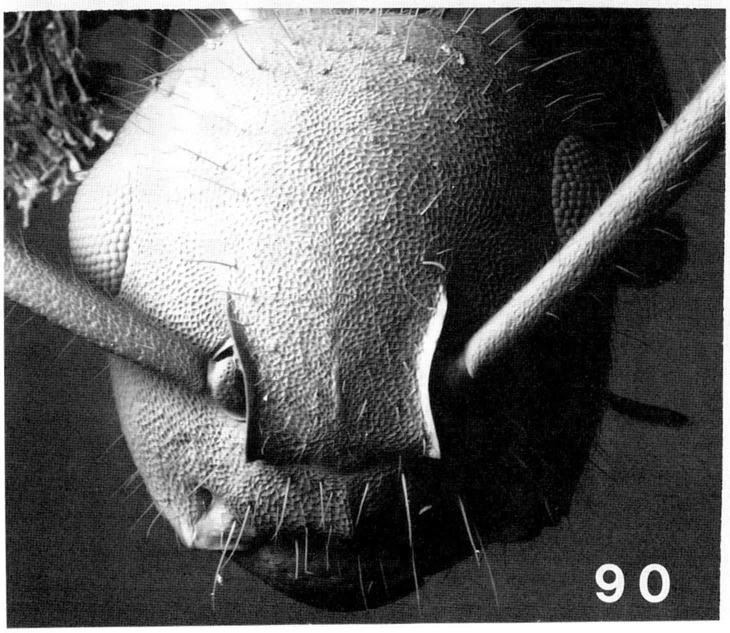
90

91

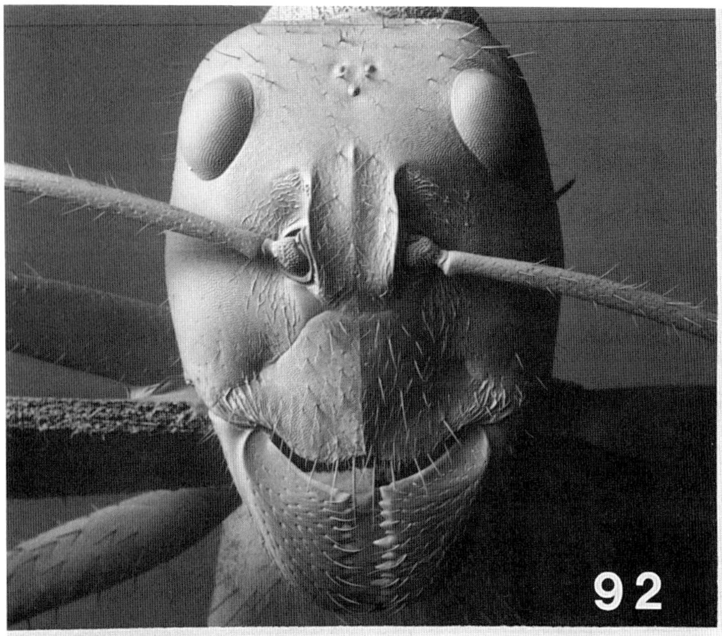
92
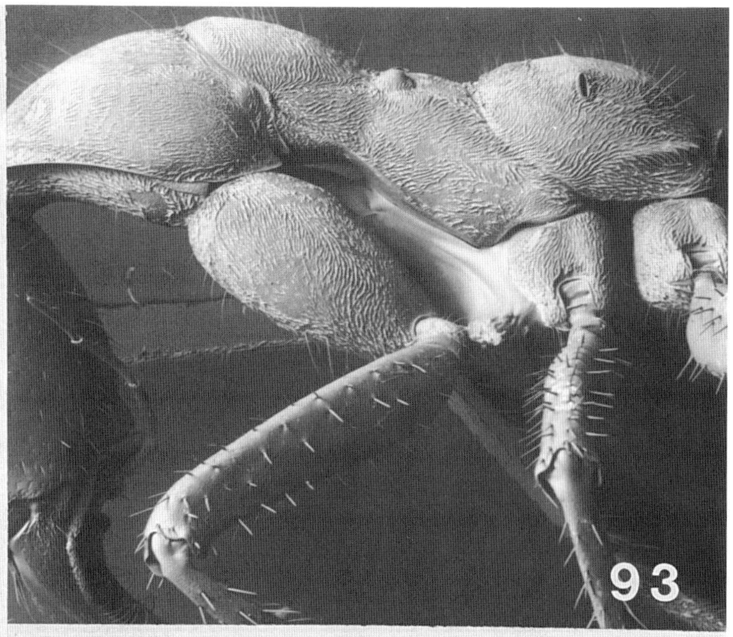
93
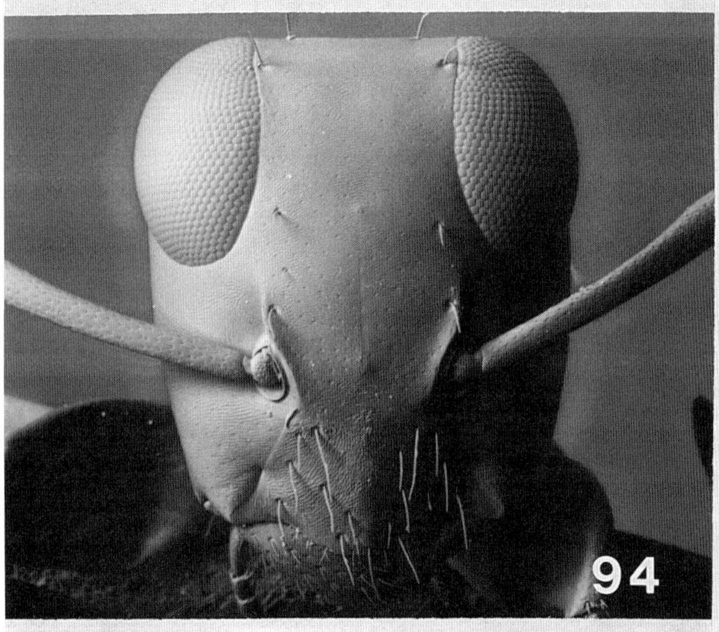
94
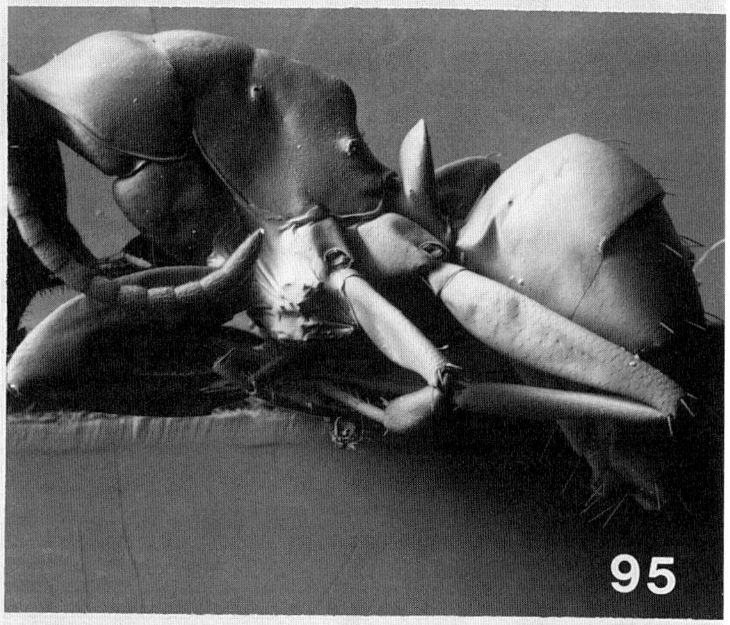
95

96

97

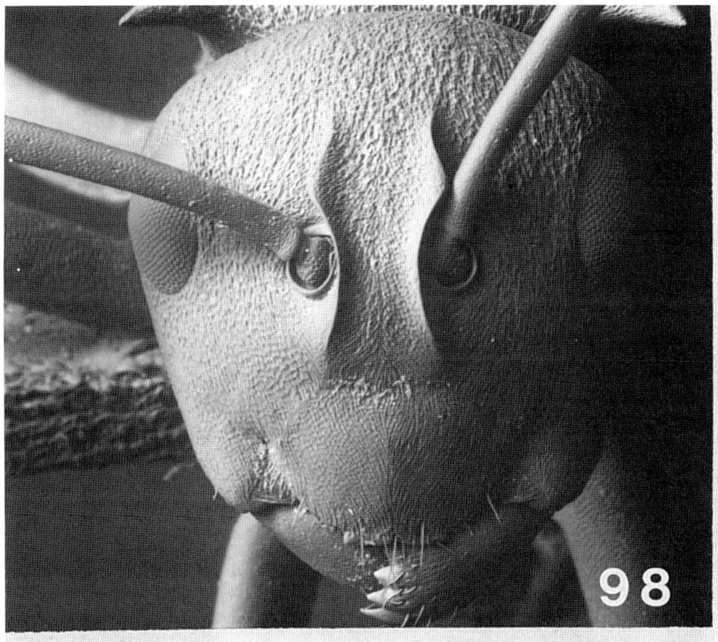
98

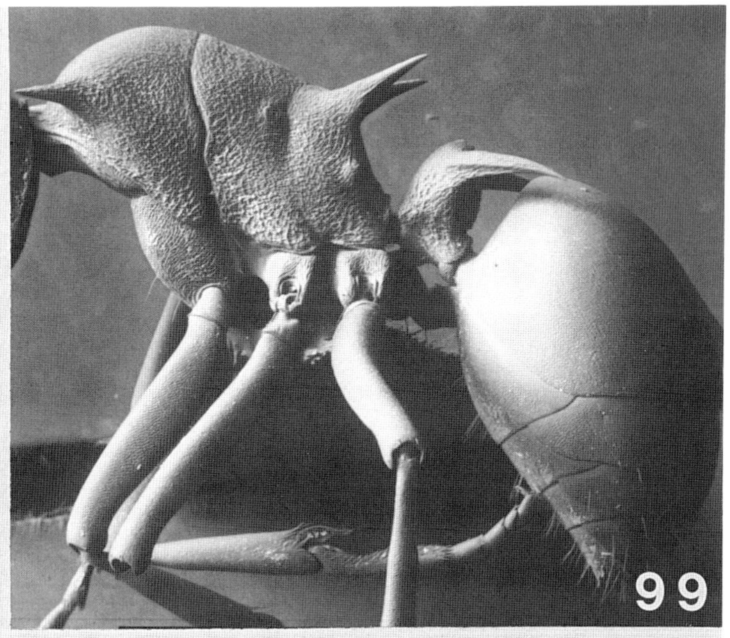
99

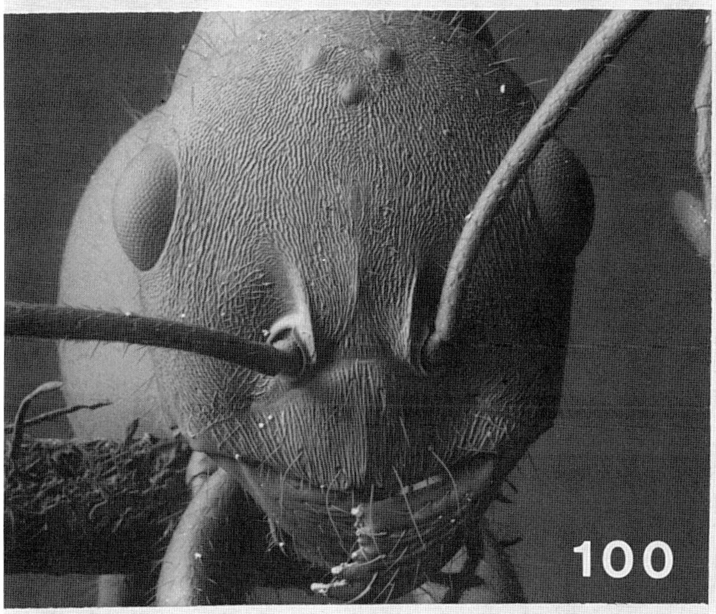
100

101

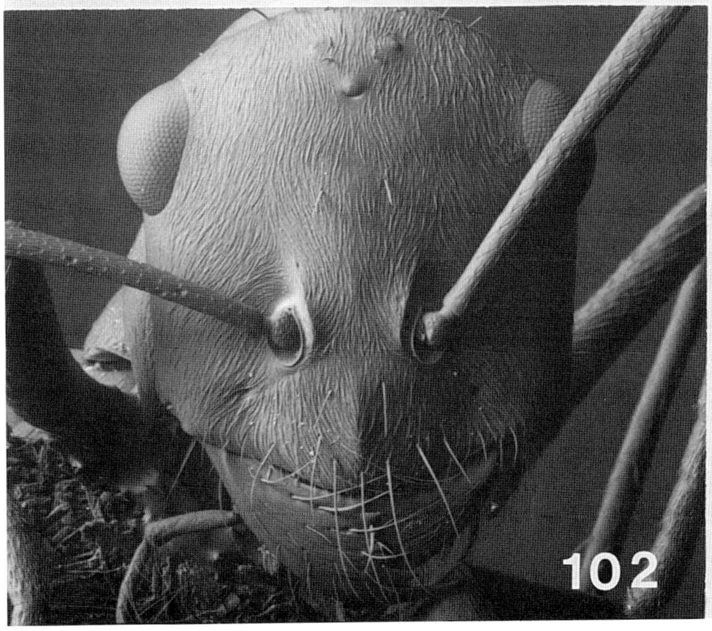
102

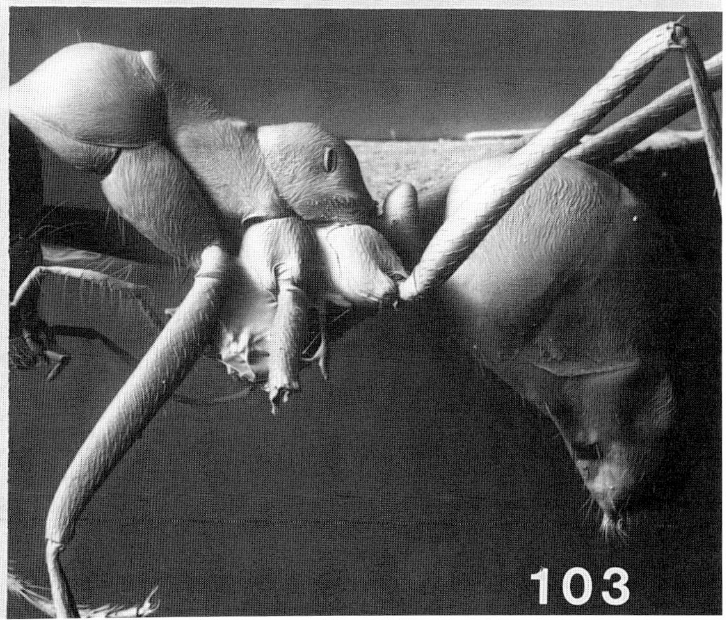
103

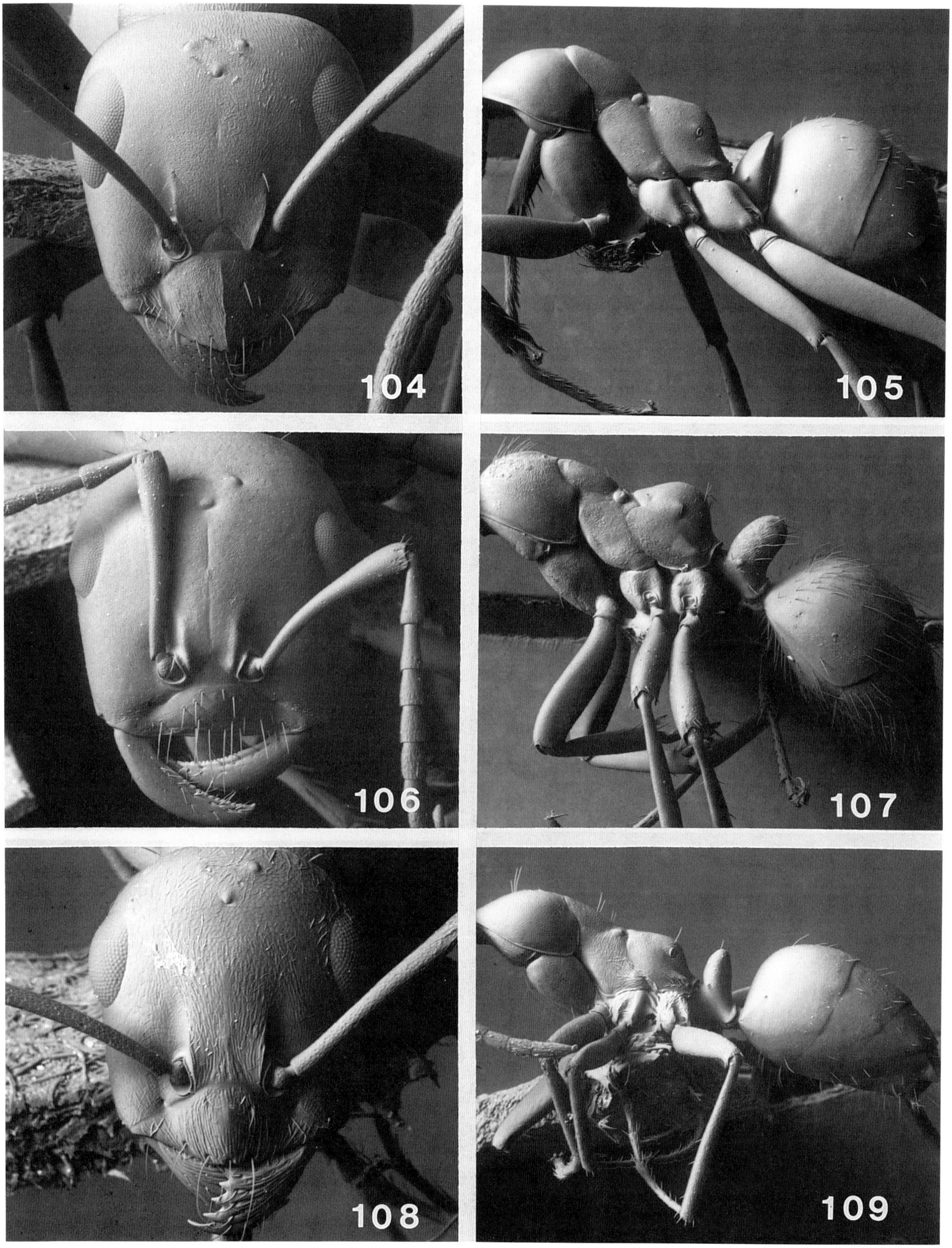
104
105
106
107
108
109

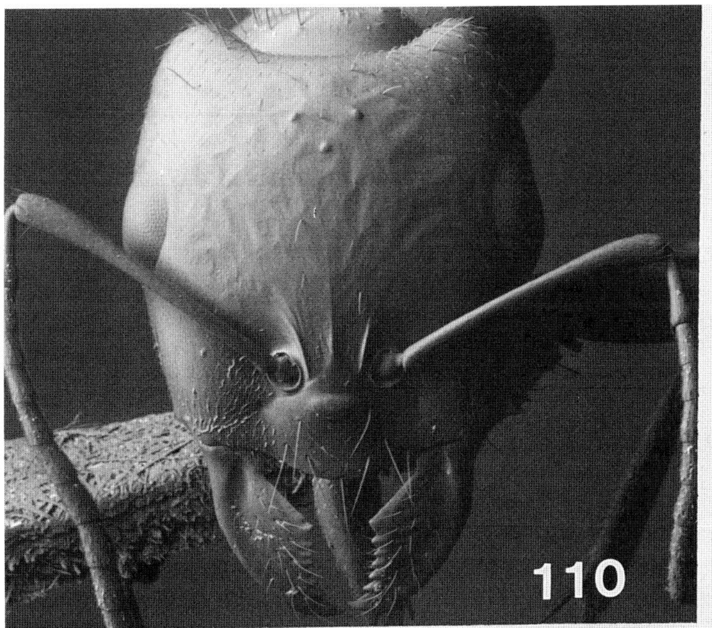
110

111

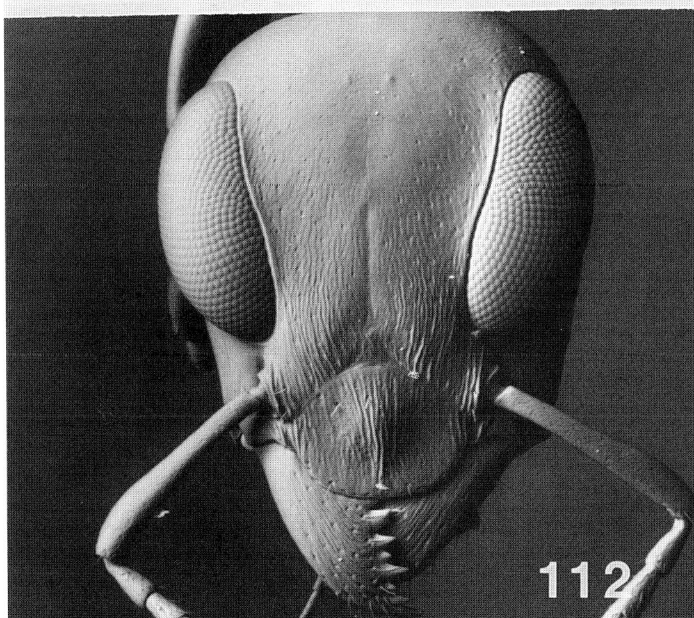
112

113

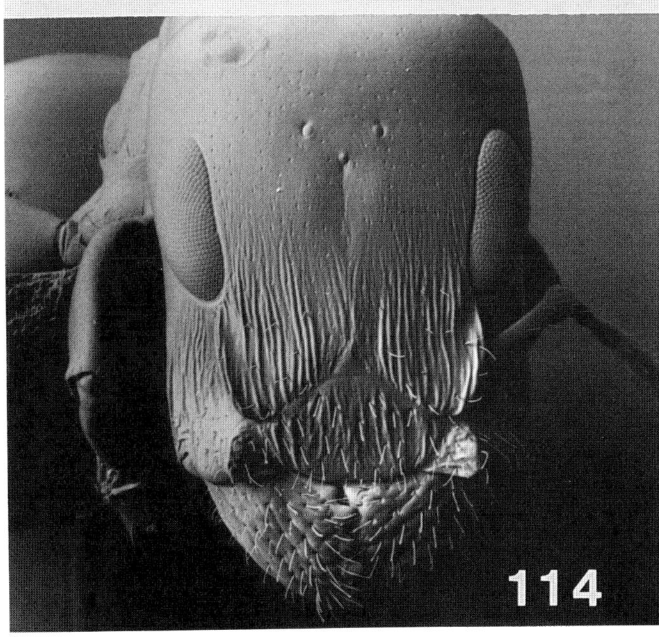
114

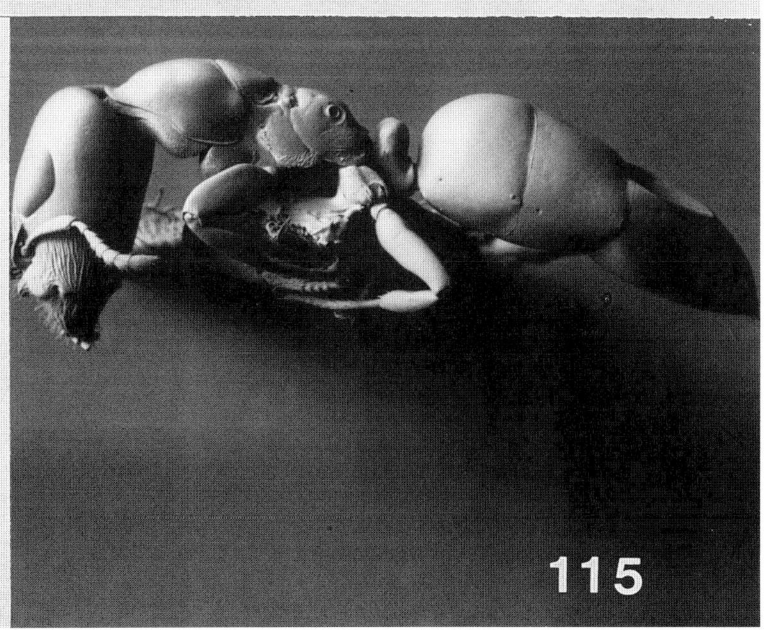
115

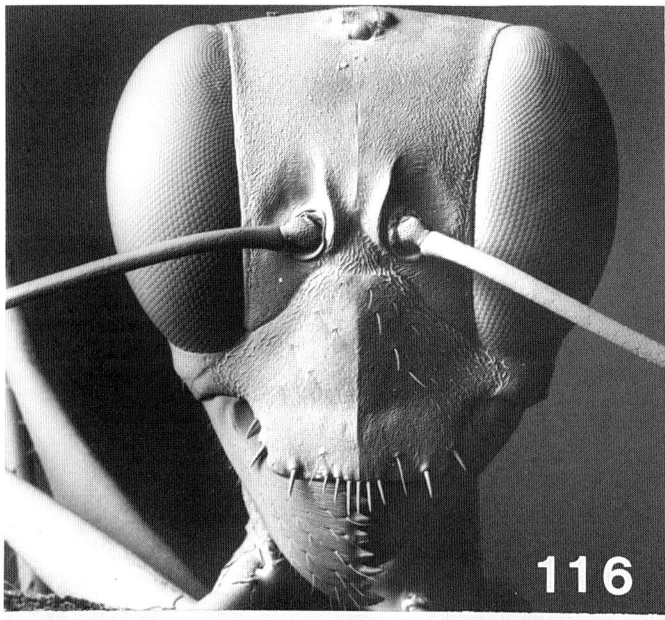
116

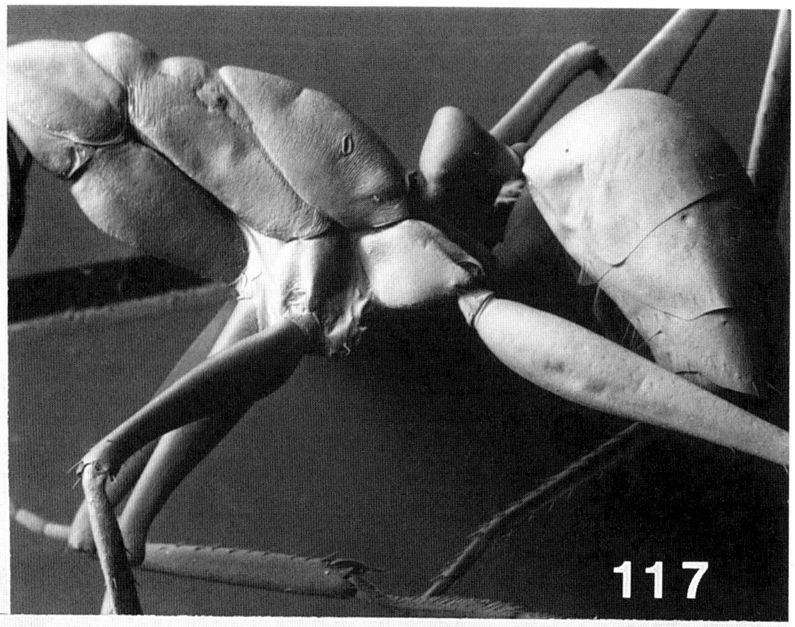
117

118

119

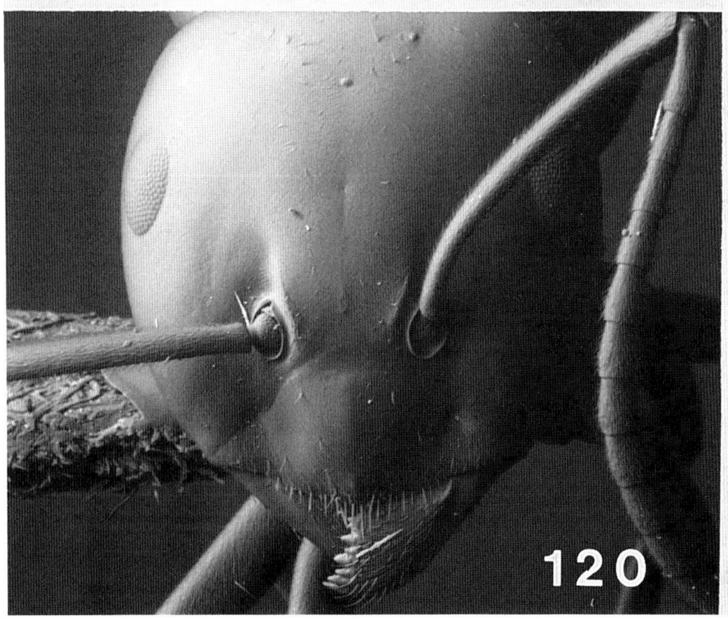
120

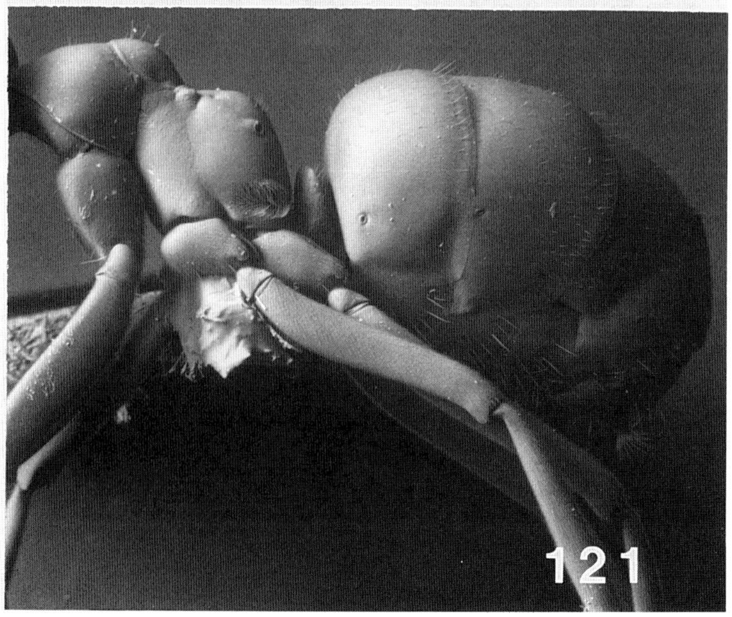
121

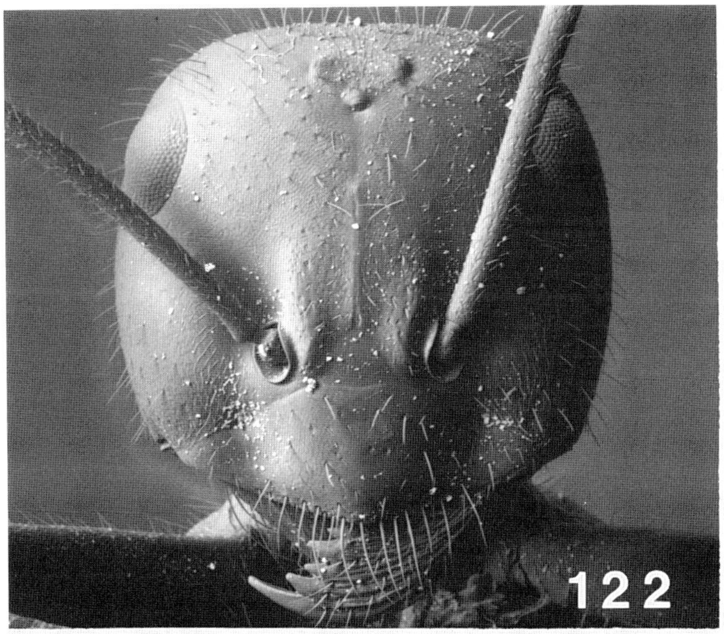
122

123

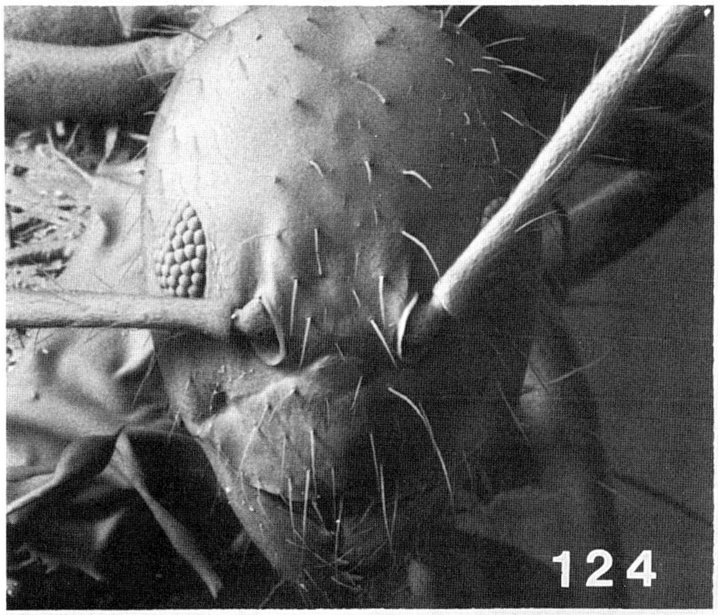
124

125

126

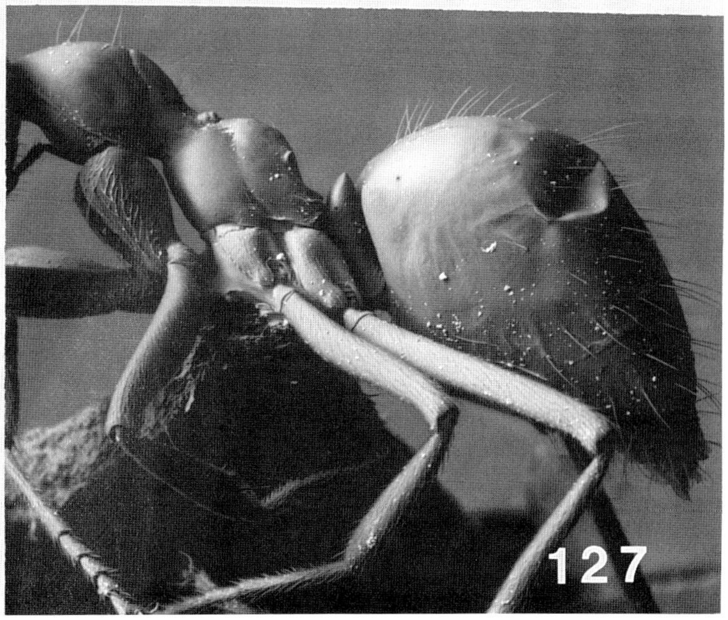
127

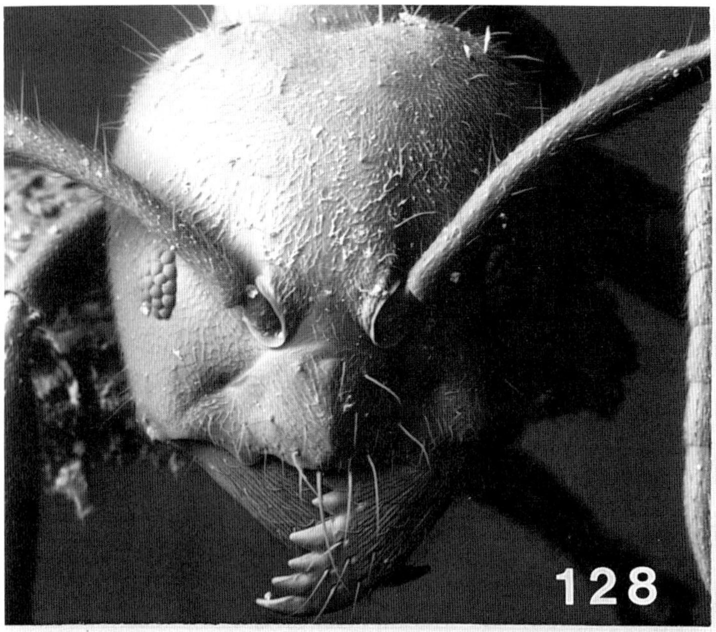
128

129

130

131

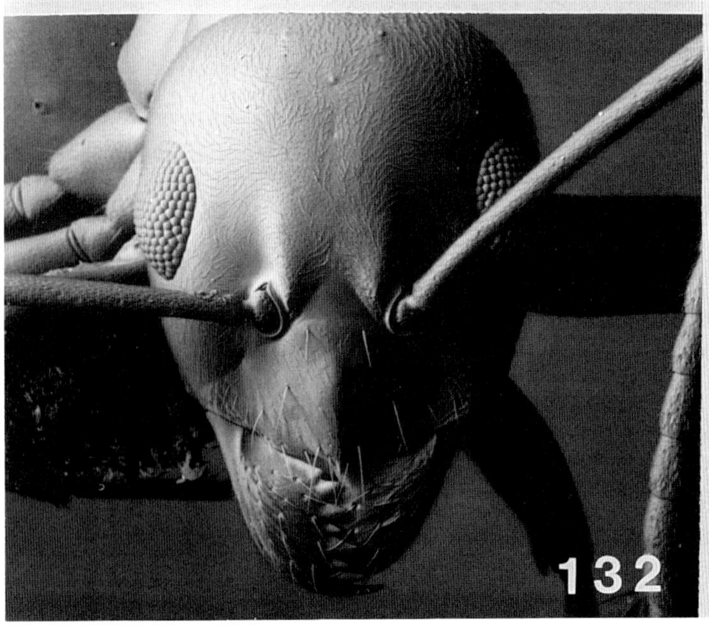
132

133

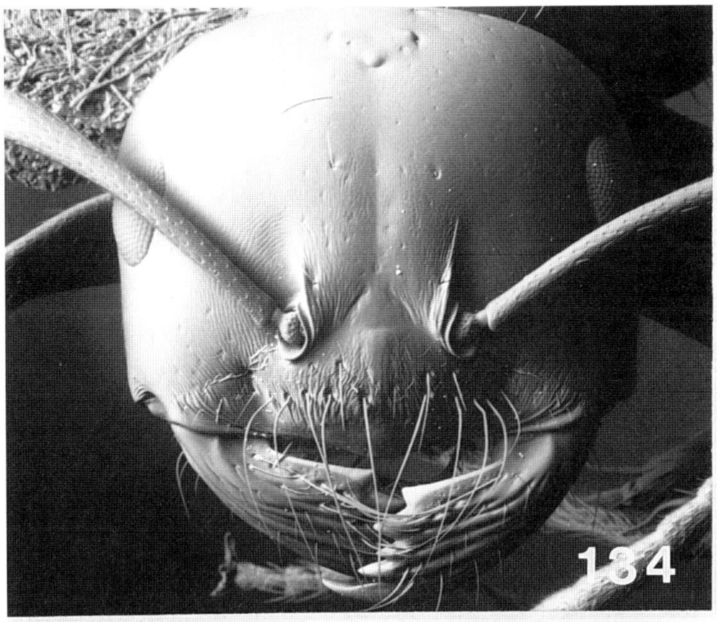
134

135

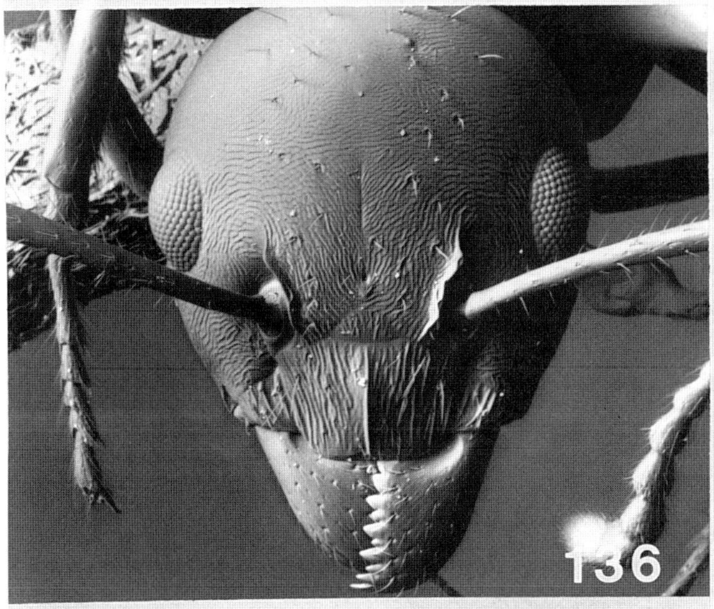
136

137

138

139

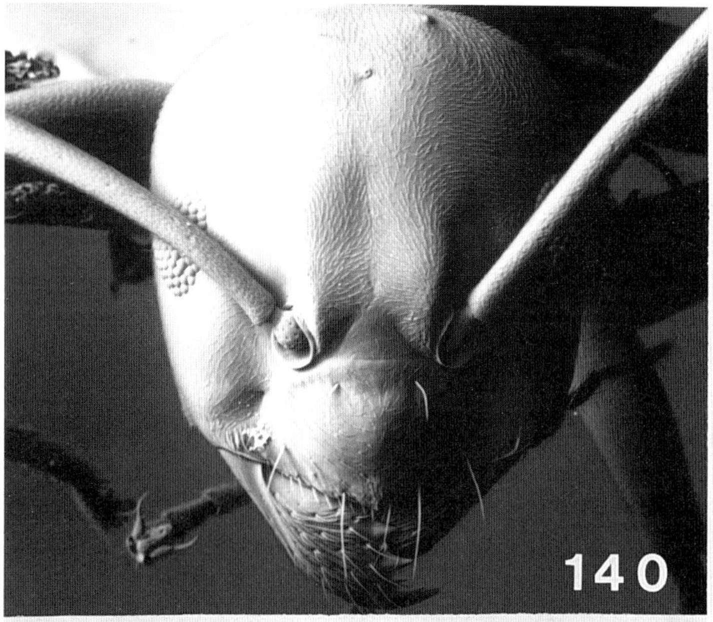
140

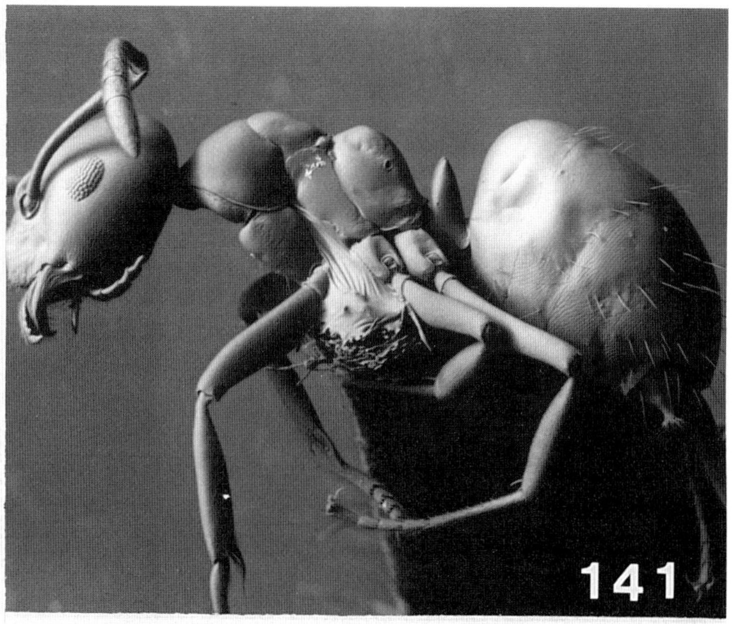
141

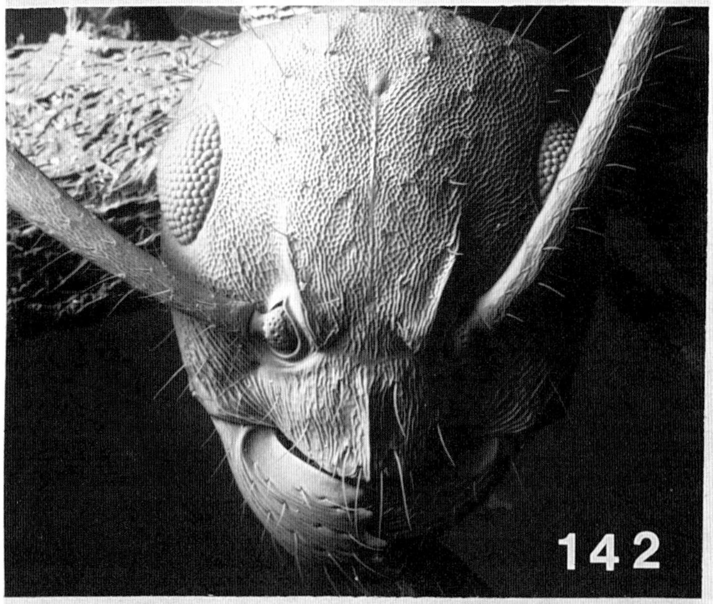
142

143

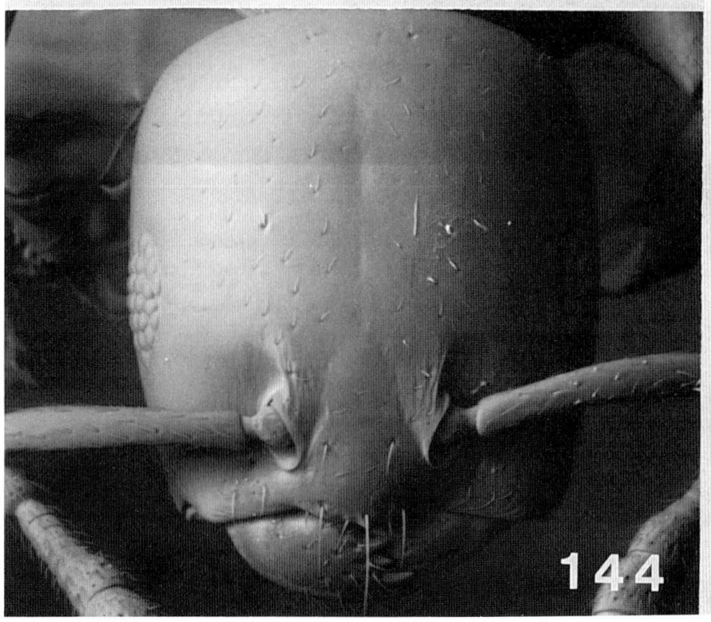
144

145

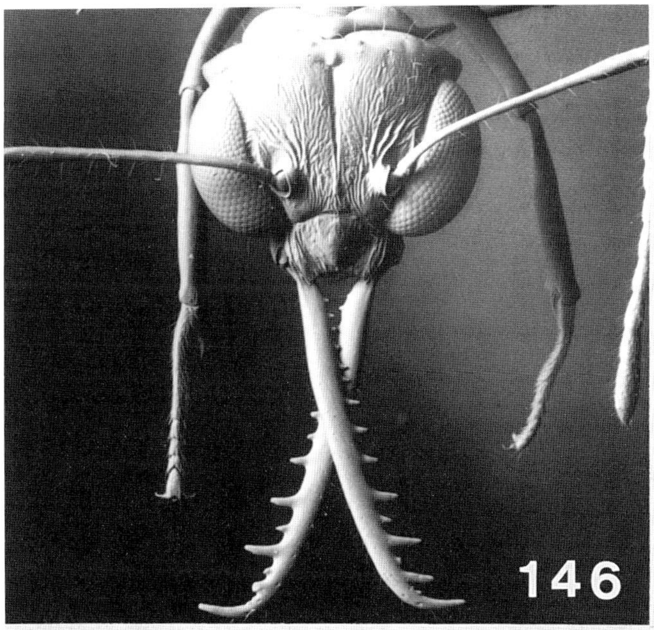
146

147

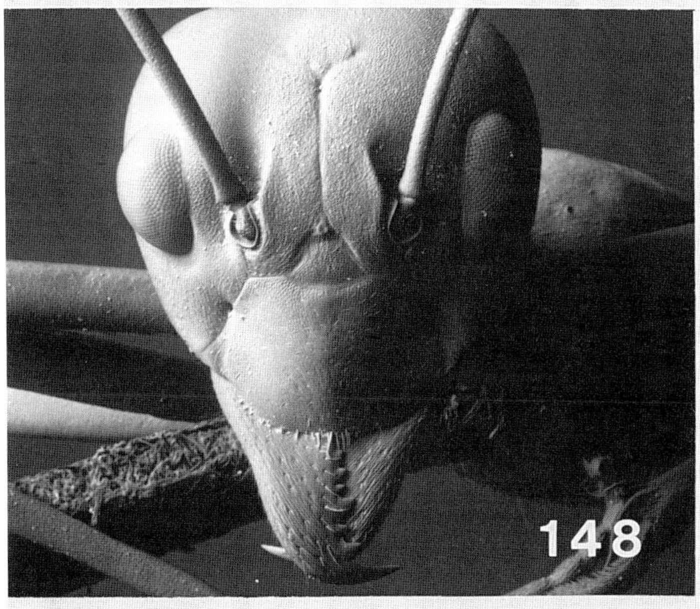
148

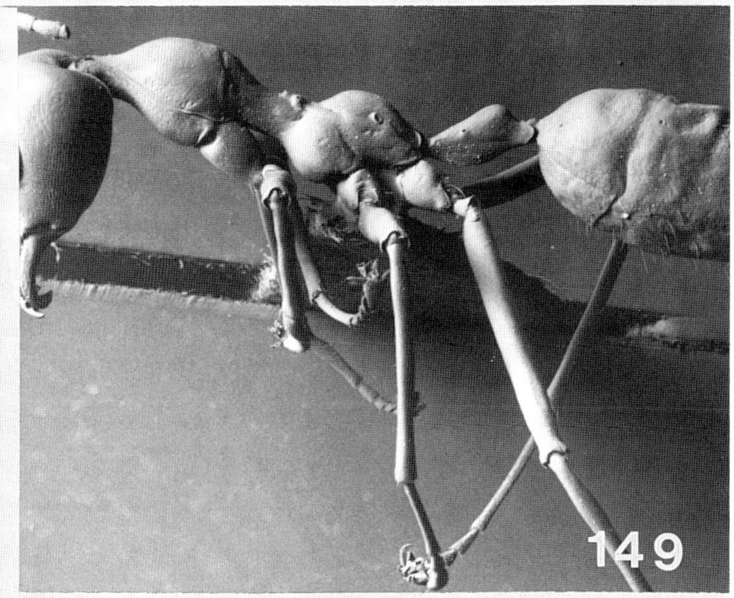
149

150

151

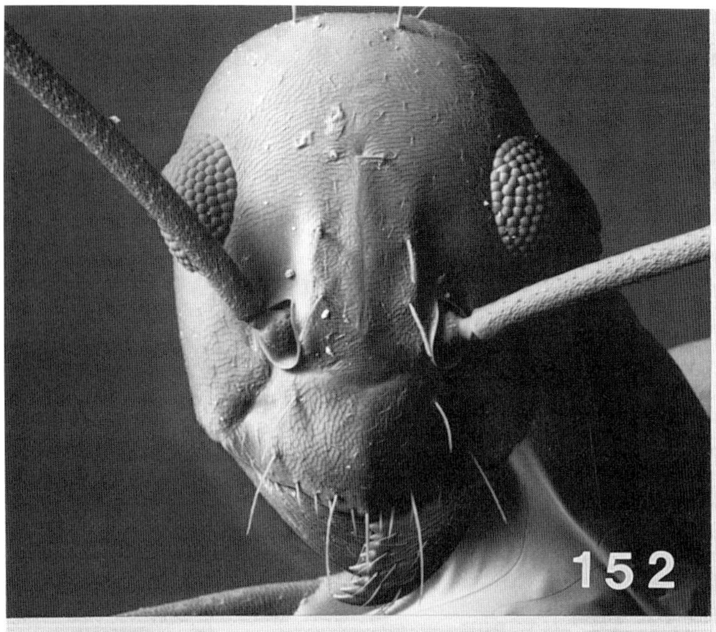
152

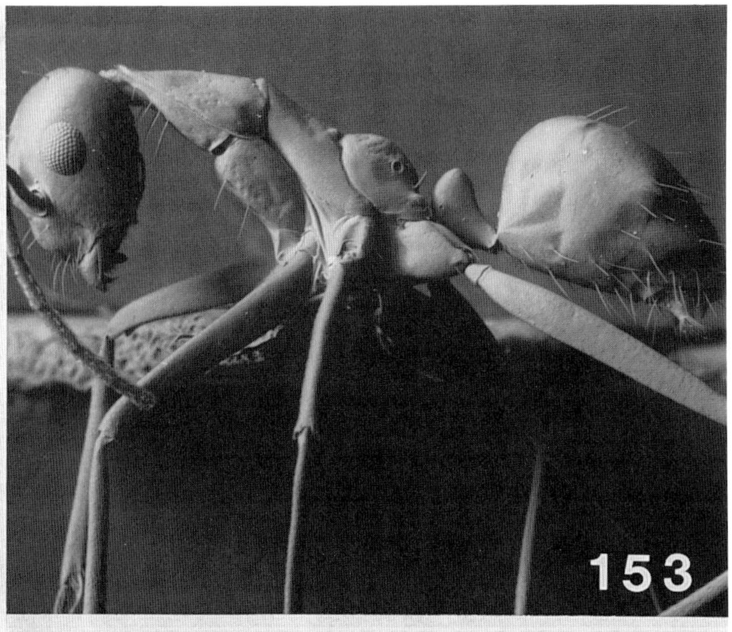
153

154

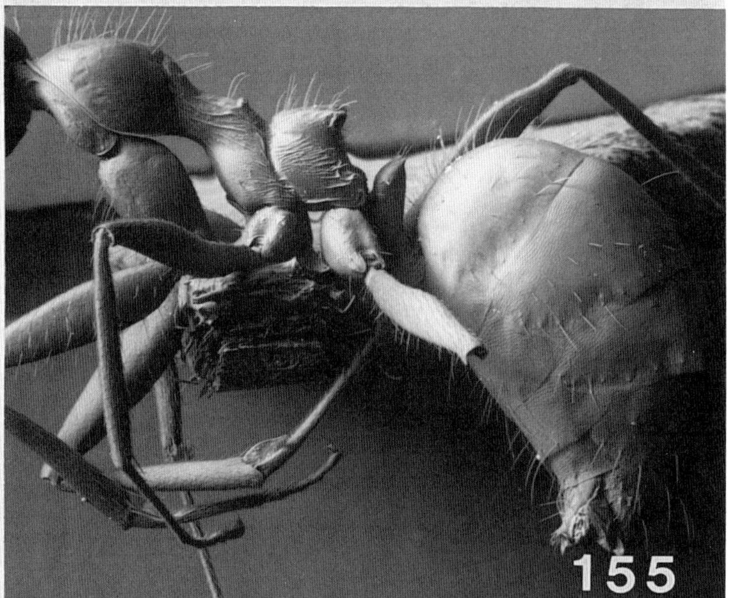
155

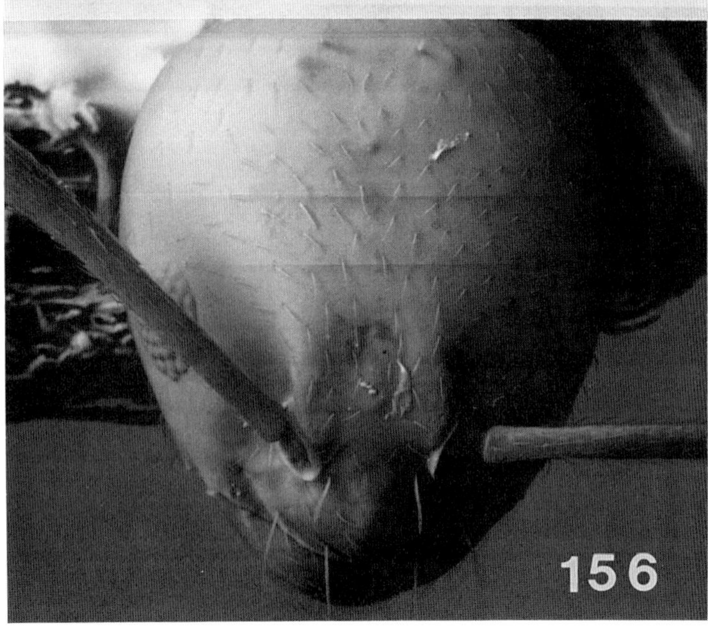
156

157

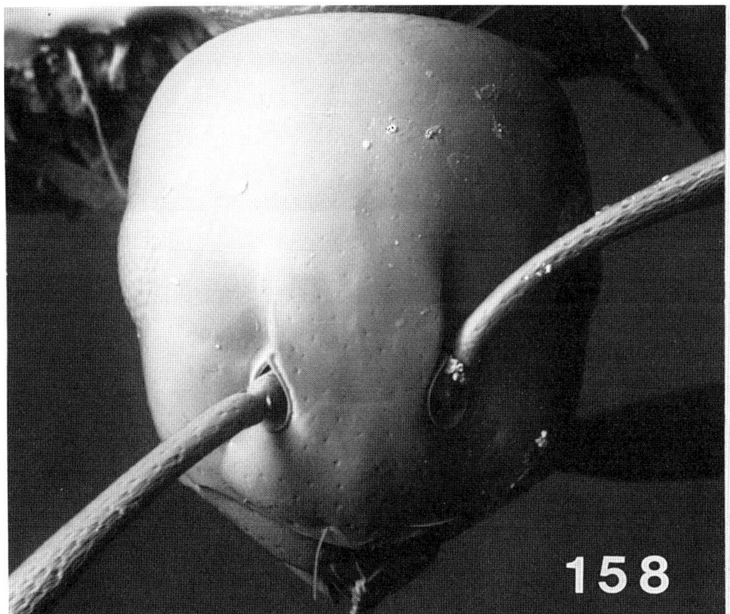
158

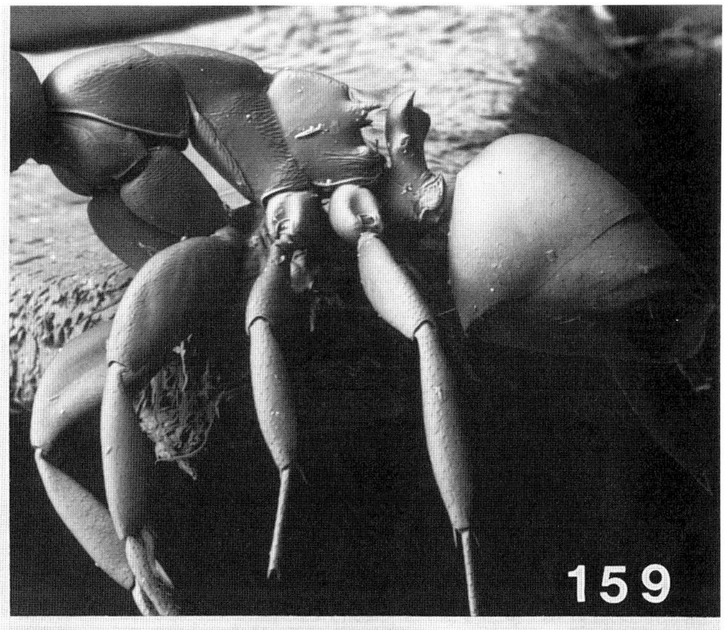
159

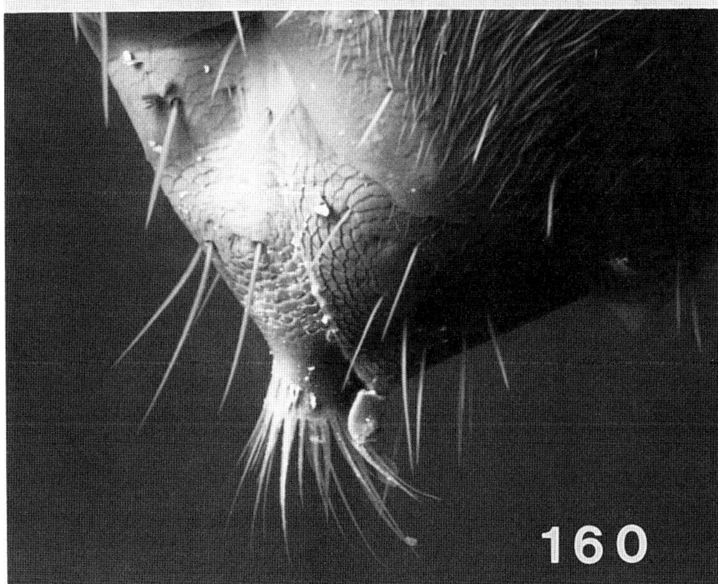
160

161

162

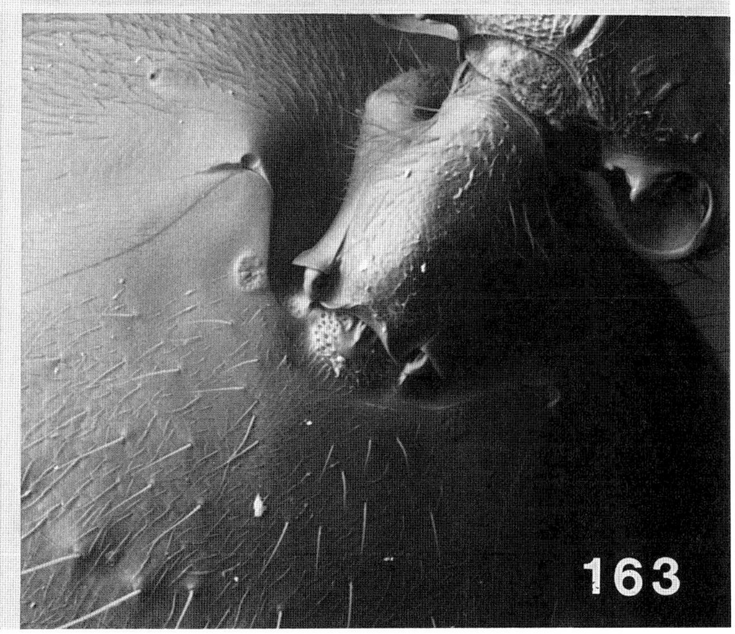
163

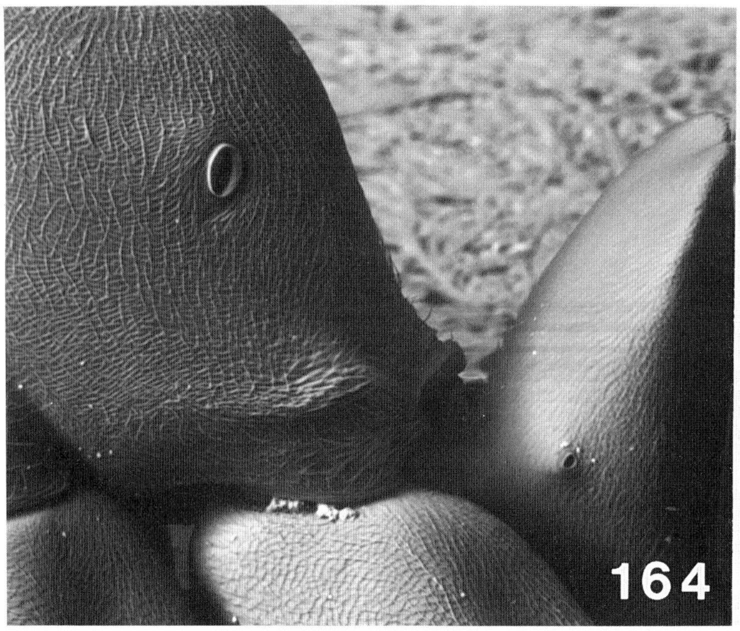
164

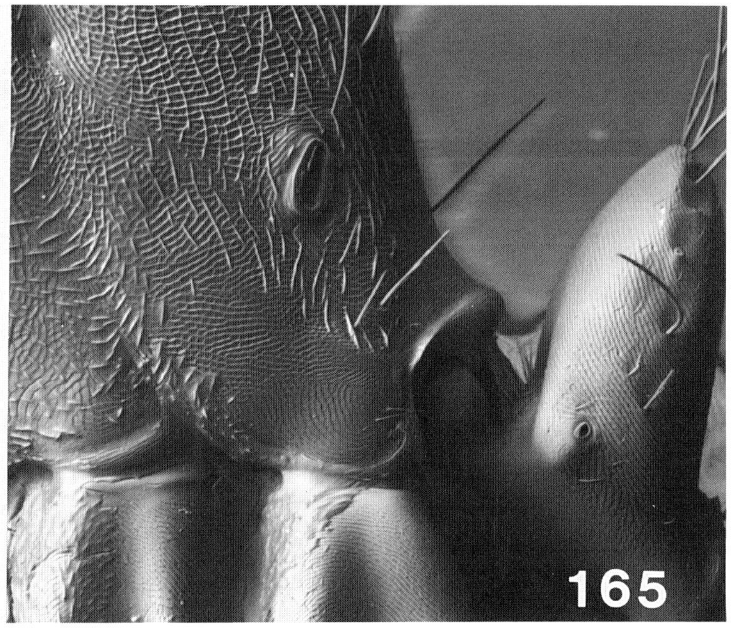
165

Subfamily LEPTANILLINAE

Diagnosis of Worker (Figs. 166–169)

Ants with the following combination of characters together.

1 Clypeus sometimes broad but usually narrow from front to back, bringing the antennal sockets close to the anterior margin of the head.

2 Antennal sockets horizontal, in the plane of the transverse axis of the head, exposed in full-face view.

3 Frontal lobes absent.

4 Narrow neck joining condylar bulb of antennal scape to shaft of scape proper straight, not sharply angled or bent downwards in frontal or full-face view.

5 Eyes absent; antenna with 12 segments.

6 Promesonotal suture present and flexible, may be deeply impressed; the pronotum capable of movement relative to the mesonotum.

7 Propodeal lobes absent.

8 Propodeal spiracle far back on sclerite, low down on the side except in a few species where the dorsum is depressed.

9 Metapleural gland orifice in lower posterior corner of metapleuron, opening laterally, the orifice concealed behind a cuticular flange or flap.

10 Metacoxal cavities closed; cuticular annulus around each cavity broad and complete, the annulus not broken medioventrally nor with a flexible suture traversing the annulus from the coxal cavity to the cavity in which the petiole articulates.

11 Waist of 2 segments, petiole plus postpetiole (= abdominal segments 2 and 3); the petiole sessile anteriorly.

12 Petiole tergite and sternite fused together, without trace of a suture.

13 Abdominal stridulatory system absent.

14 Abdominal spiracles 5–7 (= gastral spiracles 2–4) concealed by the posterior margins of the preceding segments and not visible without distension or dissection of the abdomen.

15 Helcium sternite small and retracted, concealed by the tergite, not visible in profile.

16 Abdominal segment 3 (= postpetiole) with tergosternal fusion; tergites and sternites of following abdominal segments (4–7 = gastral segments 1–4) not fused.

17 Abdominal segment 4 (= gastral segment 1) with presclerites sharply defined and differentiated from the postsclerites, the former fitting tightly within the posterior end of the third segment.

18 Pygidium (tergite of abdominal segment 7 = gastral segment 4) large and simple, unarmed, convex across and down-curved posteriorly.

19 Sting large, well developed, and functional.

Key to World LEPTANILLINAE (Workers)

1 Mandible with 3–5 teeth, all located on distal half of masticatory margin. Maxillary palp with 1 segment. Metanotal groove vestigial to absent. Bulla of metapleural gland rounded, located behind level of propodeal spiracle. Metapleural trench absent. Clypeus very narrow in front of antennal insertions, the latter very close to the anterior margin of the head (Figs. 166, 167). (Old World tropics and temperate zones) . . . . ***Leptanilla***

— Mandible with more than 5 peg-like to hook-like teeth. Maxillary palp with 4 segments. Metanotal groove present, strong, and impressed. Bulla of metapleural gland elongate and narrow, running longitudinally below the propodeal spiracle. Metapleural trench present. Clypeus relatively broad in front of antennal insertions, the latter well back from the anterior margin of the head (Figs. 168, 169) **2**

2 Mandible elongate, narrowly triangular, and down-curved, equipped with numerous peg-like teeth on inner surface (Fig. 168). (Southern Palaearctic, Oriental, Indo-Australian) . . . . ***Protanilla***

— Mandible surmounted by a large, erect lamella, which is lined internally with numerous, short, recurved cuticular teeth. (Palaearctic, Indo-Australian) . . . . ***Anomalomyrma***

Synoptic Classification

Subfamily **LEPTANILLINAE.**
Tribe **Anomalomyrmini.** Genera: ***Anomalomyrma, Protanilla*** (Figs. 168, 169).
Tribe **Leptanillini.** Genera: ***Leptanilla*** (Figs. 166, 167) (= *Leptomesites*), ***Noonilla*** (males only), ***Phaulomyrma*** (males only), ***Scyphodon*** (males only), **Yavnella** (males only).
[Material of the unavailable name Metadorylinae is in part referable to Leptanillini.]

Distribution

This subfamily is entirely absent from the New World and has not yet been discovered in the Malagasy region, though it may well prove to be present there. Otherwise the genus *Leptanilla* occurs in all Old World regions and the four genera known only from males are found in the Indo-Australian region *(Noonilla, Phaulomyrma, Scyphodon)* and the Palaearctic and Oriental regions *(Yavnella)*. The two remaining genera, *Anomalomyrma* and *Protanilla*, are centred on the Indo-Australian region, where the greatest number of species occur, but both also have species in the Palaearctic and the latter is also present in the Oriental region.

Taxonomic References

Identification of extant species
Leptanilla: Baroni Urbani (1977a). ***Yavnella:*** J. Kugler (1987).

Other taxonomic references
Leptanillinae: Petersen (1968); Baroni Urbani (1977a); Hölldobler and Wilson (1990); Bolton (1990b); Baroni Urbani, Bolton and Ward (1992).

Subfamily LEPTANILLOIDINAE

Diagnosis of Worker (Figs. 170–171)

Ants with the following combination of characters together.

1 Clypeus very reduced, narrow from front to back especially in front of the antennal insertions, bringing the antennal sockets extremely close to the anterior margin of the head.

2 Antennal sockets horizontal, in the plane of the transverse axis of the head, exposed in full-face view.

3 Frontal lobes absent; usually narrow vertical carinae present between the antennal sockets.

4 Narrow neck joining condylar bulb of antennal scape to shaft of scape proper straight, not sharply angled or bent downwards in frontal or full-face view.

5 Eyes absent; antenna with 12 segments.

6 Promesonotal suture present and flexible, the pronotum capable of movement relative to the mesonotum.

7 Metapleural gland orifice in lower posterior corner of metapleuron, opening laterally, the orifice concealed behind a ventrally directed cuticular flange or flap.

8 Propodeal spiracle situated low on side of sclerite, at or behind its midlength.

9 Metacoxal cavities closed; cuticular annulus around each cavity broad and complete, the annulus not broken medioventrally nor with a flexible suture traversing the annulus from the coxal cavity to the cavity in which the petiole articulates.

10 Waist of 2 segments, petiole and postpetiole (= abdominal segments 2 and 3).

11 Petiole sessile to subsessile, the tergite and sternite not fused; sternite of petiole with a simple posterior margin and simple articulation to the postpetiole.

12 Abdominal stridulatory system absent.

13 Abdominal spiracles 5–7 (= gastral spiracles 2–4) shifted backwards, not concealed by the posterior margins of the preceding segments and visible without distension or dissection of the abdomen.

14 Helcium sternite large and convex, bulging ventrally, visible in profile in normally mounted specimens; tergite of helcium lacking a notch or impression in its dorsal margin anteriorly.

15 Abdominal segment 3 (postpetiole) with tergosternal fusion; tergites and sternites of following abdominal segments (4–7 = gastral segments 1–4) not fused.

16 Abdominal segment 4 (= gastral segment 1) with presclerites sharply defined and differentiated from the postsclerites, the former fitting tightly within the posterior end of the third segment.

17 Gaster with deep girdling constrictions between segments 1 and 2, and between segments 2 and 3 (= abdominal segments 4, 5, and 6); strongly developed presclerites differentiated on abdominal segments 5 and 6 (= gastral segments 2 and 3).

18 Pygidium (tergite of abdominal segment 7 = gastral segment 4) extremely reduced, represented by a small, U-shaped sclerite, which is overhung by the tergite of abdominal segment 6 (= gastral segment 3).

19 Sting present but apparently reduced.

Synoptic Classification

Subfamily **LEPTANILLOIDINAE.**
Tribe **Leptanilloidini.** Genus: ***Leptanilloides*** (Figs. 170, 171).

Distribution

The single genus of this subfamily is restricted to the Neotropical region. Only two species are known, each from single collections in Bolivia and Colombia.

Taxonomic References

Leptanilloides: Brown (1975); Bolton (1990a); Baroni Urbani, Bolton and Ward (1992).

Figures 166–171 Worker heads in full-face view and bodies in profile:

166–169, LEPTANILLINAE: 166–167, **Leptanillini,** *Leptanilla;* 168–169, **Anomalomyrmini,** *Protanilla*

170–171, LEPTANILLOIDINAE, *Leptanilloides.*

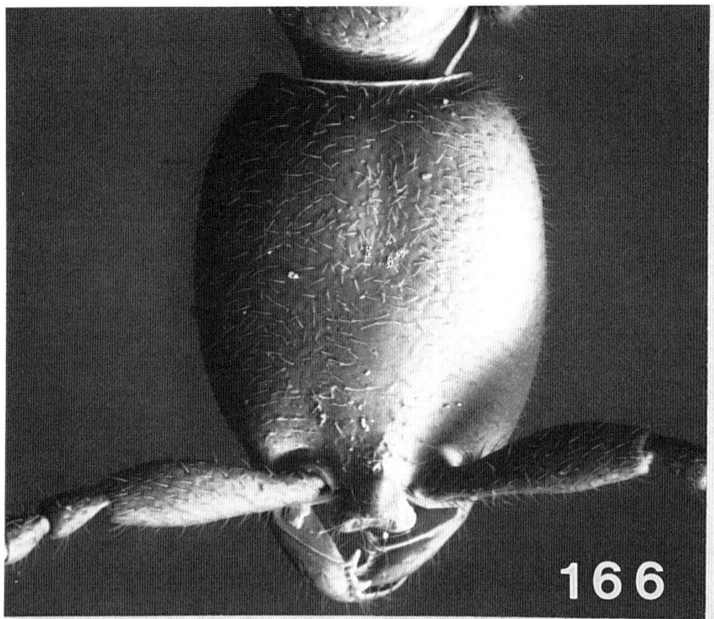

166

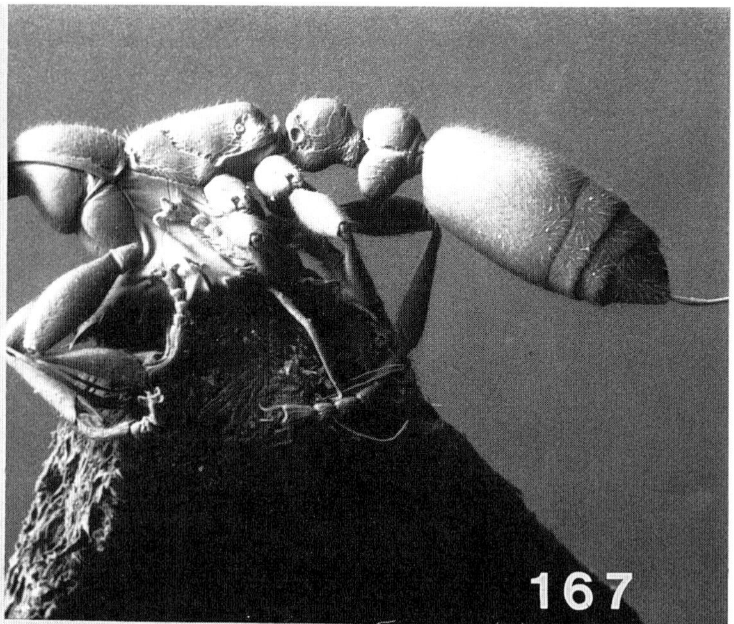

167

168

169

170

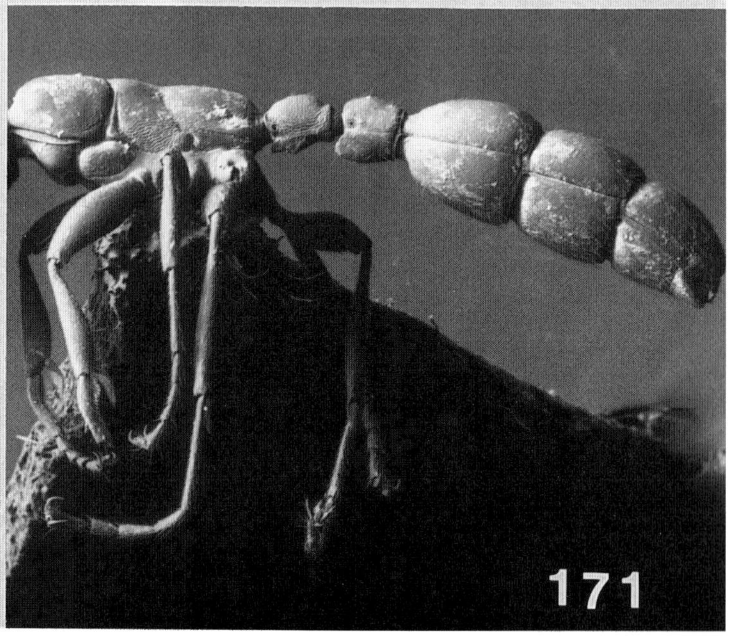

171

Subfamily MYRMECIINAE

Diagnosis of Worker (Figs. 172–173)

Ants with the following combination of characters together.

1 Clypeus broad from front to back so that antennal sockets are well back from anterior margin of head. Median portion of clypeus extended backwards between the frontal lobes.
2 Clypeo-labral hinge fully exposed and dorsum of labrum projecting anteriorly between the mandibular bases.
3 Antennal sockets inclined; their margins and section of torulus closest to the midline of the head on a higher level than the margin most distant from the midline.
4 Frontal lobes present over the antennal sockets but narrow and elevated; the lobes continued backwards for a short distance as narrow frontal carinae.
5 Narrow neck joining condylar bulb of antennal scape to shaft of scape proper straight.
6 Eyes present, large, situated anteriorly on the head with their anterior margins very close to the posterior clypeal margin. Ocelli always present. Antenna with 12 segments.
7 Promesonotal suture present and flexible, the pronotum capable of movement relative to the mesonotum.
8 Metapleural gland orifice at lower posterior corner of metapleuron, opening laterally, the orifice not concealed by a cuticular flange or flap.
9 Mesonotum distinctly defined; metanotum present on dorsal alitrunk.
10 Metacoxal cavities open; cuticular annulus around each cavity with a wide break or interruption medially so that the coxal cavity is confluent with the cavity in which the petiole articulates.
11 Propodeal lobes present.
12 Waist of 2 segments, the petiole and postpetiole (= abdominal segments 2 and 3).
13 Abdominal stridulatory system absent.
14 Abdominal segment 4 (= gastral segment 1) with sharply defined and differentiated presclerites, which fit tightly within the posterior end of the third segment.
15 Abdominal spiracles 5–7 (= gastral 2–4) concealed by posterior margins of preceding tergites, not visible without distension of the gaster.
16 Helcium sternite relatively small, retracted, concealed by the tergite, and not visible in profile.
17 Abdominal segments 2–7 (petiole to apex of gaster) without tergosternal fusion.
18 Pygidium (tergite of abdominal segment 7 = gastral segment 4) simple, biconvex, unarmed.
19 Sting present, large, and strongly developed.

Synoptic Classification

A name prefixed by * indicates an extinct taxon.
Subfamily **MYRMECIINAE.**
Tribe **Myrmeciini.** Genus: ***Myrmecia*** (Figs. 172, 173) (= *Halmamyrmecia,* = *Pristomyrmecia,* = *Promyrmecia*).
Tribe ***Prionomyrmecini.*** Genus: ****Prionomyrmex.***
Genera unplaced to tribe: ****Ameghinoia,*** ****Cariridris.***
[Material of the unavailable name Paleoponerinae is referable to Myrmeciini.]

Distribution

The single extant genus of this subfamily is restricted to the Australasian region, where it forms a fairly common and conspicuous fraction of the total ant fauna. All the extant species are from Australia except for one, which occurs naturally in New Caledonia. Fossil genera, from the Baltic Amber, Argentina, and Brazil, indicate that the subfamily was much more widely distributed in Late Cretaceous and Tertiary times.

Taxonomic References

Identification of extant species
Myrmecia: Clark (1951, out of date); Ogata and Taylor (1991).

Other taxonomic references
Myrmecia: Brown (1953b, 1954a); Hölldobler and Wilson (1990); Baroni Urbani, Bolton and Ward (1992); Ogata (1991a) [species-groups].

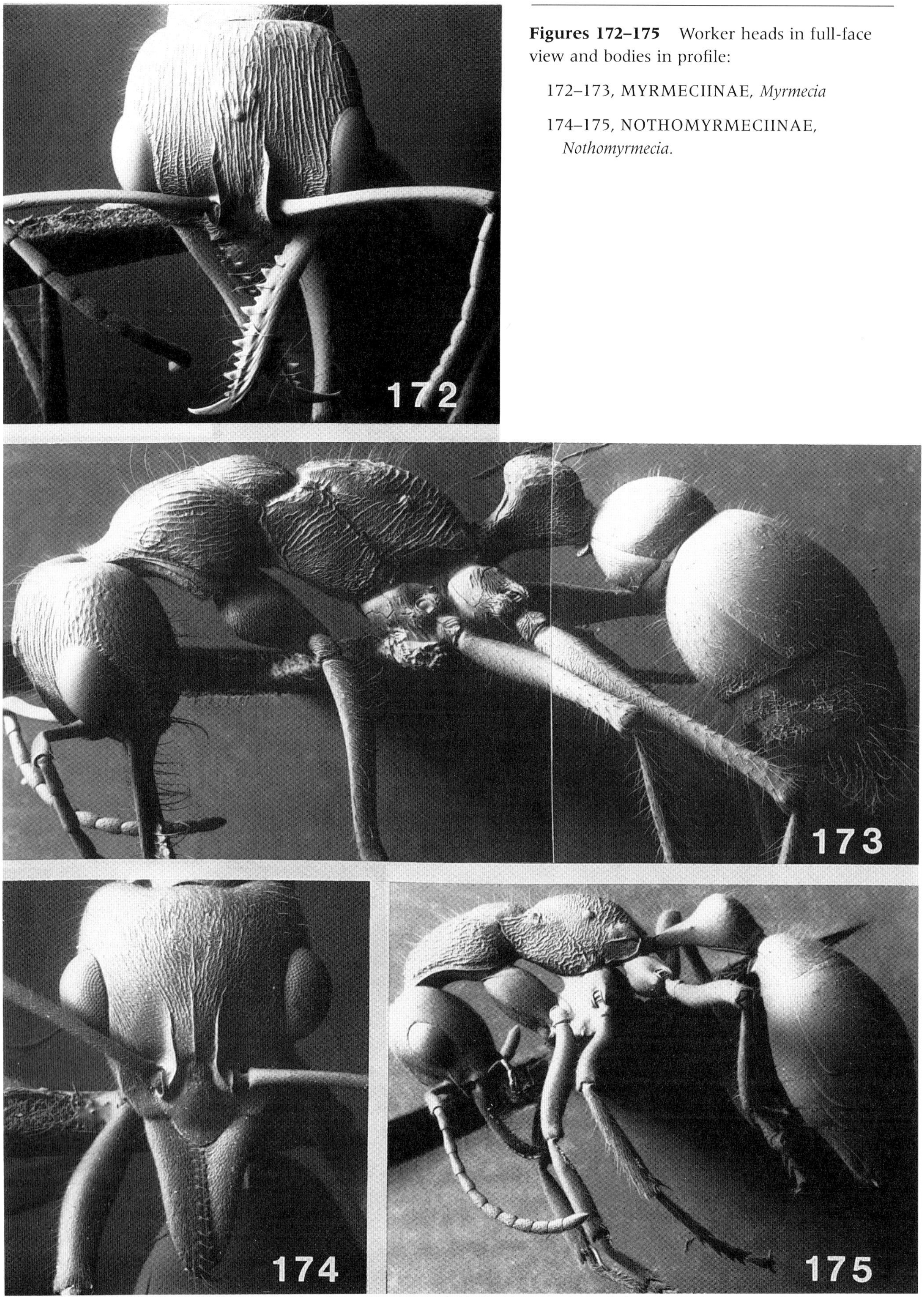

Figures 172–175 Worker heads in full-face view and bodies in profile:

172–173, MYRMECIINAE, *Myrmecia*

174–175, NOTHOMYRMECIINAE, *Nothomyrmecia*.

Subfamily MYRMICINAE

Diagnosis of Worker (Figs. 176–423)

Ants with the following combination of characters together.

1 Median portion of clypeus extended backwards between the frontal carinae.

2 Antennal sockets inclined to almost vertical; their margins and arc of torulus closest to the midline of the head on a higher level than the margin most distant from the midline, or the former almost directly above the latter.

3 Frontal lobes usually present, frequently large or very large and mostly or entirely concealing the antennal sockets; less commonly frontal lobes very small or rarely absent, leaving the antennal sockets partially to entirely exposed.

4 Narrow neck joining condylar bulb of antennal scape to shaft of scape proper usually straight, bent in only one genus.

5 Eyes usually present, less commonly vestigial or absent; antenna with 4–12 segments.

6 Promesonotal suture always absent, the pronotum and mesonotum firmly fused together and immobile with respect to each other; at most a weak line or feeble indentation may occur at the original site of the suture.

7 Metapleural gland orifice frequently invisible and probably absent; when present situated in lower posterior corner of metapleuron, opening laterally or posteriorly, and not concealed by a cuticular flange or flap.

8 Metacoxal cavities closed; cuticular annulus around each cavity broad and complete, not interrupted by a suture or gap linking the coxal cavity to the cavity in which the petiole articulates.

9 Propodeal lobes usually present; vestigial or absent in a few genera.

10 Waist of 2 segments, the petiole and postpetiole (= abdominal segments 2 and 3).

11 Abdominal stridulatory system usually present, only rarely absent; the stridulitrum situated on the pretergite of abdominal segment 4 (= gastral segment 1), the plectrum posteriorly on the preceding tergite.

12 Abdominal segment 4 (= gastral segment 1) with sharply defined and differentiated presclerites that fit tightly within the posterior end of the third segment.

13 Abdominal spiracles 5–7 (= gastral 2–4) concealed by posterior margins of preceding tergites, not visible without distension of the gaster.

14 Helcium sternite relatively large and convex, not retracted nor concealed by the tergite; helcium sternite attached across apices of tergite in frontal view.

15 Abdominal segment 2 (petiole) with tergosternal fusion; remaining segments (postpetiole to apex) with tergites and sternites not fused.

16 Pygidium (tergite of abdominal segment 7 = gastral segment 4) variable in size, simple.

17 Sting present, usually large and strongly developed, but reduced and nonfunctional as a weapon in some.

Key to Palaearctic MYRMICINAE (Workers)

1 Postpetiole articulated on dorsal surface of first gastral segment (Fig. 219); the gaster in dorsal view roughly heart-shaped and capable of reflexion over the alitrunk. Petiole dorsoventrally flattened and without a node .... ***Crematogaster***

— Postpetiole articulated on anterior surface of first gastral segment (e.g., Figs. 187, 197, 211, 255, 267, 303, 327, 337, 371, 389, 407, 417); the gaster in dorsal view not roughly heart-shaped, not capable of reflexion over the alitrunk. Petiole not dorsoventrally flattened, with a node of some form **2**

2 Apical and preapical antennal segments much larger than preceding funicular segments and forming a conspicuous, 2-segmented club (e.g., Figs. 177, 203, 220, 240, 288, 319, 347, 351, 355, 357, 391, 393, 411) **3**

— Antenna never terminating in a 2-segmented club. Either apical plus 2 preapical funicular segments of antenna enlarged and forming a conspicuous, 3-segmented club, or less commonly the club with more than 3 segments (e.g., Figs. 179, 257, 271, 272, 277, 285, 297, 309, 313, 329, 383, 389, 399, 407, 417).

Rarely the funiculus filiform and without a developed apical club **13**

3 Antenna with 4 or 6 segments **4**
— Antenna with 9–12 segments **11**

4 Mandibles elongate and linear, produced into narrow projecting blades (Figs. 234, 240). Mandibles never triangular or subtriangular, never serially multidentate or denticulate **5**
— Mandibles triangular or subtriangular, not produced into narrow projecting blades; apical (masticatory) margins usually serially multidentate or denticulate, but teeth sometimes reduced (Figs. 242, 248, 288) **7**

5 Apex of each mandibular blade either with a single, long tooth at the dorsal apex subtended by a series of minute denticles, or with a series of minute denticles only; always lacking an apical fork of 2 spiniform teeth. Labral lobes long and conical, visible between the mandibles in full-face view even when the mandibles are closed (Fig. 240) . . . . ***Epitritus***
— Apex of each mandibular blade armed with a fork of 2 spiniform teeth set in a vertical series, with or without intercalary denticles between the spiniform fork teeth. Labral lobes not long and conical, not visible between the mandibles when the latter are closed (Fig. 234) **6**

6 Antenna with 4 segments . . . . ***Quadristruma***
— Antenna with 6 segments . . . . ***Strumigenys***

7 Spongiform appendages present on petiole, postpetiole, or both (Figs. 243, 249). Frontal lobes widely separated, situated laterally on anterior half of head (Figs. 242, 248). Mandible with more than 4 teeth or denticles. Antennal scrobes present. Anterior coxae as large as or larger than the middle and hind coxae **8**
— Spongiform appendages absent from petiole and postpetiole (Fig. 289). Frontal lobes confluent, situated centrally and high on dorsum of head (Fig. 288). Mandible with 4 teeth. Antennal scrobes absent. Anterior coxae much smaller than the massively developed middle and hind coxae . . ***Melissotarsus***

8 Fully closed mandibles with a strongly defined transverse basal margin, which is separated from the anterior clypeal margin by a conspicuous impression or gap. Basal lamella of mandible situated ventral to the basalmost tooth, in a plane almost at right angle to the anterior portion of the mandible, not visible in full-face view with the mandibles open . . . . . ***Trichoscapa***
— Fully closed mandibles without a strongly defined basal margin, the basal region of the mandible overlapped by the anterior clypeal margin, the two not separated by an impression or gap. Basal lamella of mandible following basalmost tooth in the same plane (sometimes separated by a long diastema), visible in full-face view with the mandibles open **9**

9 Propodeum unarmed. Dorsal alitrunk with two distinct convexities, promesonotal and propodeal, which are separated by a transverse groove (Fig. 249) . . . . ***Kyidris***
— Propodeum armed with a pair of spines or teeth, equipped with large spongiform lamellae, or both. Dorsal alitrunk not raised into 2 separate convexities separated by a transverse groove (Fig. 243) **10**

10 Standing setae of some form usually present on head, alitrunk, or both. If setae absent from both these areas then the propodeum lacks teeth but has an extensive spongiform lamella running the height of the declivity . . . . ***Smithistruma***
— Standing setae always completely absent from head and alitrunk; propodeal teeth always present . . . . . ***Pentastruma***

11 Antenna with 12 segments. Palp formula 5,3. Mandible with 5 teeth. Lateral portions of clypeus flattened and prominent, fused to the raised, projecting median portion of the clypeus to form a shelf which projects forward over the mandibles (Fig. 258) . . . . ***Cardiocondyla*** (part)
— Antenna with 9–11 segments. Palp formula 1,2 or 2,2. Mandible with 4 or 5 teeth. Lateral portions of clypeus not flattened and prominent, not fused with median portion of clypeus nor forming a shelf projecting forward over the mandibles (Figs. 350, 352, 376) **12**

12 Anterior clypeal margin with a single, long, anteriorly projecting median seta at the midpoint of the margin (Fig. 376). Propodeum always unarmed and rounded (Fig. 377). Antenna always with 10 segments . . . . ***Solenopsis***
— Anterior clypeal margin lacking a single median seta, instead with a pair of setae straddling the midpoint of the margin (Figs. 350, 352). Propodeum with spines or teeth, or at least sharply angulate (Figs. 351, 353). Antenna with 9–11 segments . . . . ***Oligomyrmex***

13 Antenna with 11 segments **14**
— Antenna with 12 segments **24**

14 Propodeum armed with a pair of spines that curve upwards and forwards (Fig. 343). Postpetiole-gaster junction strongly dorsoventrally compressed and very narrow in profile. Basal tooth of mandible broad and with two points . . ***Recurvidris***
— Propodeum unarmed or with a pair of teeth or spines which are directed posteriorly (Figs. 251, 255, 349, 379, 417). Postpetiole-gaster junction not strongly dorsoventrally compressed. Basal tooth of mandible with a single point, or sometimes the mandible edentate **15**

15 Frontal lobes absent so that the antennal articulations are exposed and the depressed area containing the antennal sockets clearly visible (Fig. 300). Anterior clypeal margin irregular, crenulate to denticulate . . . . ***Pristomyrmex***
— Frontal lobes present, covering most or all of the antennal articulations, the antennal sockets not visible in dorsal view (Figs. 250, 252, 254, 262, 328, 348, 378, 416). Anterior clypeal margin unarmed or with a pair of small teeth . . . **16**

16 Petiole sessile to subsessile (Figs. 253, 255, 263). In profile the petiole lacking an anterior bar-like peduncle between the portion which articulates with the alitrunk and the ascending (anterior) face of the node **17**
— Petiole distinctly pedunculate (Figs. 251, 329, 349, 379). In profile the petiole with a longitudinal anterior bar-like pedun-

cle between the portion which articulates with the alitrunk and the ascending (anterior) face of the node **20**

17 Antennal scrobes present on sides of head above the eyes (Figs. 262, 263). Mandible edentate . . . . ***Harpagoxenus***

— Antennal scrobes absent (Figs. 250, 252, 254). Mandible usually with teeth, only extremely rarely edentate **18**

18 Sternite of petiole expanded ventrally into a massive process, lamella, or both (Fig. 255). Postpetiole ventrally lacking a median tooth-like projecting process. Palp formula 3,2 or 4,2 . . . . ***Epimyrma*** (part)

— Either sternite of petiole not expanded ventrally into a massive process or lamella, or postpetiole ventrally with a median tooth-like projecting process. Palp formula 4,3 or 5,3 . . . **19**

19 Postpetiole ventrally with a median tooth-like projecting process (Fig. 253). Palp formula 4,3 . . . . ***Formicoxenus***

— Postpetiole ventrally without a median tooth-like projecting process. Palp formula 5,3 . . . . ***Leptothorax*** (part)

20 Propodeum unarmed (Fig. 379). Mandible with only 3 or 4 teeth. Midpoint of anterior clypeal margin with a single, long seta (Fig. 378) . . . . ***Monomorium*** (part)

— Propodeum armed with a pair of spines or teeth (Figs. 251, 329, 349). Mandible with 5 or more teeth. Midpoint of anterior clypeal margin without a single, long seta; instead a pair of setae usually straddle the midpoint (Figs. 328, 348, 416) **21**

21 Palp formula 2,2. Propodeal lobes minute to absent (Fig. 349). Pronotal dorsum a flat plateau which is sharply marginate laterally, the marginations terminating anteriorly in projecting flat, acute, tooth-like or triangular processes, which are above and behind the true humeral angles of the pronotum . . . . ***Lophomyrmex***

— Palp formula 3,2 or more (up to 5,3). Propodeal lobes conspicuous (Figs. 251, 329, 417). Pronotal dorsum usually rounded, only rarely otherwise in which case marginations and processes as described above are absent **22**

22 Sting with an apicodorsal lamellate appendage projecting from the shaft (Fig. 417). Lateral portions of clypeus raised into a sharp ridge or shield wall on each side, in front of the antennal insertions (Fig. 416). Propodeal spiracle low on the side and distinctly behind the midlength of the sclerite . . . . ***Tetramorium*** (part)

— Sting without an apicodorsal lamellate appendage projecting from the shaft. Lateral portions of clypeus not raised into sharp ridges or shield walls in front of the antennal insertions (Figs. 250, 328). Propodeal spiracle high on the side, at or close to the midlength of the sclerite **23**

23 With head in profile the large eyes drawn out anteroventrally in a broad lobe, which runs down the side and almost onto the ventral surface of the head close to the mandibular insertion (Fig. 329) . . . . ***Oxyopomyrmex***

— With head in profile the eyes roughly oval, not drawn out anteroventrally in a broad lobe running almost onto the ventral surface of the head; anterior margin of eye a considerable distance from the mandibular insertion (Fig. 251) . . . . ***Leptothorax*** (part)

24 Palp formula 6,4. Tibial spurs of middle and hind legs usually pectinate, rarely the spurs simple or absent **25**

— Palp formula 1,2 to 5,3. Tibial spurs of middle and hind legs usually simple, sometimes absent; only rarely are the spurs pectinate **26**

25 Propodeum bidentate to bispinose (Fig. 309). Mandible with 6–10 teeth (Fig. 308). Metasternal process a closely approximated pair of raised flanges or plates, the ventral midline not visible between them . . . . ***Myrmica***

— Propodeum unarmed and rounded. Mandible with more than 12 teeth. Metasternal process a pair of crudely arched-convex, thickened lobes, the ventral midline visible between them . . . . ***Manica***

26 Sting with an apicodorsal triangular to pennant-shaped lamellate appendage projecting from the shaft (Figs. 417, 419). Lateral portions of clypeus raised into a sharp-edged ridge or shield wall on each side in front of the antennal insertions (Figs. 416, 418, 420); median portion of clypeus broadly inserted between the frontal lobes **27**

— Sting without apicodorsal triangular to pennant-shaped lamellate appendage projecting from the shaft. Lateral portions of clypeus usually not raised into sharp-edged ridges or shield walls in front of the antennal insertions (Figs. 250, 256, 254, 258, 302, 326, 334, 338, 340, 378, 388, 396), but if weakly raised then median portion of clypeus narrowly inserted between the frontal lobes **29**

27 Mandible narrow and falcate, edentate, or at most with a single, minute denticle close to the acute apex (Fig. 420) . . . . ***Strongylognathus***

— Mandible triangular or subtriangular, dentate, with 2 or 3 larger teeth distally, which are followed proximally by a row of 3 or more smaller teeth or denticles (Figs. 416, 418) **28**

28 Head heart-shaped in full-face view. Ventral margin of petiole convex and keel-like. Anterior clypeal margin strongly arcuate and prominent (Fig. 418). Eyes behind midlength of side of head. Median clypeal and median cephalic carinae vestigial or absent. Palp formula 3,2 . . . . ***Rhoptromyrmex***

— Head not heart-shaped in full-face view. Ventral margin of petiole not convex and keel-like. Anterior clypeal margin not strongly arcuate (Fig. 416). Eyes usually at or in front of midlength of sides of head, only extremely rarely otherwise. Either median clypeal carina or median cephalic carina usually present, or both present; only infrequently with both absent. Palp formula predominently 4,3, rarely reduced . . . . ***Tetramorium*** (part)

29 Sides of head with broad, deep scrobes that can accommodate the entire antenna when folded (Figs. 396, 397). Apical (masticatory) margin of mandible serially dentate from apex to base, without a long diastema . . . . ***Lordomyrma***

— Antennal scrobes absent or weakly present; if the latter then apical (masticatory) margin of mandible with an apical and preapical tooth, which are followed by a long diastema and then 1 or 2 smaller basal teeth **30**

30 Ventrolateral margin of head delineated by a sharp longitudinal carina on each side. The carina starts close to the inner-ventral mandibular base, runs the length of the head below the eye, and ascends the occipital surface posteriorly (Fig. 303) . . . . ***Myrmecina***

— Ventrolateral margin of head without a longitudinal carina on each side (Figs. 251, 257, 295, 335, 339, 341, 389) **31**

31 With head in profile the large eyes drawn out anteroventrally in a broad lobe, which runs down the side and almost onto the ventral surface of the head close to the mandibular insertions . . . . ***Goniomma***

— With head in profile the eyes not drawn out anteroventrally in a broad lobe running almost onto the ventral surface of the head; anterior margin of eye a considerable distance from the mandibular insertions **32**

32 Median portion of clypeus (in front of level of frontal lobes) sharply raised and with a relatively narrow dorsal area, the raised portion with a pair of longitudinal carinae, which usually arise between the frontal lobes and which usually diverge anteriorly (Figs. 294, 378, 388). Median portion of clypeus never with a single, median longitudinal carina **33**

— Median portion of clypeus (in front of level of frontal lobes) broad, evenly convex to more or less flat, not sharply raised, without carinae or frequently with an unpaired median carina (Figs. 250, 254, 256, 258, 332, 334, 338, 340). Never with a median pair of anteriorly divergent carinae upon a sharply raised narrow dorsal area **36**

33 Petiole sessile, lacking an anterior peduncle. Petiole node ventrally with a large and strongly projecting plate-like process (Fig. 295) . . . . ***Vollenhovia***

— Petiole pedunculate. Petiole node ventrally without a large plate-like process (usually a small anteroventral process present on the peduncle) (Figs. 379, 389) **34**

34 Dorsum of pronotum medially with a pair of blunt raised tubercles; mesonotal dorsum posteriorly with a single median tubercle. Anterior clypeal margin lacking long, projecting setae. Lateral portions of clypeus raised into blunt crests in front of the antennal insertions. Dorsum of head with a shallow longitudinal median furrow . . . . ***Dacatria*** [Description in preparation, by F. Rigato, University of Milan; name not available here.]

— Dorsum of pronotum without a pair of tubercles; mesonotal dorsum without a median tubercle. Anterior clypeal margin with long projecting setae present at least close to the midpoint. Lateral portions of clypeus not raised into crests in front of the antennal insertions. Dorsum of head lacking a longitudinal median furrow **35**

35 Propodeum unarmed and rounded (Fig. 379). Mandible with 3–5 (usually 4) teeth or denticles in total. Petiolar spiracle at the node or on the peduncle very close to the node. Midpoint of anterior clypeal margin with a single, long seta, which projects forward over the mandibles (Fig. 378) . . . . ***Monomorium*** (part)

— Propodeum bidentate (Fig. 389). Mandible with 6 or more (usually more) teeth or denticles in total, though sometimes the basalmost teeth poorly defined. Petiolar spiracle on the peduncle very close to the articulation with the alitrunk. Midpoint of anterior clypeal margin straddled by a pair of setae, without an unpaired, long median seta (Fig. 388) . . . . ***Stenamma***

36 Petiole sessile; in profile the petiole lacking a longitudinal anterior bar-like peduncle between the section which articulates with the alitrunk and the ascending (anterior) face of the node **37**

— Petiole pedunculate; in profile the petiole with a longitudinal anterior bar-like peduncle between the section which articulates with the alitrunk and the ascending (anterior) face of the node **38**

37 Sternite of petiole expanded ventrally into a massive process (Fig. 255) . . . . ***Epimyrma*** (part)

— Sternite of petiole at most with a small, anteroventral tooth . . . . ***Leptothorax*** (part)

38 Either the apical (masticatory) margin of the mandible with more than 5 teeth or denticles in total (dental count usually more than 7) (Figs. 334, 338), or the teeth variously worn down to blunt stubs (Fig. 340), or the margin functionally edentate. In the last two cases the original basal outline of each tooth may or may not be discernible **39**

— Apical (masticatory) margin of mandible with 3–5 well-defined teeth or denticles in total (Figs. 250, 256, 258, 332); apical margin not showing the remains of numerous teeth worn down to stubs nor functionally edentate **41**

39 Apical (masticatory) margin of mandible with the third tooth (counting from the apex) smaller than the fourth tooth, or the reduced third tooth followed by a minute denticle before the larger fourth tooth (Fig. 334). Palp formula 2,2 or 3,2. Antennal funiculi terminating in a strongly defined, 3-segmented club . . . . ***Pheidole*** (part)

— Apical (masticatory) margin of mandible with the third tooth (counting from the apex) larger than the fourth tooth; or teeth worn to stubs or the margin edentate (Figs. 338, 340). Palp formula 4,3 or 5,3. Antennal funiculi terminating in a weakly defined club of 4 segments or without a differentiated club, the segments gradually increasing in size to the apex **40**

40 Metasternal process large or very large. Head massive and broad in media and major workers (CI > 90) (Fig. 340). Mandibles short and powerful, massively constructed, their outer margins strongly curved toward the midline. Mostly polymorphic species, some of which may have the mandibular teeth much worn down . . . . **_Messor_**

— Metasternal process minute to absent. Head narrow in all workers (CI 90 at maximum, usually much less) (Fig. 338). Mandibles elongate-triangular and not massively constructed, their outer margins not strongly curved toward the midline. Monomorphic species . . . . **_Aphaenogaster_**

41 Mandible powerfully constructed, armed with 2 large apical teeth followed by a long diastema and then with 1 or 2 (rarely 3) basal teeth (Fig. 332). Two to 4 hypostomal teeth usually present on the posterior margin of the buccal cavity. Palp formula 2,2 or 3,2 . . . . **_Pheidole_** (part)

— Mandible delicately constructed, armed with 5 teeth, serially dentate, and the teeth decreasing in size from apex to base, not arranged as above (Figs. 250, 256, 258). Hypostomal teeth absent from posterior margin of buccal cavity. Palp formula 5,3 . . . . **42**

42 Midpoint of anterior clypeal margin with an unpaired, long seta, which projects forward over the mandibles. Lateral portions of clypeus flattened and projecting over the mandibles, sometimes the lateral portions projecting farther forward than the median portion of the clypeus (Fig. 258) . . . . **_Cardiocondyla_** (part)

— Midpoint of anterior clypeal margin without an unpaired, long seta; instead, the midpoint usually straddled by a pair of setae. Lateral portions of clypeus not flattened and projecting over the mandibles, never projecting farther forward than the median portion of the clypeus (Figs. 250, 256) **43**

43 Frontal carinae absent (Fig. 250) . . . . **_Leptothorax_** (part)

— Frontal carinae present but usually weak, running back from the frontal lobes as a pair of fine raised ridges or lines (Fig. 256) . . . . **_Chalepoxenus_**

Key to Afrotropical and Malagasy MYRMICINAE (Workers)

1 Postpetiole articulated on dorsal surface of first gastral segment (Fig. 219); the gaster in dorsal view roughly heart-shaped and capable of reflexion over the alitrunk. Petiole dorsoventrally flattened and without a node . . . . **_Crematogaster_**

— Postpetiole articulated on anterior face of first gastral segment (e.g., Figs. 187, 197, 211, 255, 267, 303, 327, 337, 371, 389, 407, 417); the gaster in dorsal view not roughly heart-shaped, not capable of reflexion over the alitrunk. Petiole not dorsoventrally flattened, with a node of some form **2**

2 Apical and preapical antennal segments much larger than preceding funicular segments and forming a conspicuous and usually very distinctive 2-segmented club (e.g., Figs. 177, 203, 220, 240, 288, 319, 347, 351, 355, 357, 391, 393, 411) . . **3**

— Antenna never terminating in a conspicuous, 2-segmented club. Either apical plus 2 preapical segments of antenna enlarged and forming a conspicuous, 3-segmented club, or less commonly the club with more than 3 segments (e.g., Figs. 179, 257, 271, 272, 277, 285, 297, 309, 313, 329, 383, 389, 399, 407, 417). Rarely the funiculus filiform and without a developed apical club **24**

3 Mandibles elongate and linear, produced into narrow projecting blades, each one of which is much longer than broad (Figs. 226, 234, 240). Mandibles never triangular or subtriangular, never serially multidentate or denticulate **4**

— Mandibles triangular or subtriangular, not produced into narrow projecting blades, (Figs. 242, 244, 246, 258, 288, 346, 350, 412); apical (masticatory) margins usually serially multidentate or denticulate, but teeth sometimes reduced . . . **8**

4 Apex of each mandibular blade armed with a fork of 2 or 3 spiniform teeth set in a more or less vertical series, with or without intercalary teeth between the spiniform fork teeth **5**

— Apex of each mandibular blade either with a single, long tooth at the dorsal apex subtended by a series of minute denticles, or with a series of minute denticles only; always lacking an apical fork of 2 or 3 spiniform teeth **7**

5 Apical fork of mandible with 3 spiniform teeth; blade of mandible without preapical teeth (Fig. 226). Maxillary palp 3-segmented. Antennal scrobes absent, the eyes dorsolateral. Petiole node with a pair of teeth or short spines (Fig. 227), postpetiole with lateral lamellate appendages . . . . **_Microdaceton_**

— Apical fork of mandible with 2 spiniform teeth; blade of mandible usually with preapical teeth (Figs. 234, 240). Maxillary palp 1-segmented. Antennal scrobes present, the eyes ventrolateral. Petiole node unarmed (Figs. 235, 241), postpetiole with spongiform appendages **6**

6 Antenna with 4 segments . . . . **_Quadristruma_**

— Antenna with 6 segments . . . . **_Strumigenys_**

7 Antennal scape with a broad anteriorly projecting subbasal lobe (Fig. 240). Clypeal margin with spatulate or strap-like projecting setae. Head with large orbicular setae present; the head broad, wider than long, CI > 100 . . . . **_Epitritus_**

— Antennal scape linear, without a projecting lobe (as in Fig. 246). Clypeal margin without spatulate or strap-like projecting setae. Head only with simple setae present and the head longer than wide, CI < 100 . . . . **_Cladarogenys_**

8 Antenna with 4–6 segments **9**

— Antenna with 8–12 segments **14**

9 Spongiform or lamellate appendages absent from petiole and postpetiole (Fig. 289). Frontal lobes confluent, situated centrally and high on dorsum of head (Fig. 288). Mandible with 4 teeth. Antennal scrobes absent. Anterior coxae much smaller than the massively developed middle and hind coxae (Fig. 289) . . . . **_Melissotarsus_**

— Spongiform or lamellate appendages present on petiole, postpetiole, or both (Figs. 243, 245, 247, 249). Frontal lobes widely separated, situated laterally on anterior half of head (Figs. 242, 244, 246, 248). Mandible with more than 4 teeth. Antennal scrobes present. Anterior coxae as large as or larger than the middle and hind coxae (Figs. 243, 247) **10**

10 Differentiated, prominent basal lamella of mandible absent. Apical (masticatory) margin of mandible with more than 20 denticles (Fig. 246), the basal 4–8 of which may be enlarged. Mandible relatively long, MI > 25 .... ***Serrastruma***

— Differentiated, prominent basal lamella of mandible present. Apical (masticatory) margin of mandible with 17 or fewer teeth or denticles of varying size (Figs. 242, 244, 248). Mandible relatively short, MI < 25 .... **11**

11 Fully closed mandibles with a strongly defined transverse basal margin, which is separated from the anterior clypeal margin by a conspicuous impression or gap. Basal lamella of mandible situated ventral to the basalmost tooth, in a plane almost at right angle to the anterior portion of the mandible, not visible in full-face view with the mandibles open ..... ***Trichoscapa***

— Fully closed mandibles without a strongly defined basal margin, the basal region of the mandible contiguous with or overlapped by the anterior clypeal margin, the two not separated by an impression or gap. Basal lamella of mandible following basalmost tooth in the same plane, visible in full-face view with the mandibles open **12**

12 Propodeum unarmed. Dorsal alitrunk with two distinct convexities, promesonotal and propodeal, which are separated by a transverse groove (Fig. 249) .... ***Kyidris***

— Propodeum armed with a pair of spines or teeth, which may or may not be incorporated in spongiform tissue. Dorsal alitrunk not raised into two convexities separated by a transverse groove (Figs. 243, 245) **13**

13 With the head in profile the mandible increasing in width from base to apex and the distal portion of the blade passing into a strongly down-curved arc so that part or most of the apical margin is at right angle to the long axis of the head (Fig. 245). Apical (masticatory) margin of mandible armed with a basal lamella plus 8–11 teeth, the basal 5–8 of which may be very strong (Fig. 244) .... ***Glamyromyrmex***

— With the head in profile the mandible with its upper and lower margins approximately parallel for most of its length or evenly tapering anteriorly (Fig. 243). At most the extreme tip of the mandible down-curved, without the major part of the apical margin at right angle to the long axis of the head. Apical (masticatory) margin of mandible armed with a basal lamella plus 12–17 teeth or denticles, the apicalmost group of which are minute .... ***Smithistruma***

14 Mandible with 7 large teeth which increase in size from apex to base; between each tooth is a minute denticle. Mesopleuron with a depressed, circular organ filled with fine, radially arranged hairs. Antennal scrobes present; antenna with 8 segments .... ***Pilotrochus***

— Mandible with 4–6 teeth which decrease in size from apex to base, or with only 2 teeth apically which are followed by an edentate oblique margin; without denticles between the teeth. Mesopleuron lacking a hair-filled circular organ. Antennal scrobes usually absent and antenna generally with 9–12 segments, only extremely rarely with 8 .... **15**

15 Antenna with 12 segments **16**

— Antenna with 8–11 segments **17**

16 Palp formula 5,3. Frontal lobes separated and median portion of clypeus broadly inserted between them (Fig. 258). Lateral portions of clypeus flattened and prominent, fused to the raised projecting median portion of the clypeus to form a shelf, which projects forward over the mandibles (Fig. 258). Propodeal lobes low and rounded, not connected to propodeal spines (when present) by broad projecting lamellae (Fig. 259) .... ***Cardiocondyla*** (part)

— Palp formula 2,2. Frontal lobes closely approximated and median portion of clypeus reduced to an extremely narrow strip between them (Fig. 412). Lateral portions of clypeus not prominent, not fused to median portion, and not forming a shelf; instead median portion of clypeus sharply raised centrally and in the form of a narrow longitudinal ridge (Fig. 412). Propodeal lobes large and prominent, connected to propodeal spines by broad, conspicuous lamellae (Fig. 413) .... ***Baracidris***

17 Head with strongly developed, sinuate frontal carinae and with broad, deep antennal scrobes (Fig. 208). Lateral portions of clypeus forming a sharp transverse ridge in front of the antennal insertions .... ***Wasmannia*** (part)

— Head without frontal carinae and without antennal scrobes (Figs. 344, 346, 350, 352, 354, 356, 376). Lateral portions of clypeus not forming a sharp transverse ridge in front of the antennal insertions **18**

18 Anterior clypeal margin with a single, long, anteriorly projecting median seta at the midpoint of the margin (Fig. 376). Propodeum always unarmed and rounded (Fig. 377). Antenna always with 10 segments .... ***Solenopsis***

— Anterior clypeal margin lacking a single, median seta; instead a pair of setae usually straddle the midpoint of the margin (Figs. 346, 352, 356). Propodeum sometimes unarmed and rounded (Fig. 357) but usually with spines or teeth (Figs. 345, 347, 351, 353), or sharply angulate (Fig. 355). Antenna with 8–11 segments **19**

19 Antenna with 8 or 9 segments **20**

— Antenna with 10 or 11 segments **22**

20 Propodeum bidentate, bispinose, or sharply angulate in profile. Worker caste dimorphic, without intermediates .... ***Oligomyrmex*** (part)

— Propodeum unarmed. Worker caste monomorphic **21**

21 Eyes absent. Mandible with 5 or 6 teeth. Promesonotum not marginate laterally (Fig. 357) .... ***Carebara***

— Eyes present. Mandible with 4 or 5 teeth. Promesonotum marginate laterally (Fig. 355) .... ***Paedalgus***

22 Mandible with 2 teeth apically, the teeth followed by an elongate, very oblique, edentate margin, which ends at the acute basal angle. With mandibles at full closure there is a distinct gap between their inner borders and the anterior clypeal margin . . . . ***Afroxyidris***
[Note that the mandibles are the same as those described for the Dominican Amber fossil **Oxyidris.*]

— Mandible with 4 or more teeth, which occupy the entire apical margin; without an elongate, oblique, edentate margin behind the two apical teeth (Figs. 344, 346, 350, 352). With mandibles at full closure without a gap between their inner borders and the anterior clypeal margin **23**

23 Clypeus longitudinally bicarinate on median portion (Figs. 350, 352). Worker caste dimorphic, without intermediates . . . . ***Oligomyrmex*** (part)

— Clypeus not bicarinate on median portion (Figs. 344, 346). Worker caste polymorphic with a graded series of intermediates connecting minor to major workers . . . . ***Pheidologeton***

24 Antenna with 7 segments (Figs. 306, 307) . . . . . . ***Myrmicaria***

— Antenna with 9–12 segments **25**

25 Antenna with 9 segments. Petiole sessile, without an anterior peduncle (Fig. 305). Pronotum and mesonotum fused into a laterally projecting shield, which overhangs the sides of the alitrunk on each side and sometimes also overhangs the propodeum posteriorly . . . . ***Meranoplus***

— Antenna with 10–12 segments. Petiole usually with an elongate anterior peduncle; if not then the pronotum and mesonotum do not form a shield overhanging sides of the alitrunk . . **26**

26 Median portion of clypeus vertical, with a conspicuous anteriorly projecting bilobed appendage above (the clypeal fork), which projects over the mandibles from about the same level as the frontal lobes (Figs. 404–407) **27**

— Median portion of clypeus not vertical, without a bilobed appendage projecting over the mandibles from about the same level as the frontal lobes (e.g., Figs. 208–211, 272–279, 330–335, 364–369, 394, 395, 412–421) **28**

27 Antenna with 11 segments. Peduncle of petiole short and very thick in profile (Fig. 407). All body setae simple, without bizarre pilosity . . . . ***Dicroaspis***

— Antenna with 12 segments. Peduncle of petiole elongate and narrow in profile (Fig. 405). Body setae bizarre; either spatulate, squamate, clavate, star-shaped, or very short, thick, and stubbly with abruptly tapered points . . . . . . ***Calyptomyrmex***

28 Propodeal spiracle long and narrow, its orifice slit-like (Fig. 331). Mesothoracic spiracles opening on dorsum of alitrunk. Mandible with at least the third tooth from the apex, and usually the third and fourth teeth, double-ranked . . . . ***Ocymyrmex***

— Propodeal spiracle circular to subcircular, very rarely oval but never long and narrow with a slit-like orifice. Mesothoracic spiracles concealed by a pronotal flap on the sides of the alitrunk. Without double-ranked mandibular teeth **29**

Afrotropical and Malagasy MYRMICINAE *(continued)*

29 Antenna with 10 segments **30**

— Antenna with 11 or 12 segments **31**

30 Sting shaft with an apicodorsal, triangular to pennant-shaped lamelliform appendage. Propodeal spiracle low on side and behind midlength, abutting the metapleural gland, widely separated from the dorsal outline in profile (Fig. 415). Petiole with a long anterior peduncle. Palp formula 4,3 . . . . ***Decamorium***

— Sting shaft without an apicodorsal lamelliform appendage. Propodeal spiracle very high on side and slightly in front of the midlength, widely separated from the metapleural gland, very close to the dorsal outline in profile (Fig. 273). Petiole subsessile, with an extremely short, inconspicuous anterior peduncle. Palp formula 5,3 . . . . ***Atopomyrmex***

31 Antenna with 11 segments **32**

— Antenna with 12 segments **42**

32 Antennal scrobes present, running below the eyes (Figs. 210, 211). Dorsum of gaster consisting entirely of the expanded first tergite, the remaining tergites visible in profile below the posterior margin of the first . . . . ***Cataulacus***

— Antennal scrobes either absent or present but running above the eyes (Figs. 208, 209, 250, 251, 300, 301, 368, 369, 378, 379, 394, 395, 416, 417). Dorsum of gaster not consisting entirely of the first tergite, the remaining tergites continuing the line of the first and visible in dorsal view **33**

33 Frontal lobes vestigial or absent so that the antennal articulations are exposed and the depressed area containing the antennal sockets clearly visible (Fig. 300). Anterior clypeal margin armed with denticles . . . . ***Pristomyrmex***

— Frontal lobes present, covering most or all of the antennal articulations, the antennal sockets not fully visible in dorsal view (Figs. 208, 250, 363, 368, 378, 394, 416). Anterior clypeal margin unarmed or with a pair of small teeth . . . **34**

34 Eyes located behind midlength of sides of head. Median portion of clypeus raised and produced forward as a large, shield-like lobe projecting strongly over the mandibles. Tibiae and basitarsi of middle and hind legs terminating in a number of peg-like, stout spines . . . . ***Metapone***

— Eyes located at or in front of the midlength of the sides of the head, or sometimes absent. Median portion of clypeus not produced forward as a large, shield-like lobe projecting strongly over the mandibles. Tibiae and basitarsi of middle and hind legs not terminating in peg-like, stout spines . . **35**

35 Maxillary palp with 1 or 2 segments. Propodeum rounded to angulate, never armed with differentiated teeth or spines (Figs. 365, 369, 379). Anterior clypeal margin with a single median seta (Figs. 368, 378). Antennal scrobes always absent and mandible with only 4 teeth **36**

— Maxillary palp with 3–5 segments. Propodeum bidentate or bispinose (Figs. 209, 251, 395, 417). Anterior clypeal margin without a single median seta, usually with a pair of setae that straddle the midline (Figs. 250, 394, 416). Antennal scrobes frequently, but not always, present and mandible usually with 5 or more teeth **38**

36 Eyes absent. Propodeal spiracle enormously enlarged, circular (Fig. 369). Frontal lobes closely approximated and median portion of clypeus narrow posteriorly between the lobes (Fig. 368) . . . . ***Bondroitia***

— Eyes present. Propodeal spiracle small, usually pinhole-like (Figs. 365, 379). Frontal lobes widely separated and median portion of clypeus broad posteriorly between the lobes (Figs. 364, 378) **37**

37 Median portion of clypeus distinctly raised, strongly to weakly longitudinally bicarinate (Fig. 378). Postpetiole node less voluminous than petiole node in profile and narrowly attached to the gaster . . . . ***Monomorium*** (part)

— Median portion of clypeus evenly transversely convex, not distinctly raised nor longitudinally bicarinate (Fig. 364). Postpetiole node much more voluminous than petiole node in profile and very broadly attached to gaster (Fig. 365) . . . . ***Diplomorium***

38 Mandible with 4 or 5 teeth or denticles in total, the basal tooth generally concealed by the anterior clypeal margin. Sting acute apically, not terminating in a lamellate, spatulate, or dentiform appendage **39**

— Mandible with 6 or more teeth or denticles, usually with at least 7. Sting terminating in an apical or apicodorsal lamellate, spatulate, or dentiform appendage **40**

39 Frontal carinae and antennal scrobes present (Fig. 208, 209). Palp formula 3,2. Petiole with an anterior peduncle. Nodes of petiole and postpetiole without spines or tubercles . . . . ***Wasmannia***

— Frontal carinae and antennal scrobes absent. Palp formula 5,3. Petiole sessile, without an anterior peduncle. Nodes of petiole and postpetiole with spines or tubercles . . . . ***Leptothorax*** (part)

40 Lateral portions of clypeus not raised into a narrow ridge or wall in front of the antennal insertions. Median portion of clypeus narrow and bicarinate (Fig. 394), narrowly inserted between frontal lobes. Mandible armed with 10–14 teeth which decrease in size from apex to base. Promesonotum in profile with a swollen and dome-like outline (Fig. 395) . . . . ***Cyphoidris***

— Lateral portions of clypeus raised into a narrow ridge or wall in front of the antennal insertions. Median portion of clypeus broad, not bicarinate (Figs. 416, 418), broadly inserted between frontal lobes. Mandible armed with 2 or 3 enlarged teeth apically, followed by a row of at least 4 smaller denticles, sometimes more. Promesonotum in profile without a swollen and dome-like outline (Figs. 417, 419) **41**

41 Palp formula 3,2. Head heart-shaped in full-face view. Median portion of clypeus with a prominent, arcuate anterior margin, which overlaps the basal angle of the mandible. Antennal scrobes always absent. Ventral margin of petiole keel-like. Eyes behind midlength of sides of head (Figs. 418, 419) . . . . ***Rhoptromyrmex*** (part)

— Palp formula usually 4,3, only very rarely reduced. Head not heart-shaped in full-face view. Median portion of clypeus without a prominent, arcuate anterior margin. Antennal scrobes usually, but not always, present. Ventral margin of petiole not keel-like. Eyes at, or somewhat in front of, midlength of sides of head (Figs. 416, 417) . . ***Tetramorium***

42 Petiole sessile, lacking an anterior peduncle, and the petiole node ventrally with a large and very strongly projecting plate-like process (Fig. 295). Median portion of clypeus longitudinally bicarinate (Fig. 294) . . . . ***Vollenhovia***

— Petiole distinctly pedunculate to subsessile, in either case the petiole node ventrally without a large plate-like process (usually a small anteroventral process present on the peduncle). If petiole subsessile then median portion of clypeus not bicarinate **43**

43 Dorsum of petiole node armed with a pair of sharp spines (Figs. 273, 275, 285) **44**

— Dorsum of petiole node unarmed or indented medially, lacking sharp spines **46**

44 All of visible portion of gaster consisting of the first tergite, which is massively enlarged and subglobose, ball-like but with an anteroventral orifice within which the remaining gastral segments are telescoped (Fig. 285). Eyes at extreme posterior corners of head (Fig. 284). Clypeus projecting far forward and almost concealing the mandibles . . . . ***Ankylomyrma***

— Gaster composed of 4 visible tergites and sternites which decrease in size posteriorly, the gaster with the first tergite not massively enlarged and ball-like (Figs. 273, 275). Eyes not at extreme posterior corners of head (Figs. 272, 274). Clypeus not projecting far forward over the mandibles **45**

45 Occipital corners of head evenly broadly rounded in full-face view (Fig. 272). Ventral surface of alitrunk with a very deep, broad pit between the hind coxae. Ventral margin of sides of metapleuron eroded in front of the metapleural gland bulla (Fig. 273). Polymorphic species, the propodeum armed with a pair of long spines . . . . ***Atopomyrmex*** (part)

— Occipital corners of head angulate to denticulate in full-face view (Fig. 274). Ventral surface of alitrunk without a deep, broad pit between the hind coxae. Ventral margin of sides of metapleuron not eroded in front of metapleural gland bulla but with a conspicuous, broad groove running forward to the mesopleuron (Fig. 275). Monomorphic species, the propodeum bidentate to unarmed . . . . ***Teratauer*** (part)

46 Lateral portions of clypeus raised up into a sharp-edged ridge or shield wall in front of the antennal insertions (Figs. 314, 416, 418) **47**

— Lateral portions of clypeus not raised up into a sharp-edged ridge or shield wall in front of the antennal insertions (Figs. 250, 258, 274, 278, 332, 334, 338, 340, 378) **49**

47 Sting lacking a spatulate or dentiform lamellate appendage dorsally at or close to the apex of the shaft. Anterior clypeal margin with a small triangular point medially (Fig. 314) . . . . ***Eutetramorium***

— Sting with a spatulate or dentiform lamellate appendage dorsally at or close to the apex of the shaft (Figs. 417, 419). Anterior clypeal margin without a small triangular point medially (Figs. 416, 418) **48**

48 Head heart-shaped in full-face view. Ventral margin of petiole convex and keel-like. Anterior clypeal margin strongly arcuate and prominent (Fig. 418). Eyes behind midlength of sides of head and propodeum unarmed . . ***Rhoptromyrmex*** (part)

— Head not heart-shaped in full-face view. Ventral margin of petiole never convex and keel-like. Anterior clypeal margin not strongly convex nor prominent (Fig. 416). Eyes only rarely behind midlength of sides of head and propodeum usually armed with a pair of spines or teeth . . . ***Tetramorium*** (part)

49 Occipital corners of head distinctly tuberculate to sharply denticulate (Fig. 275). Pronotum marginate laterally. Frontal carinae present (Fig. 274) . . . . ***Terataner*** (part)

— Occipital corners of head usually rounded, rarely angular, but never tuberculate or denticulate (Figs. 251, 259, 279, 341, 333, 335). Pronotum usually lacking lateral margination. Frontal carinae absent (Figs. 250, 258, 278, 332, 334, 338, 340, 378 . . . . **50**

50 Apical (masticatory) margin of mandible with more than 5 teeth or denticles in total, usually with 7 or more altogether; or very rarely the apical margin may be worn down and entirely edentate (Figs. 334, 338, 340) **51**

— Apical (masticatory) margin of mandible with 3–5 teeth or denticles altogether, never with more; apical margin never entirely edentate (Figs. 250, 258, 278, 378) **54**

51 With alitrunk in profile the dorsal outline simple, more or less evenly shallowly convex from front to back, without breaks in the outline and lacking a metanotal groove. Pronotum and mesonotum indistinguishable . . . . ***Undescribed genus*** [University of California, Davis, and Natural History Museum, London]

— With alitrunk in profile the dorsal outline complex; the pronotum or pronotum plus anterior mesonotum forming a high dome-like or markedly convex arc. Behind this the mesonotum may or may not form a second eminence before sloping steeply to the metanotal groove. Propodeum forming a separate convexity or flat plateau behind the metanotal groove (Figs. 335, 339, 341). Pronotum and mesonotum usually distinguishable **52**

52 Apical (masticatory) margin of mandible with the third tooth (counting from the apex) smaller than the fourth, or the reduced third tooth followed by a minute denticle before the larger fourth tooth (Fig. 334); margin never edentate. Antennal funiculi terminating in a strongly defined 3-segmented club. Palp formula 2,2 or more rarely 3,2 . . ***Pheidole*** (part)

Afrotropical and Malagasy MYRMICINAE *(continued)*

— Apical (masticatory) margin of mandible with the third tooth (counting from the apex) larger than the fourth, or margin edentate. Antennal funiculi terminating in a weakly defined 4-segmented club or without a differentiated club, the segments gradually increasing in size toward the apex. Palp formula 4,3 or 5,3 . . . . **53**

53 Ventral surface of head with a psammophore. Head massive and broad, CI > 90 (Fig. 340). Mandibles massive, their outer margins strongly curved toward the midline; sometimes edentate. Metasternal process large or very large, conspicuous. Polymorphic species . . . . ***Messor***

— Ventral surface of head without a psammophore. Head narrow, CI 90 at maximum but usually less (Fig. 338). Mandibles elongate-triangular, not massive, their outer margins not curved toward the midline; never edentate. Metasternal process minute to absent. Monomorphic species . . . . ***Aphaenogaster***

54 Mandible powerfully constructed, armed with 2 large apical teeth followed by a long diastema and then 1 or 2 (rarely 3) basal teeth (Fig. 332). Two to 4 hypostomal teeth usually present on posterior margin of buccal cavity. Palp formula 2,2 or 3,2 and clypeus lacking a long, unpaired median seta on the anterior margin (Fig. 332) . . . . ***Pheidole*** (part)

— Mandible delicately constructed, armed with 3–5 teeth, serially dentate and decreasing in size from apex to base, not arranged as above (Figs. 250, 258, 278, 378). Hypostomal teeth absent from posterior margin of buccal cavity. If palp formula 2,2 then clypeus with a long, unpaired median seta on the anterior margin **55**

55 Midpoint of anterior clypeal margin with a single, elongate seta, which projects forward over the mandibles and is usually very conspicuous (Figs. 258, 378) **56**

— Midpoint of anterior clypeal margin without a single, elongate seta; instead usually with a pair of short setae, one on each side of the midpoint (Figs. 250, 278) **57**

56 Median portion of clypeus concave to prominent anteriorly, usually overhanging the mandibles, weakly to acutely bicarinate; lateral portions of clypeus not expanded forward nor fused with the median portion to form a broad, projecting shelf (Fig. 378). Maxillary palps usually with 1 or 2 segments, rarely more. Mandible with 3–5 teeth (usually 4). Propodeum usually unarmed (Fig. 379), rarely angulate or bidenticulate . . . . ***Monomorium*** (part)

— Median portion of clypeus raised and projecting, not bicarinate, fused to the flattened prominent lateral portions of the clypeus to form a shelf, which projects forward over the mandibles (Fig. 258). Maxillary palps 5-segmented. Mandible with 5 teeth. Propodeum usually bidentate or bispinose (Fig. 259), only extremely rarely unarmed . . . . . ***Cardiocondyla*** (part)

57 With the alitrunk in profile the anterior margin of the mesonotum suddenly and very steeply raised above the level of the pronotum (Fig. 279); mesonotal free anterior face nearly vertical and somewhat concave, projecting and prominent (known only from Mauritius) . . . . ***Ireneopone***

— With the alitrunk in profile the mesonotum following the line of the pronotum, not suddenly and steeply raised above the level of the pronotum (Fig. 251); mesonotum without a near-vertical, somewhat concave free anterior face . . ***Leptothorax***

Key to Oriental and Indo-Australian MYRMICINAE (Workers)

1 Antennal scrobes present which run below the eye (Figs. 197, 203, 211). Eye usually distinct but rarely may be minute and situated on the underside of the upper scrobe margin, apparently absent in full-face view (Figs. 196, 202, 210, 220) . **2**

— Either antennal scrobes absent (e.g., Figs. 218, 258, 266, 283, 306, 338), or present but running above the eye (e.g., Figs. 235, 243, 245, 249, 265, 277, 396, 417); in some genera both eyes and scrobes absent (Figs. 290, 356, 366) **5**

2 Antenna with 11 segments. Petiole sessile, without an anterior peduncle. Dorsum of gaster consisting entirely of the first tergite, the remaining tergites visible in profile below the posterior margin of the first (Fig. 211) . . . . ***Cataulacus***

— Antenna with 6 or 7 segments (Figs. 202, 220). Petiole with an anterior peduncle. Dorsum of gaster not consisting entirely of the first tergite, the remaining tergites continuing the line of the first and visible in dorsal view (Figs. 197, 203, 221) . **3**

3 Antenna with 6 segments (Fig. 220). Petiole and postpetiole with extensive foliaceous or membranous outgrowths (Fig. 221). Palp formula 5,3 . . . . ***Colobostruma***

— Antenna with 7 segments (Fig. 202). Petiole and postpetiole without foliaceous or membranous outgrowths (Figs. 197, 203). Palp formula 2,2 or less **4**

4 Mandibles triangular, their whole serially dentate apical (masticatory) margins engaging directly at full closure (Fig. 196) . . . . ***Eurhopalothrix***

— Mandibles linear, their insertions remote so that their apical (masticatory) margins cross or engage only near their apices (Fig. 202) . . . . ***Rhopalothrix***

5 Postpetiole articulated on dorsal surface of first gastral segment (Fig. 219); the gaster in dorsal view roughly heart-shaped and capable of reflexion over the alitrunk. Petiole dorsoventrally flattened and without a node. Eyes present . . ***Crematogaster***

— Postpetiole articulated on anterior surface of first gastral segment (e.g., Figs. 187, 197, 211, 255, 267, 303, 327, 337, 371, 389, 407, 417); the gaster in dorsal view not roughly heart-shaped, not capable of reflexion over the alitrunk. Petiole not dorsoventrally flattened, with a node of some form. If postpetiole articulated high on anterior face of first gastral segment, then petiole with a node and eyes absent (Fig. 366, 367) **6**

6 Apical and preapical antennal segments much larger than preceding funicular segments and forming a conspicuous club of 2 segments (e.g., Figs. 177, 203, 220, 240, 288, 319, 347, 351, 355, 357, 391, 393, 411); or the apical and preapical segments preceded by an elongate, bar-like fusion segment (Figs. 232, 233) **7**

— Antenna never terminating in a conspicuous, 2-segmented club. Either apical plus two preapical funicular segments of antenna enlarged and forming a conspicuous, 3-segmented club, or less commonly the club with more than 3 segments (e.g., Figs. 179, 257, 271, 272, 277, 285, 297, 309, 313, 329, 383, 389, 399, 407, 417). Rarely the funiculus filiform and without a developed apical club **31**

7 Antenna with 4–6 segments **8**

— Antenna with 8–12 segments **18**

8 Mandibles elongate and linear, produced into narrow projecting blades which are much longer than broad and which always lack a cluster of 3 stout teeth near the midlength of each blade (Figs. 232–235, 240, 241). Mandible never triangular or subtriangular, never serially multidentate or denticulate **9**

— Mandibles triangular to subtriangular (Figs. 238, 239, 242–245, 248, 249) or with a cluster of 3 stout teeth near the midlength of each blade. Mandible usually serially multidentate or denticulate but sometimes with diastemata **12**

9 Apex of each mandibular blade either with a single, long tooth at the dorsal apex subtended by a series of minute denticles, or with a series of minute denticles only; always lacking an apical fork of 2 or 3 spiniform teeth. Labral lobes long and conical, visible between the mandibles in full-face view when the latter are closed (Fig. 240) . . . . ***Epitritus***

— Apex of each mandibular blade armed with a fork of 2 or 3 spiniform teeth set in a more or less vertical series; with or without intercalary denticles between the fork teeth. Labral lobes not long and conical, not visible between the mandibles in full-face view when the latter are closed (Figs. 232, 234) **10**

10 Antennal scrobes absent. Antenna with 5 segments; of the 4 funicular segments the second is bar-like and elongate (Figs. 232, 233). Eyes lateral. Palp formula 5,3 . . . ***Orectognathus***

— Antennal scrobes present though reduced in some. Antenna with 4 or 6 segments, never with 5; if antenna 6-segmented then second funicular segment never bar-like nor elongate (Figs. 234, 235). Eyes ventrolateral, placed on or near the ventral margin of the scrobe. Palp formula 1,1 . . . . **11**

11 Antenna with 4 segments . . . . ***Quadristruma***

— Antenna with 6 segments . . . . ***Strumigenys***

12 Closed mandibles with a strongly defined transverse basal margin, which is separated from the anterior clypeal margin by a conspicuous impression or gap (Fig. 238) **13**

— Closed mandibles without a strongly defined basal margin, the basal region of the mandible contiguous with or overlapped by the anterior clypeal margin, the two not separated by an impression or gap (Figs. 242, 244, 248) **14**

13 Mandible with 12 teeth or denticles in total. At full closure the entire lengths of the apical (masticatory) margins engaging directly, from basal to apical tooth .... ***Trichoscapa***

— Mandible with more than 20 teeth or denticles in total. At full closure only the distal halves of the apical (masticatory) margins engaging, with a distinct gap between the basal portions of the margins when the distal halves are engaged (Fig. 238) .... ***Dysedrognathus***

14 Propodeum unarmed. Dorsal alitrunk with 2 distinct convexities, promesonotal and propodeal, which are separated by a transverse groove (Fig. 249) .... ***Kyidris***

— Propodeum armed with a pair of spines or teeth, or with extensive spongiform tissue, or both. Dorsal alitrunk not raised into 2 convexities separated by a transverse groove (Figs. 243, 245) **15**

15 Apical (masticatory) margin of mandible with a cluster of 3 stout teeth near the midlength. Between this cluster and the clypeal margin the mandibular margin is unarmed. Distal to the 3 stout teeth are a diastema and a single small tooth before the apical series of small teeth and minute denticles .... ***Asketogenys***

— Apical (masticatory) margin of mandible serially dentate or denticulate, lacking a cluster of 3 stout teeth near the midlength (Figs. 242, 244). Diastema between basal tooth and clypeal margin variably developed but usually absent **16**

16 Mandible short, powerful, and bear-trap-like, armed with relatively few teeth in total (usually 8 or less), most or all of which are large and strongly developed (Fig. 244) .... ***Glamyromyrmex***

— Mandible triangular to elongate-triangular, not bear trap-like, armed with many teeth or denticles in total (10 or more), all of which are short and triangular to peg-like **17**

17 Standing setae of some form usually present on the head, alitrunk, or both. If setae absent from both these areas then the propodeum lacks teeth but has an extensive spongiform lamella running the height of the declivity .. ***Smithistruma***

— Standing setae always completely absent from head and alitrunk; propodeal teeth always present ..... ***Pentastruma***

18 Antenna with 12 segments **19**

— Antenna with 8–11 segments **21**

19 Dorsal alitrunk with a series of pairs of conical tubercles or prominences (Fig. 403). Frontal carinae present (Fig. 402) .... ***Proatta*** (part)

— Dorsal alitrunk without conical tubercles or prominences. Frontal carinae absent **20**

20 Frontal lobes closely approximated; median portion of clypeus, where it is inserted between the lobes, narrower than either of the frontal lobes (Fig. 410). Basal margin of mandible with a tooth close to or behind its midlength (Fig. 410). Palp formula 2,2 or less .... ***Adelomyrmex***

— Frontal lobes widely separated; median portion of clypeus, where it is inserted between the lobes, much broader than either of the frontal lobes (Fig. 258). Basal margin of mandible unarmed. Palp formula 5,3 ...... ***Cardiocondyla*** (part)

21 Antennal scrobes present, varying from long, broad, but shallow indentations bounded above by the frontal carinae, to extensive excavations in the sides of the head above the eyes (Figs. 208, 209, 318, 319, 408, 409) **22**

— Antennal scrobes and frontal carinae absent (Figs. 286, 287, 344–347, 350–357, 376, 377, 390, 391) **24**

22 Mandible with 12 teeth, which alternate in size and which become larger basally. Antenna with 9 segments (Fig. 319). In full-face view anterolateral corners of head formed by the frontal lobes (Fig. 318) .... ***Ishakidris***

— Mandible with 4 or 5 teeth, which decrease in size from the apex. Antenna with 10 or 11 segments (Figs. 209, 409). In full-face view anterolateral corners of head formed by the clypeus (Figs. 208, 408) **23**

23 Median portion of clypeus with a near-vertical anterior face and forming a bilobed or bidentate process above, which projects forward over the mandibles (Fig. 408). Antenna 10-segmented .... ***Mayriella***

— Median portion of clypeus convex but lacking a near-vertical anterior face and not produced into a bilobed or bidentate process projecting forward over the mandibles (Fig. 208). Antenna with 11 segments .... ***Wasmannia*** (part)

24 Anterior clypeal margin with a single, long, anteriorly projecting median seta at the midpoint of the margin (Fig. 376) .... ***Solenopsis***

— Anterior clypeal margin lacking a single median seta, instead usually with a pair of setae that straddle the midpoint (Figs. 286, 346, 352, 356) **25**

25 Antenna with 8–10 segments **26**

— Antenna with 11 segments **29**

26 Postpetiole very broadly attached to gaster (Fig. 287). Frontal lobes very closely approximated, touching or separated only by an extremely narrow impression (Fig. 286). Alitrunk box-like, the flattened dorsum finely and densely longitudinally striate .... ***Rhopalomastix***

— Postpetiole narrowly attached to gaster (Figs. 347, 353, 355, 357). Frontal lobes separated by median portion of clypeus (Figs. 344, 346, 350, 352, 354, 356, 390). Alitrunk usually not box-like but if so then the dorsum not finely and densely longitudinally striate **27**

27 Propodeum bidentate, bispinose (Figs. 351, 353), or sharply angulate in profile. Worker caste dimorphic, without intermediates .... ***Oligomyrmex*** (part)

— Propodeum unarmed (Figs. 355, 357). Worker caste monomorphic **28**

28 Eyes absent. Mandible with 4–6 teeth. Promesonotum not marginate laterally (Figs. 356, 357) . . . . ***Carebara***
— Eyes present. Mandible with 4 teeth. Promesonotum marginate laterally (Figs. 354, 355) . . . . ***Paedalgus***

29 Clypeus posteriorly very narrowly inserted between broad frontal lobes (Fig. 390). Ventral surfaces of petiole and postpetiole with diffuse spongiform appendages; a pad of loose spongiform tissue also present ventrally at base of first gastral sternite (Fig. 391) . . . . ***Tetheamyrma***
— Clypeus posteriorly more broadly inserted between frontal lobes (Figs. 344, 346, 350, 352). Ventral surfaces of petiole and postpetiole without spongiform tissue; base of first gastral sternite without spongiform pad (Figs. 351, 353, 345, 347) **30**

30 Median portion of clypeus longitudinally bicarinate (Figs. 350, 352). Workers dimorphic without intermediates . . . . ***Oligomyrmex*** (part)
— Median portion of clypeus not longitudinally bicarinate (Figs. 344, 346). Workers polymorphic with a graded series of intermediates between minors and majors . . . . ***Pheidologeton***

31 Antenna with 7 segments (Figs. 306, 307) . . . . . . ***Myrmicaria***
— Antenna with 9–12 segments **32**

32 Apical (masticatory) margin of mandible with a long, edentate edge apically and 4 small teeth basally, the edentate portion of the margin longer than the dentate section (Fig. 422). Entirety of the gastral dorsum formed by the first tergite, which curves strongly downwards posteriorly so that tergites 2–4 are on the ventral surface (Fig. 423). Eyes vestigial . . . . ***Secostruma***
— Apical (masticatory) margin of mandible not divided into a long, edentate apical portion and a shorter 4-dentate basal section. First gastral tergite not forming entire dorsum nor downcurved posteriorly; tergites 2–4 following the first and not on the ventral surface. Eyes present (even if only of a single ommatidium) or completely absent **33**

33 Eyes completely absent **34**
— Eyes present, varying from large and conspicuous to a single ommatidium **35**

34 Frontal lobes very closely approximated, the posterior portion of the clypeus, which passes between them, narrower than the width of either lobe (Fig. 366). Petiole without an anteroventral process. Postpetiole articulated at top of anterior face of first gastral segment (Fig. 367) . . . . ***Anillomyrma***
— Frontal lobes widely separated, the posterior portion of the clypeus, which passes between them, much broader than the width of either lobe (Fig. 290). Petiole with a large to very large anteroventral process. Postpetiole articulated close to center of anterior face of first gastral segment (Fig. 291) . . . . ***Liomyrmex***

35 Median portion of clypeus vertical, with a conspicuous, anteriorly projecting, bilobed appendage above (the clypeal fork), which projects over the mandibles from about the same level as the frontal lobes (Figs. 404, 405) . . . . ***Calyptomyrmex***

— Median portion of clypeus not vertical, without a bilobed appendage projecting over the mandibles from about the same level as the frontal lobes **36**

36 Antenna with 9 segments **37**
— Antenna with 10–12 segments **39**

37 Antennal scrobes present, deep and extensive. Frontal lobes present, the antennal sockets not fully exposed in full-face view. Apical (masticatory) margin of mandible with 4–6 teeth, the basal margin unarmed **38**
— Antennal scrobes absent. Frontal lobes absent, the antennal sockets fully exposed in full-face view (as in Fig. 300). Apical (masticatory) margin of mandible with 3 teeth, the basal margin with a tooth at about the midlength . . . . ***Perissomyrmex***

38 Petiole sessile, without an elongate, bar-like anterior peduncle between the articulation with the alitrunk and the node (Fig. 305). Promesonotum sharply marginate laterally, the margins usually expanded and often equipped with spines, lobes, foliaceous processes, or other outgrowths. Frontal lobes widely separated, the median portion of the clypeus where it projects between them broad and broadly convex in full-face view (Fig. 304) . . . . ***Meranoplus***
— Petiole pedunculate, with an elongate, bar-like anterior peduncle between the articulation with the alitrunk and the node. Promesonotum rounded laterally, lacking projections of any kind. Frontal lobes very closely approximated, the median portion of the clypeus where it projects between them reduced to an extremely narrow strip in full-face view . . . . ***Rostromyrmex***
[Description in preparation, by K. Rosciszewski, Staatliches Museum für Naturkunde, Karlsruhe]

39 Antenna with 10 segments . . . . ***Tetramorium*** (part)
— Antenna with 11–12 segments **40**

40 Antenna with 11 segments **41**
— Antenna with 12 segments **55**

41 Frontal lobes absent or reduced and elevated so that the antennal articulations are exposed (Fig. 300); in either case the anterior clypeal margin denticulate or sharply crenulate. Distal portion of mandible suddenly broadened, broader than proximal portion. Long axis of mandible rotated so that at full closure the masticatory margin is vertical or near-vertical below the anterior clypeal margin . . . . ***Pristomyrmex***
— Frontal lobes present and covering most or all of antennal articulations (Figs. 258, 282, 294, 342, 348, 378, 398, 416). Anterior clypeal margin at most with a pair of teeth, usually unarmed. Distal portion of mandible not suddenly broadened. Long axis of mandible not rotated, at full closure the masticatory margin not vertical below the anterior clypeal margin **42**

42 Sting either with a lamellate appendage projecting dorsally, close to the sting apex but at an angle to the shaft (Fig. 417), or rarely the appendage continuing the line of the shaft and up-curved at its apex. Mandible with 7 teeth, consisting of 3 larger teeth apically, followed by 4 smaller teeth . . . . ***Tetramorium*** (part)

— Sting shaft usually simple and lacking a lamellate appendage, or rarely the sting straight-spatulate apically. Mandible usually with fewer than 7 teeth, but if 7 or more present they are not arranged as above **43**

43 Apical (masticatory) margin of mandible with 8 or more teeth, denticles, or crenulations in total (Figs. 398, 400) **44**

— Apical (masticatory) margin of mandible with 4–6 teeth or denticles in total (Figs. 258, 270, 294, 348, 378) **46**

44 With alitrunk in profile the propodeal spiracle at or extremely close to the margin of the declivity (Figs. 399, 401). Antennal scrobes present, varying from weakly defined, broad, shallow, longitudinal impressions to deep excavations (Figs. 398, 400) **45**

— With alitrunk in profile the propodeal spiracle well in front of the margin of the declivity (as in Fig. 349). Antennal scrobes completely absent (as in Fig. 348) . . . . ***Lophomyrmex***

45 Spongiform tissue present ventrally on petiole, postpetiole, and base of first gastral sternite (Fig. 399). Petiole without a differentiated peduncle between node and articulation with alitrunk . . . . ***Dacetinops***

— Spongiform tissue absent from petiole, postpetiole, and first gastral sternite (Fig. 401). Petiole with a peduncle between node and articulation with alitrunk . . . . ***Indomyrma***

46 Narrow but deep antennal scrobes present which are bordered above by very broad, horizontal, laterally directed or down-curved frontal carinae, the frontal carinae extensively overhanging the scrobe and concealing it from dorsal view. Eyes situated below posterior ends of the scrobal impressions . . . . ***Metapone***

— Antennal scrobes completely absent (Figs. 258, 282, 294, 342, 348, 378) or shallowly present (Fig. 208). If the latter then the scrobal impressions bounded above by narrow frontal carinae, which do not extensively overhang and conceal the scrobe, and the eyes not situated below the posterior ends of the scrobal impressions **47**

47 Maxillary palp with 4 or 5 segments **48**

— Maxillary palp with 1–3 segments **51**

48 Petiole subsessile, with an extremely short and very broad anterior peduncle (Fig. 283). Petiole node high and thickly scale-like . . . . ***Stereomyrmex***

— Petiole either with an elongate, narrow anterior peduncle or the petiole subcylindrical and armed above with a single tooth; in either case the node not high and thickly scale-like (Figs. 259, 271, 343) **49**

49 Propodeum armed with a pair of spines which curve upwards and forwards (Fig. 343). Junction of postpetiole and gaster strongly dorsoventrally compressed and very narrow in profile. Basal tooth of mandible broad and with 2 points, or basal margin of mandible with a tooth . . . . ***Recurvidris***

— Propodeum unarmed or with a pair of teeth or spines which are more or less straight and directed posteriorly or posterodorsally (Figs. 259, 271). Junction of postpetiole and gaster not strongly dorsoventrally compressed. Basal tooth of mandible with a single point and basal margin of mandible edentate **50**

50 Femora of middle and hind legs grossly swollen medially. Standing pilosity present on the dorsal alitrunk. Lateral portions of clypeus not flattened, not fused with median portion, not projecting as a shelf over the mandibles (Fig. 270). Anterior clypeal margin usually with a pair of setae that straddle the midpoint (Fig. 270) . . . . ***Podomyrma***

— Femora of middle and hind legs slender, not grossly swollen medially. Standing pilosity absent from the dorsal alitrunk. Lateral portions of clypeus flattened, fused with median portion, and projecting as a shelf over the mandibles (Fig. 258). Anterior clypeal margin with an unpaired median seta (Fig. 258) . . . . ***Cardiocondyla*** (part)

51 Broad but shallow antennal scrobes present on sides of head above the eyes, the scrobes running almost the length of the sides of the head capsule (Figs. 208, 209) . . . . ***Wasmannia*** (part)

— Antennal scrobes completely absent (Figs. 294, 295, 248, 349, 378, 379) **52**

52 Petiole sessile to subsessile and with a large to very large ventral process (Fig. 295). Petiole not enormously more voluminous than postpetiole in dorsal view and in profile . . . . ***Vollenhovia*** (part)

— Petiole pedunculate and at most with a small, tooth-like ventral process on the peduncle (Figs. 349, 379). If peduncle short and stout then petiole enormously more voluminous than postpetiole in dorsal view and in profile **53**

53 Pronotal dorsum a flat plateau which is sharply marginate laterally, the marginations terminating anteriorly in projecting flat, acute, tooth-like or triangular processes above and behind the true humeral angles of the pronotum . . . . ***Lophomyrmex*** (part)

— Pronotal dorsum convex, without lateral marginations and lacking acute, tooth-like or triangular processes above and behind the true humeral angles of the pronotum **54**

54 Propodeum unarmed, evenly rounded . . . ***Monomorium*** (part)

— Propodeum with a pair of stout, posteriorly directed spines . . . . ***Willowsiella***

55 Palp formula 6,4. Spurs on posterior tibiae usually pectinate . . . . ***Myrmica***

— Palp formula less than 6,4 (up to 5,3). Spurs on posterior tibiae usually simple or absent, only rarely pectinate **56**

56 Sting with an apicodorsal triangular to pennant-shaped lamellate appendage projecting from the shaft at an angle to its long axis (Figs. 417, 419). Lateral portions of clypeus raised into a sharp-edged ridge or shield wall on each side, in front of the antennal insertions (Figs. 416, 418) **57**

— Sting without an apicodorsal triangular to pennant-shaped lamellate appendage projecting from the shaft at an angle to its long axis, though sometimes the sting apex may be straight-spatulate. Lateral portions of clypeus not raised into a sharp-edged ridge or shield wall on each side in front of the antennal insertions **58**

57 Head heart-shaped in full-face view. Ventral margin of petiole convex and keel-like. Anterior clypeal margin strongly arcuate and prominent. Eyes behind midlength of sides of head. Median clypeal and median cephalic carinae vestigial or absent (Figs. 418, 419). Palp formula 3,2 . . . ***Rhoptromyrmex***

— Head not heart-shaped in full-face view. Ventral margin of petiole not convex and keel-like. Anterior clypeal margin not strongly arcuate. Eyes usually at or in front of the midlength of the sides of the head, only extemely rarely otherwise. Either median clypeal carina or median cephalic carina usually present, or both present; only infrequently with both absent. Palp formula usually 4,3, rarely reduced (Figs. 416, 417) . . . . ***Tetramorium*** (part)

58 Mandible very strongly falcate, sharply pointed apically, and with the apical third recurved and hook-like (Fig. 324). Strongly arched-concave masticatory margin of mandibular blade edentate except for a single, large, triangular tooth near the base, which is mostly concealed by the clypeus when the mandibles are fully closed . . . . ***Chimaeridris***

— Mandible triangular to elongate-triangular, edentate to multidentate but never strongly falcate with a recurved and hook-like apical third. Masticatory margin of mandible not arched-concave, only very rarely edentate, always lacking an isolated, large, triangular tooth near the base **59**

59 Ventrolateral margin of head delineated by a sharp longitudinal carina on each side. The carina starts close to the inner-ventral mandibular base, runs the length of the head below the eye, and ascends the occipital surface posteriorly (Fig. 302). Petiole sessile, without an anterior peduncle (Fig. 303) . . . . ***Myrmecina***

— Ventrolateral margin of head without a longitudinal carina on each side. Rarely a narrow, sinuate, longitudinal groove may run the length of the head immediately below the eye (e.g., Fig. 267), or a weak ridge may be present; if the latter then the petiole has a long anterior peduncle **60**

60 Alitrunk armed with spines at the pronotal humeri and the propodeal angles; besides this the dorsum with 4 pairs of elongate conical to thickly spiniform prominences on the promesonotum, and the propodeum with a single anteromedian spine (Fig. 403). Occipital region of head with 3 pairs of similar prominences (Fig. 402) . . . . ***Proatta*** (part)

— Alitrunk at most with spines at pronotal humeri and propodeal angles; pronotum usually unarmed and propodeum sometimes unarmed; in either case the dorsal alitrunk never with 4 pairs of elongate conical to thickly spiniform prominences on the promesonotum and never with 3 pairs of similar prominences on the occipital region of the head **61**

Oriental and Indo-Australian MYRMICINAE *(continued)*

61 Mandible edentate. Either structurally so, with a sharp-edged, toothless apical (masticatory) margin from apex to base (Fig. 296); or functionally so, the teeth having been worn down to nothing from a previously multidentate condition. [Major workers of some dimorphic or polymorphic species] **62**

— Mandible dentate. The apical (masticatory) margin distinctly with 3 or more teeth or denticles present in total **64**

62 Tergite of first gastral segment medially overlapping onto the anteroventral surface, the suture between tergite and sternite of the first gastral segment basally in the form of a rounded M-shape and the postpetiole articulated in the base of the M. In profile the postpetiole attached on the apparent anteroventral surface of the gaster (Fig. 297) . . . . ***Acanthomyrmex***

— Tergite of first gastral segment medially not overlapping onto the anteroventral surface, the suture between tergite and sternite of the first gastral segment basally transverse and not a rounded M-shape; postpetiole articulated in the middle of the anterior surface. In profile the postpetiole not attached anteroventrally on the gaster (as in Figs. 333, 341) **63**

63 With the head in full-face view the eyes usually behind the midlength of the sides, or more rarely at the midlength (Fig. 340). Head generally roughly transversely rectangular. Maxillary palp with 4 or 5 segments. Metasternal process large to massive . . . . ***Messor*** (part)

— With the head in full-face view the eyes in front of the midlength of the sides, usually very obviously so (Fig. 332). Head not roughly transversely rectangular. Maxillary palp with 2 or 3 segments. Metasternal process vestigial to absent . . . . ***Pheidole*** (part)

64 Apical (masticatory) margin of mandible with only 3–6 teeth or denticles in total, the dentition usually sharply defined and the teeth decreasing in size from the apical (Figs. 258, 264, 266, 268, 294, 332, 336, 370, 378, 386). Masticatory margin never with a series of ill-defined crenulations or semi-effaced denticles near the basal angle, nor with teeth radically alternating in size along the length of the margin **65**

— Apical (masticatory) margin of mandible with 7 or more teeth or denticles in total (Figs. 298, 334, 338, 340, 388, 396), the dentition sometimes decreasing in size from the apex but often the masticatory margin with ill-defined crenulations or denticles between the main teeth, or with a series of ill-defined crenulations or denticles near the basal angle. Sometimes teeth alternating in size along the length of the margin **80**

65 Petiole sessile or subsessile, without a roughly horizontal anterior peduncle between the portion which articulates with the alitrunk and the ascending anterior face of the strongly developed node (Figs. 265, 295) **66**

— Petiole pedunculate, with a roughly horizontal anterior peduncle between the portion which articulates with the alitrunk and the ascending anterior face of the node (Figs. 251, 259, 267, 333, 337, 341, 371, 379, 387), or the entire petiole roughly cylindrical to claviform and lacking a sharply defined node (Fig. 277) **67**

66 Elongate frontal carinae and shallow antennal scrobes present (Figs. 264, 265). Palp formula 5,3. Ventral process of petiole a minute denticle. Propodeum armed with a pair of stout, long spines . . . . ***Romblonella***

— Frontal carinae and antennal scrobes absent (Figs. 294, 295). Palp formula 2,2. Ventral process of petiole large and angular to plate-like. Propodeum unarmed or at most with a pair of small triangular teeth . . . . ***Vollenhovia*** (part)

67 Basal margin of mandible with 2 posteriorly directed, broad, rounded lobes, the first lobe close to the basalmost tooth of the 5 on the apical (masticatory) margin (Fig. 370), the second lobe near the trulleum . . . . ***Epelysidris***

— Basal margin of mandible without 2 posteriorly directed lobes; basal margin usually unarmed but rarely a single small tooth may occur **68**

68 Midpoint of anterior clypeal margin with an unpaired median seta, which is usually elongate and stout and which projects forward over the mandibles (Figs. 258, 276, 378) **69**

— Midpoint of anterior clypeal margin without an unpaired median seta; instead either with a pair of setae that straddle the midpoint (Figs. 266, 268, 294, 336), or with an unbroken row of long, strong setae (Fig. 340), or hairless **72**

69 With head in full-face view the occipital corners acutely angulate to dentate. Frontal carinae and antennal scrobes present (Figs. 276, 277) . . . . ***Dilobocondyla***

— With head in full-face view the occipital corners broadly to narrowly rounded. Frontal carinae and antennal scrobes absent (Figs. 258, 259, 378, 379, 386, 387) **70**

70 Propodeum unarmed and rounded (Fig. 379) or at most with minute denticles; if the latter then the eyes are only of a single ommatidium . . . . ***Monomorium*** (part)

— Propodeum armed with a pair of teeth or spines (Figs. 259, 387); eyes always with many ommatidia **71**

71 Maxillary palp with 5 segments. Lateral portions of clypeus dorsoventrally flattened and thin (Fig. 258), strongly prominent over the mandibles, and sometimes projecting farther than the median clypeal portion. Median portion of clypeus not longitudinally bicarinate (Fig. 258) . . . . ***Cardiocondyla*** (part)

— Maxillary palp with fewer than 5 segments (usually 2 or 3). Lateral portions of clypeus not dorsoventrally flattened nor projecting over the mandibles (Fig. 386), not projecting as far forward as the median clypeal portion. Median portion of clypeus longitudinally bicarinate (Fig. 386) . . . . ***Rogeria***

72 With alitrunk in profile the dorsal outline simple, more or less flat to evenly shallowly convex from front to back, without breaks in the outline or at most with the metanotal groove

impressed (Figs. 251, 267, 269, 295). Pronotum and mesonotum usually indistinguishable **73**

— With alitrunk in profile the dorsal outline complex; the pronotum or pronotum plus anterior mesonotum forming a high, dome-like or markedly convex arc. Behind this the mesonotum may or may not form a second eminence before sloping steeply to the metanotal groove. Propodeum forming a separate convexity or plateau behind the metanotal groove (Figs. 333, 337, 341). Pronotum and mesonotum usually distinguishable **77**

73 Median portion of clypeus longitudinally bicarinate (Fig. 294). Clypeus posteriorly narrowly inserted between frontal lobes, the clypeus at most only as broad as one of the lobes where it passes between them. Propodeum unarmed. Palp formula less than 5,3 (usually 2,2) . . . . ***Vollenhovia*** (part)

— Median portion of clypeus not longitudinally bicarinate (Figs. 250, 266, 268). Clypeus posteriorly broadly inserted between frontal lobes, the clypeus much broader than one of the lobes where it passes between them. Propodeum armed with a pair of spines or teeth. Palp formula 5,3 . . . . **74**

74 Propodeal lobes elongate, sharply, narrowly triangular, and directed almost vertically . . . . ***Poecilomyrma***

— Propodeal lobes short and rounded, not projecting almost vertically as sharp, narrow triangles (Figs. 251, 267, 269) . . **75**

75 Frontal carinae weakly present, running as a pair of feeble or interrupted ridges from the frontal lobes to the occipital margin (Fig. 268). Propodeal spiracle large, low on the side, and shifted backwards, in profile the spiracle situated close to the margin of the declivity and immediately above (almost touching) the metapleural gland bulla (Fig. 269) . . . ***Rotastruma***

— Frontal carinae absent (Figs. 250, 266). Propodeal spiracle small, high on the side and far forward, in profile the spiracle situated far in front of the margin of the propodeal declivity and widely separated from the metapleural gland bulla (Figs. 251, 267) **76**

76 Mandible with 5 (or extremely rarely 6) teeth, which are regularly spaced and which lack a long diastema between teeth 3 and 4 (Fig. 250). Side of head below eye never with a sinuate longitudinal groove running part or all of its length (Fig. 251) . . . . ***Leptothorax***

— Mandible with 5 teeth, which are arranged as a group of 3 apically, followed by a long diastema, and terminating in a basal group of 2 smaller teeth (Fig. 266). Side of head below eye usually with a sinuate longitudinal groove, the groove generally running the length of the head (Fig. 267) but sometimes terminating at the eye (absent in one species) . . . . ***Vombisidris***

77 With the petiole in profile the spiracle situated at the node, behind the level of the anterior sloping face of the node . . . . ***Undescribed genus*** [University of California, Davis, and Natural History Museum, London]

— With the petiole in profile the spiracle situated on the peduncle, in front of the anterior sloping face of the node **78**

78 With head in full-face view the eyes in front of the midlength of the sides (Fig. 332). Apical (masticatory) margin of mandible with 2 teeth apically, a long diastema, and 0–3 teeth basally (usually 2). Maxillary palp with 2 or 3 segments . . . . ***Pheidole*** (part)

— With head in full-face view the eyes at or more usually behind the midlength of the sides (Figs. 336, 340). Apical (masticatory) margin of mandible with up to 6 teeth arranged along its length, without a long central diastema. Maxillary palp with 4 or 5 segments **79**

79 Metasternal process large to massive. Median portion of clypeus more or less flat both longitudinally and transversely (Fig. 340); not forming a distinctly convex outward bulge in profile. Vertex of head in profile or in full-face view without a marked depression of the surface behind the level of the eyes (Fig. 341) . . . . ***Messor*** (part)

— Metasternal process vestigial to absent. Median portion of clypeus strongly swollen, strikingly convex both longitudinally and transversely (Fig. 336); forming a distinctly convex outward bulge in profile. Vertex of head in profile or in full-face view with a marked depression of the surface behind the level of the eyes (Fig. 337) . . . . ***Kartidris***

80 Tergite of first gastral segment medially overlapping onto the anteroventral surface, the suture between tergite and sternite of the first gastral segment basally in the form of a rounded M-shape and the postpetiole articulated in the base of the M. In profile the postpetiole attached on the apparent anteroventral surface of the gaster (Fig. 299) . . . . ***Acanthomyrmex***

— Tergite of first gastral segment medially not overlapping onto the anteroventral surface, the suture between tergite and sternite of the first gastral segment basally transverse and not a rounded M-shape; postpetiole articulated in the middle of the anterior surface. In profile the postpetiole not attached anteroventrally on the gaster (Figs. 281, 335, 339, 387, 389, 397) **81**

81 Petiole armed dorsally with a pair of narrow, acute spines directed posteriorly . . . . ***Ancyridris***

— Petiole unarmed dorsally or with a single tooth or spine; sometimes dorsum of node emarginate but never with a pair of posteriorly directed narrow spines **82**

82 Median portion of clypeus narrow and longitudinally bicarinate (Figs. 294, 386, 388, 396), the surface between the two carinae usually transversely concave. Frontal lobes relatively close together so that the posteromedian portion of the clypeus, where it projects between the frontal lobes, is at most only slightly broader than one of the lobes. Frontal lobes themselves usually flat and transverse, not sharply elevated **83**

— Median portion of clypeus broad and not longitudinally bicarinate (Figs. 280, 334, 338, 340). Frontal lobes relatively far apart so that the posteromedian portion of the clypeus, where it projects between the frontal lobes, is usually very much broader than one of the lobes. If posteromedian clypeus not distinctly much broader than one of the lobes (very rare) then the frontal lobes are markedly elevated **86**

83 Elongate frontal carinae present, running back from the posteriormost points of the frontal lobes (Fig. 396). Antennal scrobes variously developed, ranging from broad but shallow impressions to extensive excavations in the sides of the head (Fig. 397) . . . . ***Lordomyrma*** (part)

— Frontal carinae absent (Figs. 294, 386, 388). Antennal scrobes absent (Figs. 295, 387, 389) **84**

84 Petiole sessile to subsessile, without a roughly horizontal anterior peduncle between the portion which articulates with the alitrunk and the anterior ascending face of the node. Petiole with a large to enormous ventral process (Fig. 295) . . . . ***Vollenhovia***

— Petiole pedunculate, with a roughly horizontal anterior peduncle between the portion which articulates with the alitrunk and the anterior ascending face of the node. Petiole at most with a small, dentiform anteroventral process (Figs. 387, 389) **85**

85 Petiole with a very long anterior peduncle, much longer than the height of the node; petiole node small, low, and conical to subconical in profile. Antennal club of 4 segments (Fig. 389) . . . . ***Stenamma***

— Petiole with a shorter anterior peduncle, shorter than or at most about equal to the height of the node; petiole node not low and conical in profile. Antennal club of 3 segments (Fig. 387) . . . . ***Rogeria*** (part)

86 With alitrunk in profile the dorsum of the promesonotum flat or forming a single, very shallowly convex curve from front to back (Fig. 281). Dorsal surface of propodeum on approximately the same level as the promesonotum, at most only fractionally lower. Petiole node rectangular or rounded-rectangular, blocky . . . . ***Paratopula***

— With alitrunk in profile the pronotum or pronotum plus anterior mesonotum forming a high, dome-like or markedly convex arc (Figs. 335, 339, 341). Behind this the mesonotum may or may not form a second eminence before sloping steeply, and sometimes sinuously, to the metanotal groove. Dorsal surface of propodeum depressed below the level of the promesonotum, usually considerably so. Petiole node usually conical to subconical, only rarely otherwise **87**

87 Palp formula 2,2 or 3,2. Apical (masticatory) margin of mandible with the third tooth (counting from the apex) smaller than the fourth, or the reduced third tooth followed by a minute denticle before the larger fourth tooth (Fig. 334) [all minor workers]; or mandible with 2 large apical and 1 or 2 enlarged basal teeth, the margin between these groups of teeth irregularly crenulate or bluntly dentate [major workers of a few species] . . . . ***Pheidole*** (part)

— Palp formula 4,3 or 5,3. Apical (masticatory) margin of mandible with the third tooth (counting from the apex) larger than the fourth (Figs. 338, 340). Mandible never with dentition as described for major workers above **88**

88 Metasternal process large or very large. Head massive and broad in media and major workers (Fig. 340). Mandibles short and powerful, massively constructed, their outer margins strongly curved toward the midline. Mostly polymorphic species . . . . ***Messor*** (part)

— Metasternal process minute to absent. Head elongate and narrow in all workers (Fig. 338). Mandibles elongate-triangular and not massively constructed, their outer margins not strongly curved toward the midline. Monomorphic species . . . . ***Aphaenogaster***

Key to Australasian MYRMICINAE (Workers)

1 Either antennal scrobes present which run below the eye (Figs. 197, 203, 225), or the scape, when laid back in its normal resting position, passes below the eye or across the lower margin of the eye in profile (the latter when the eye relatively large) (Figs. 220–223). Eye usually distinct but rarely may be minute and situated on the underside of the upper scrobe margin **2**

— Either antennal scrobes entirely absent (e.g., Figs. 218, 258, 266, 283, 306, 338) or present but running conspicuously above the eye (e.g., Figs. 235, 243, 245, 249, 265, 277, 396, 417); in some instances both eyes and scrobes absent (e.g., Figs. 290, 356, 366) but when eyes present the scape, when laid back in its normal resting position, always passing distinctly above the eye **6**

2 Antenna with 4–6 segments (Figs. 220, 222, 224). Palp formula 5,3 . . . . **3**

— Antenna with 7 segments (Figs. 196, 202). Palp formula 2,2 or less **5**

3 Mandibles elongate and linear, their insertions remote so that their masticatory margins at full closure engage only near their apices and leave a broad gap between the blades for the remainder of their length (Fig. 224) . . . . ***Epopostruma***

— Mandibles triangular to elongate-triangular, their masticatory margins at full closure engaging throughout their length, without a broad gap between the blades at full closure (Figs. 220, 222) **4**

4 With petiole in dorsal view or profile the node equipped with laterally projecting, aliform cuticular prominences (Fig. 221) . . . . ***Colobostruma***

— With petiole in dorsal view or profile the node lacking laterally projecting, aliform cuticular prominences (Fig. 223) . . . . ***Mesostruma***

5 Mandibles triangular, their whole serially dentate masticatory margins engaging directly at full closure (Fig. 196) . . . . ***Eurhopalothrix***

— Mandibles linear, their insertions remote so that the masticatory margins cross or engage only near their apices (Fig. 202) . . . . ***Rhopalothrix***

6 Postpetiole articulated on dorsal surface of first gastral segment (Fig. 219); the gaster in dorsal view roughly heart-shaped and capable of reflexion over the alitrunk. Petiole dorsoventrally flattened and without a node . . . . ***Crematogaster***

— Postpetiole articulated on anterior surface of first gastral segment (e.g., Figs. 187, 197, 211, 255, 267, 303, 327, 337, 371, 389, 407, 417); the gaster in dorsal view not roughly heart-shaped, not capable of reflexion over the alitrunk. Petiole not dorsoventrally flattened, usually with a node of some form but sometimes roughly cylindrical or claviform **7**

7 Apical and preapical antennal segments much larger than preceding funicular segments and forming a conspicuous 2-segmented club (e.g., Figs. 177, 203, 220, 240, 288, 319, 347, 351, 355, 357, 391, 393, 411), or the apical and preapical segments preceded by an elongate, bar-like fusion segment (Fig. 230) **8**

— Antenna never terminating in a conspicuous, 2-segmented club. Either apical plus two preapical funicular segments of antenna enlarged and forming a conspicuous, 3-segmented club, or less commonly the club with more than 3 segments (e.g., Figs. 179, 257, 271, 272, 277, 285, 297, 309, 313, 329, 383, 399, 407, 417). Rarely the funiculus filiform and without a developed apical club **20**

8 Antenna with 4–6 segments **9**

— Antenna with 8–12 segments **13**

9 Mandibles short, triangular to subtriangular, serially multidentate (Figs. 242, 244), and not terminating in an apical fork of 2 or 3 spiniform teeth in a vertical series **10**

— Mandibles elongate and linear, produced into narrow, projecting blades, which are much longer than broad and each of which terminates in an apical fork of 2 or 3 spiniform teeth arranged in a vertical series **11**

10 With the head in profile the mandible increasing in width from base to apex and the distal portion of the blade passing into a strongly down-curved arc so that part or most of the apical margin is at right angle to the long axis of the head (Fig. 245). Apical (masticatory) margin of mandible armed with a basal lamella plus 8–11 teeth, the basal 5–8 of which may be very strong (Fig. 244) . . . . ***Glamyromyrmex***

— With the head in profile the mandible with its upper and lower margins approximately parallel for most of its length or evenly tapering anteriorly. At most the extreme tip of the mandible down-curved, without the major part of the apical margin at right angle to the long axis of the head (Fig. 243). Apical (masticatory) margin of mandible armed with a basal lamella plus 12–17 teeth or denticles, the apicalmost group of which are minute (Fig. 242) .... ***Smithistruma***

11 Antenna with 5 segments; of the 4 funicular segments the second is elongate and bar-like (Figs. 232, 233). Palp formula 5,3. Eyes situated laterally on head .... ***Orectognathus***

— Antenna with 4 or 6 segments, never with 5; the second funicular segment never elongate and bar-like (Figs. 234, 235). Palp formula 1,1. Eyes situated ventrolaterally on head, on or near the ventral scrobe margin **12**

12 Antenna with 4 segments .... ***Quadristruma***

— Antenna with 6 segments .... ***Strumigenys***

13 Antenna with 12 segments **14**

— Antenna with 8–11 segments **15**

14 Frontal lobes closely approximated; median portion of clypeus, where it is inserted between the lobes, narrower than either of the frontal lobes (Fig. 410). Basal margin of mandible with a tooth close to or behind its midlength. Palp formula 2,2 or less .... ***Adelomyrmex***

— Frontal lobes widely separated; median portion of clypeus, where it is inserted between the lobes, much broader than either of the frontal lobes (Fig. 258). Basal margin of mandible unarmed. Palp formula 5,3 .... ***Cardiocondyla*** (part)

15 Antennal scrobes present, varying from long, broad, but shallow indentations bounded above by the frontal carinae, to extensive excavations in the sides of the head above the eyes (Figs. 208, 209, 408, 409) **16**

— Antennal scrobes and frontal carinae completely absent (Figs. 286, 287, 344–347, 350–353, 376, 377) **17**

16 Median portion of clypeus with a near-vertical anterior face and forming a bilobed or bidentate process, which projects forward over the mandibles (Figs. 408, 409). Antenna 10-segmented .... ***Mayriella***

— Median portion of clypeus convex, lacking a near-vertical anterior face, and not produced into a bilobed or bidentate process projecting forward over the mandibles (Figs. 208, 209). Antenna 11-segmented .... ***Wasmannia*** (part)

17 Anterior clypeal margin with a single, long, anteriorly projecting median seta at the midpoint of the margin (Fig. 376) .... ***Solenopsis***

— Anterior clypeal margin lacking a single, long median seta, instead usually with a pair of setae that straddle the midpoint (Figs. 286, 346, 350, 352) **18**

18 Median portion of clypeus longitudinally bicarinate (Figs. 350, 352). Worker caste dimorphic, without intermediates .... ***Oligomyrmex***

— Median portion of clypeus not longitudinally bicarinate (Figs. 286, 344, 346). Worker caste either monomorphic or polymorphic with a graded series of intermediates between largest and smallest **19**

19 Antenna with 10 segments. Postpetiole very broadly attached to the gaster (Fig. 287). Frontal lobes very closely approximated, touching or separated only by an extremely narrow impression (Fig. 286). Alitrunk box-like, the dorsum flattened and finely longitudinally striate. Monomorphic . . ***Rhopalomastix***

— Antenna with 11 segments. Postpetiole narrowly attached to gaster (Fig. 347). Frontal lobes separated by median portion of clypeus (Figs. 344, 346). Alitrunk not box-like, the dorsum not flattened and not longitudinally striate. Polymorphic .... ***Pheidologeton***

20 Median portion of clypeus vertical, with a conspicuous, anteriorly projecting, broad, bilobed appendage above (the clypeal fork), which projects over the mandibles from about the same level as the frontal lobes and which overhangs the vertical median clypeus (Figs. 404, 405) .... ***Calyptomyrmex***

— Median portion of clypeus not vertical, without a bilobed appendage above which projects out over the mandibles from about the same level as the frontal lobes **21**

21 Antenna with 9 segments. Promesonotum forming a fused dorsal shield with sharp lateral margins that usually project laterally (Fig. 305) .... ***Meranoplus***

— Antenna with 10–12 segments. Promesonotum rounded to laterally marginate but never forming a fused shield with sharp lateral margins **22**

22 Antenna with 10 segments .... ***Monomorium*** (part)

— Antenna with 11–12 segments **23**

23 Antenna with 11 segments **24**

— Antenna with 12 segments **36**

24 Frontal lobes entirely absent or vestigial, the antennal insertions fully exposed in full-face view (Fig. 300) **25**

— Frontal lobes present and distinctive, covering most or all of the antennal insertions in full-face view (Figs. 208, 270, 316, 360, 378, 416) **26**

25 Anterior clypeal margin denticulate or sharply crenulate (Fig. 300). Humeral angles of pronotum bispinose; petiole node unarmed dorsally (Fig. 301). Long axis of mandible rotated so that at full closure the masticatory margin is vertical or near-vertical below the anterior clypeal margin. Mandible tridentate, with a pair of teeth apically, followed by a diastema and a single basal tooth .... ***Pristomyrmex***

— Anterior clypeal margin unarmed. Humeral angles of pronotum unarmed; petiole node bidenticulate to bidentate dorsally. Long axis of mandible not rotated, at full closure the masticatory margin projecting, not vertical below the anterior clypeal margin. Mandible with 5 teeth, without diastemata .... ***Podomyrma*** (part)

26 Sting either with a lamellate appendage which projects dorsally, close to the sting apex but at an angle to the shaft (Fig. 417), or more rarely the lamellate appendage continuing the line of the shaft but up-curved at its apex . . ***Tetramorium*** (part)

— Sting shaft usually simple, always lacking a lamellate appendage; rarely the sting shaft itself may be straight-spatulate apically **27**

27 Antennal scrobes present; varying from narrow and deep, through moderately developed (Figs. 208, 209), to very feeble (Fig. 270); scrobes bounded above by conspicuous or interrupted frontal carinae which run back from the frontal lobes (Figs. 208, 270) **28**

— Antennal scrobes and frontal carinae both completely absent (Figs. 316, 360, 378) **31**

28 Antennal scrobes narrow and deep, bordered above by very broad, horizontal, laterally directed or down-curved frontal carinae, the frontal carinae extensively overhanging the scrobes and concealing them in full-face view. Eyes, which may be very small, situated below the posterior ends of the scrobal impressions. Median portion of clypeus a prominent truncated lobe, which projects forward over the mandibles . . . . ***Metapone***

— Antennal scrobes broad and shallow, bordered above by narrow frontal carinae which are not directed laterally and do not overhang and conceal the scrobes in full-face view. Eyes not situated below posterior ends of scrobal impressions. Median portion of clypeus shallowly arcuate anteriorly, not forming a truncated lobe projecting forward over the mandibles . . **29**

29 Palp formula 3,2. Lateral portions of clypeus raised into a sharp ridge in front of the antennal insertions. Hind femora narrow, not strongly inflated in the middle third or more of their length . . . . ***Wasmannia*** (part)

— Palp formula 4,3 or 5,3. Lateral portions of clypeus not raised into a sharp ridge in front of the antennal insertions. Hind femora inflated in the middle third or more of their length, often grossly so **30**

30 Petiole and postpetiole in profile both very high and narrow, conical and pointed apically, the points directed slightly posteriorly . . . . ***Peronomyrmex***

— Petiole in profile low and rounded, subcylindrical, or armed dorsally with 1–3 teeth or spines; not high and conical. Postpetiole not high and conical, nor pointed apically . . . . ***Podomyrma*** (part)

31 Anterior clypeal margin with a single, long, anteriorly projecting median seta at the midpoint of the margin (Fig. 378) . . . . ***Monomorium*** (part)

— Anterior clypeal margin lacking a single, long median seta; instead usually with a pair of setae that straddle the midpoint (Figs. 270, 316, 360) **32**

32 Maxillary palp with 2 or 3 segments **33**

— Maxillary palp with 4 or 5 segments **34**

33 Petiole in dorsal view enormously larger than postpetiole. Median portion of clypeus abruptly raised, the raised portion sharply marginate on each side. Alitrunk in profile with the dorsal outline evenly convex, metanotal groove absent. Monomorphic . . . . ***Willowsiella***

— Petiole in dorsal view smaller than postpetiole. Median portion of clypeus not abruptly raised, not marginate on each side. Alitrunk in profile with dorsal outline not evenly convex, metanotal groove present. Polymorphic . . . . . ***Machomyrma***

34 Femora of hind legs grossly swollen medially. Petiole in profile lacking a strongly differentiated node. Ascending anterior face of node short or inconspicuous, sometimes terminating in a tooth or short spine. Dorsal and posterior faces behind this confluent, cylindrical to subcylindrical (Fig. 271) . . . . ***Podomyrma*** (part)

— Femora of hind legs somewhat broader medially than at apex or base, but not grossly swollen. Petiole in profile with a strongly differentiated node, the node having a conspicuous ascending anterior face, a summit, and a free descending posterior face (Fig. 317) **35**

35 Palp formula 5,3. Antennal club weakly 4-segmented. Propodeum bispinose (New Zealand only) . . . . ***Huberia***

— Palp formula 4,3. Antennal club very strongly 3-segmented. Propodeum unarmed . . . . ***Adlerzia***

36 Sting with an apicodorsal triangular to pennant-shaped lamellate appendage projecting from the shaft at an angle to its long axis (Figs. 417, 419). Lateral portions of clypeus raised into a sharp-edged ridge or shield wall on each side, in front of the antennal insertions (Figs. 416, 418) **37**

— Sting without an apicodorsal triangular to pennant-shaped lamellate appendage projecting from the shaft at an angle to its long axis, though sometimes the sting apex may be straight-spatulate. Lateral portions of clypeus not raised into a sharp-edged ridge or shield wall on each side in front of the antennal insertions (Figs. 258, 264, 266, 276, 294, 332, 334, 338, 358, 378, 396) **38**

37 Head heart-shaped in full-face view. Ventral margin of petiole convex and keel-like. Anterior clypeal margin strongly arcuate and prominent. Eyes behind midlength of sides of head. Median clypeal and median cephalic carinae vestigial or absent. Palp formula 3,2 . . . . ***Rhoptromyrmex***

— Head not heart-shaped in full-face view. Ventral margin of petiole not convex and keel-like. Anterior clypeal margin not strongly arcuate. Eyes usually at or in front of midlength of sides of head, only very rarely otherwise. Either median clypeal carina or median cephalic carina usually present, or both present; only rarely both absent. Palp formula usually 4,3, rarely reduced . . . . ***Tetramorium*** (part)

38 Ventrolateral margin of head delineated by a sharp longitudinal carina on each side. The carina starts close to the inner-ventral mandibular base, runs the length of the head below the eye, and ascends the occipital surface posteriorly (Fig. 303) . . . . ***Myrmecina***

— Ventrolateral margin of head without a longitudinal carina on each side (Figs. 259, 295, 333, 335, 359, 379, 397), though sometimes a longitudinal narrow groove may run the length of the head immediately below the eye (Fig. 267) **39**

39 Apical (masticatory) margin of mandible with only 3–6 teeth or denticles in total, the dentition usually sharply defined and the teeth decreasing in size from the apical (Figs. 258, 264, 266, 294, 332, 358, 378). Apical (masticatory) margin never with a series of ill-defined crenulations or semi-effaced denticles near the basal angle nor with teeth radically alternating in size along the length of the margin **40**

— Apical (masticatory) margin of mandible with 7 or more teeth or denticles in total (Figs. 334, 338, 396), the dentition sometimes decreasing in size from the apex but often the masticatory margin with ill-defined crenulations or denticles between the main teeth, or with a series of ill-defined crenulations or denticles near the basal angle. Sometimes teeth alternating in size along the length of the margin **47**

40 Petiole sessile or subsessile, without a roughly horizontal anterior peduncle between the portion which articulates with the alitrunk and the ascending anterior face of the strongly developed node (Figs. 265, 295) **41**

— Petiole pedunculate, with a roughly horizontal anterior peduncle between the portion which articulates with the alitrunk and the ascending anterior face of the node (Figs. 259, 267, 333, 359, 379), or the entire petiole roughly cylindrical to claviform and lacking a sharply defined node (Fig. 277) **42**

41 Elongate frontal carinae and shallow antennal scrobes present (Figs. 264, 265). Median portion of clypeus not longitudinally bicarinate. Palp formula 5,3. Ventral process of petiole a minute denticle. Propodeum armed with a pair of long, stout spines . . . . ***Romblonella***

— Frontal carinae and antennal scrobes absent (Figs. 294, 295). Median portion of clypeus longitudinally bicarinate. Palp formula 2,2. Ventral process of petiole large and angular to plate-like. Propodeum unarmed or at most with a pair of small, triangular teeth . . . . ***Vollenhovia***

42 With head in full-face view the occipital corners acutely angulate to dentate (Fig. 276). Frontal carinae and antennal scrobes present, propodeum unarmed, and petiole usually subcylindrical to claviform (Figs. 276, 277) . . ***Dilobocondyla***

— With head in full-face view the occipital corners broadly to narrowly rounded. If the latter then either frontal carinae and antennal scrobes are absent, or the propodeum is armed with a pair of spines or teeth, or the petiole has a definitive node; or sometimes all of these **43**

43 Midpoint of anterior clypeal margin with a long, unpaired median seta, which projects forward over the mandibles (Figs. 258, 378) **44**

— Midpoint of anterior clypeal margin without a long, unpaired median seta; instead either with a pair of setae that straddle the midpoint, or with an unbroken row of long, strong setae, or hairless **45**

44 Maxillary palp with 5 segments. Lateral portions of clypeus dorsoventrally flattened and thin, strongly prominent over the mandibles (Fig. 258). Median portion of clypeus not longitudinally bicarinate . . . . ***Cardiocondyla*** (part)

— Maxillary palp with 1 or 2 segments. Lateral portions of clypeus not dorsoventrally flattened, not projecting over the mandibles (Fig. 378). Median portion of clypeus longitudinally bicarinate, usually strongly so . . . . ***Monomorium*** (part)

45 Sides of head immediately below eye with a fine longitudinal groove bounded by a pair of minute carinulae; this structure originates close to the lateral mandibular base, runs to the anteroventral margin of the eye, then curves to run below the eye to the latero-occipital margin (Fig. 267). Mandible armed with 3 teeth apically, these followed by a long diastema and a basal pair of teeth (Fig. 266). Monomorphic . . . . ***Vombisidris***

— Sides of head below eye without a longitudinal groove bounded by a pair of minute carinulae (Figs. 333, 359). Dentition not as above; if a diastema present then it is preceded by 2 apical teeth only. Dimorphic or polymorphic **46**

46 With alitrunk in profile the pronotum plus the anterior portion of the mesonotum forming a high, dome-like eminence or markedly convex arc (Fig 333). Behind this the posterior portion of the mesonotum slopes steeply to the metanotal groove, sometimes with a second eminence before reaching the groove. Mandible always with 2 teeth apically, plus a long diastema, plus a basal pair of teeth (Fig. 332) (major workers) . . . . ***Pheidole*** (part)

— With alitrunk in profile the pronotal dorsum shallowly convex and the mesonotal dorsum more or less flat and horizontal (Fig. 359), the mesonotum on a slightly lower level than the pronotum; mesonotum posteriorly angled suddenly downwards just before the metanotal groove. Mandibular dentition variable but usually not of 2 apical teeth plus a long diastema plus 2 basal teeth (Fig. 358) (larger workers) . . . . ***Anisopheidole*** (part)

47 Median portion of clypeus narrow and longitudinally bicarinate (Fig. 396), the surface between the two carinae usually flat to transversely concave. Frontal lobes relatively close together so that the posteromedian portion of the clypeus, where it projects between the frontal lobes, is at most only slightly broader than one of the lobes (Fig. 396) . . . . ***Lordomyrma***

— Median portion of clypeus broad and not longitudinally bicarinate (Figs. 334, 338). Frontal lobes relatively far apart so that the posteromedian portion of the clypeus, where it projects between the frontal lobes, is usually very much broader than one of the lobes. If posteromedian clypeus is not distinctly broader than one of the lobes (very rare) then the frontal lobes are markedly elevated **48**

48 With alitrunk in profile the promesonotal dorsum forming a single, evenly convex curve from front to back. Dorsal surface of propodeum on approximately the same level as the promesonotum or continuing the slope of the promesonotal outline; dorsal surface of propodeum not depressed far below the promesonotum (minor workers) . . ***Anisopheidole*** (part)

— With alitrunk in profile the pronotum plus anterior mesonotum forming a high, dome-like or markedly convex arc (Figs. 335,

339). Behind this the mesonotum may or may not form a second eminence before sloping steeply, and sometimes sinuously, to the metanotal groove. Dorsal surface of propodeum depressed far below the promesonotum and never continuing the slope of the promesonotal outline **49**

49 Antennal club of 3 segments. Palp formula 2,2 or 3,2 (minor workers) . . . . ***Pheidole*** (part)

— Antennal club of 4 segments. Palp formula 4,3 . . . . ***Aphaenogaster***

Key to Nearctic MYRMICINAE (Workers)

1 Antennal scrobes present which run below the eye. Antenna with 7 segments (Figs. 196, 197) . . . . ***Eurhopalothrix***

— Antennal scrobes usually absent (e.g., Figs. 180, 218, 250, 258, 292, 308, 310, 334, 338, 376), if present running above the eye (e.g., Figs. 235, 243, 245, 261, 393, 417) or in front of the eye (e.g., Figs. 177, 215, 217); sometimes both eyes and scrobes absent (e.g., Figs. 290, 356, 366). Antennal segment count variable but never with 7 segments **2**

2 Postpetiole articulated on dorsal surface of first gastral segment (Fig. 219); the gaster in dorsal view roughly heart-shaped and capable of reflexion over the alitrunk. Petiole dorsoventrally flattened and without a node . . . . ***Crematogaster***

— Postpetiole articulated on anterior face of first gastral segment (e.g., Figs. 187, 197, 211, 255, 267, 303, 327, 337, 371, 389, 407, 417); the gaster in dorsal view not roughly heart-shaped, not capable of reflexion over the alitrunk. Petiole usually with a node but if reduced then not dorsoventrally flattened . . **3**

3 Apical and preapical antennal segments forming a strong and conspicuously differentiated club of 2 segments (e.g., Figs. 177, 203, 220, 240, 288, 319, 347, 351, 355, 357, 391, 393, 411) **4**

— Antenna never terminating in a conspicuously differentiated, 2-segmented club. Either apical plus 2 preapical antennal segments forming a conspicuous, 3-segmented club, or club of more than 3 segments (e.g., Figs. 179, 257, 271, 272, 277, 285, 297, 309, 313, 329, 383, 389, 399, 407, 417), or the antenna without a distinctly differentiated apical club . . **11**

4 Antenna with 4 or 6 segments. Petiole and postpetiole with spongiform tissue present (Figs. 235, 243) **5**

— Antenna with 9–11 segments. Petiole and postpetiole without spongiform tissue (Figs. 209, 293, 351, 353, 377) **8**

5 Mandible elongate and linear, produced into a narrow, projecting blade (Fig. 234). Apex of each mandible with a fork of 2 spiniform teeth, arranged one above the other. Mandible never triangular or subtriangular, never serially dentate or denticulate **6**

— Mandible triangular or subtriangular, not produced into a narrow, projecting blade, and without an apical fork of spiniform teeth. Apical (masticatory) margin of mandible usually serially dentate or denticulate but sometimes with a long, basal diastema **7**

6 Antenna with 4 segments . . . . ***Quadristruma***

— Antenna with 6 segments . . . . ***Strumigenys***

7 Fully closed mandibles with a strongly defined basal margin, which is separated from the anterior clypeal margin by a conspicuous impression or gap. Basal lamella of mandible situated ventral to the basalmost tooth, in a plane almost at right angle to the anterior portion of the mandible, not visible in full-face view with the mandibles open . . . . ***Trichoscapa***

— Fully closed mandibles without a strongly defined basal margin, the basal region of the mandible overlapped by the anterior clypeal margin, the two not separated by an impression or gap. Basal lamella of mandible following basalmost tooth in the same plane (sometimes the two separated by a long diastema), visible in full-face view with the mandibles open . . . . ***Smithistruma***

8 Frontal carinae present, sinuate, running back almost to the occipital margin. Broad but shallow antennal scrobes present (Fig. 209). Median portion of clypeus without a pair of longitudinal carinae (Fig. 208). Propodeum with a pair of long spines . . . . ***Wasmannia*** (part)

— Frontal carinae absent. Antennal scrobes absent (Figs. 292, 376, 351). Median portion of clypeus with a pair of longitudinal carinae medially or at its lateral edges (Figs. 350, 352, 376). Propodeum unarmed or at most with a pair of short teeth **9**

9 Palp formula 4,2. Mandible with 5 or 6 teeth. Petiole subsessile, lacking a distinct anterior peduncle; petiole subcylindrical in shape and without a strongly differentiated node (Fig. 293) . . . . ***Xenomyrmex*** (part)

— Palp formula 2,2 or 1,2. Mandible with 4 teeth. Petiole with a distinct and well-developed anterior peduncle; petiole with a strongly differentiated node (Figs. 351, 353, 377) **10**

10 Propodeum unarmed and rounded (Fig. 377). Anterior clypeal margin with a single, anteriorly projecting median seta (Fig. 376). Antenna with 10 segments . . . . ***Solenopsis***

— Propodeum armed with a pair of short teeth (Figs. 351, 353). Anterior clypeal margin without an unpaired median seta, instead with a pair of setae that straddle the midpoint of the margin (Figs. 350, 352). Antenna with 11 segments . . . . ***Oligomyrmex***

11 Antenna with 11 segments **12**

— Antenna with 12 segments **25**

12 With the head in profile the eye situated close to the posterior margin and at the apex of a deep antennal scrobe (Fig. 217); the scrobe running forward from the anterior margin of the eye and bounded above by a massively expanded frontal carina (Fig. 216) . . . . ***Zacryptocerus***

— With the head in profile either the eyes absent, or present and situated close to or at the midlength (Figs. 180, 184, 208, 250, 262, 292, 416), not close to the posterior margin. Antennal scrobe absent or present; if the latter then the scrobe running above the eye and extending back beyond the level of the eye (Figs. 185, 209, 261, 263, 417), the eye not located at the apex of the scrobe **13**

13 Lateral portions of clypeus forming a raised, sharp-edged rim or wall in front of the antennal insertions (Figs. 208, 416) **14**

— Lateral portions of clypeus not forming a raised, sharp-edged rim or wall in front of the antennal insertions (Figs. 178, 180, 182, 184, 250, 260, 262, 292) **15**

14 Mandible with 7 teeth. Palp formula 4,3. Sting apicodorsally with a lamellate appendage which projects at an angle from the long axis of the shaft. Metasternal process large and conspicuous . . . . ***Tetramorium*** (part)

— Mandible with 5 teeth. Palp formula 3,2. Sting without a lamellate appendage projecting from the shaft apicodorsally. Metasternal process absent . . . . ***Wasmannia*** (part)

15 With head in profile a diagonal superocular carina present, which runs forward from above the eye down toward the mandibular insertion (Figs. 178, 180–183, 185); superocular carina independent of and distinct from any other sculpture that may be present. Promesonotal dorsum equipped with numerous prominences, tubercles, teeth, or spines (Figs. 179, 181, 183, 185) **16**

— With head in profile a superocular carina absent (Figs. 250–253, 260–263, 292). Longitudinal components of the sculpture may run from above the eye toward the mandibular insertion but none forms a sharply differentiated carina. Promesonotal dorsum smooth to coarsely sculptured but not equipped with prominences, tubercles, teeth, or spines (Figs. 251, 253, 261, 263, 293) **20**

16 Promesonotum with blunt tubercles (Fig. 185) or short teeth. With the head in full-face view the frontal lobes projecting far forward, anteriorly reaching or overlapping the anterior margins of the lateral portions of the clypeus (Fig. 184). Mandible with 5–7 teeth. Propodeum angular to bidentate **17**

— Promesonotum with elongate, sharp spines (Figs. 179, 181) or teeth (Fig. 183). With head in full-face view the frontal lobes not reaching the anterior margins of the lateral portions of the clypeus (Figs. 178, 180, 182). Mandible usually with more than 7 teeth. Propodeum bispinose **18**

17 Body without erect setae. Frontal lobes extensively expanded laterally, in full-face view overhanging and concealing the sides of the head in front of the eyes and the mandibular insertions (Fig. 184). Promesonotum dorsally with blunt tubercles, lacking short teeth . . . . ***Cyphomyrmex***

— Body with erect setae. Frontal lobes large but not so massively expanded, in full-face view the sides of the head in front of the eyes and the mandibular insertions visible. Promesonotum with short teeth . . . . ***Mycetosoritis***

Nearctic MYRMICINAE *(continued)*

18 Promesonotal dorsum with only 2 pairs of spines. First gastral tergite without raised cuticular tubercles (Fig. 181) . . . . ***Atta***

— Promesonotal dorsum with 3 or more pairs of spines. First gastral tergite with numerous raised cuticular tubercles (Figs. 179, 183) **19**

19 Frontal carinae present, sharply developed, and conspicuous, reaching back almost to the occipital margin (Fig. 182). Antennal scrobes present, their lower margins bounded by an extended superocular carina, which runs backward well beyond the eye . . . . ***Trachymyrmex***

— Frontal carinae vestigial to absent, at best scarcely distinguished from the surrounding sculpture and fading out well in front of the occipital margin. Antennal scrobes absent, the superocular carina curving mesad and petering out at about the level of the eye . . . . ***Acromyrmex***

20 Propodeum unarmed, the dorsum and declivity rounding together in profile (Fig. 293) **21**

— Propodeum in profile armed with a pair of teeth or spines (Figs. 251, 253, 261, 263) **22**

21 Eyes absent. Maxillary palp with 2 segments. Mandible with 4 teeth on a strongly oblique apical (masticatory) margin. Node of petiole well developed . . . . ***Undescribed genus*** [Museum of Comparative Zoology, Harvard, and Natural History Museum, London]

— Eyes present. Maxillary palp with 4 segments. Mandible with 5 or 6 teeth on a perpendicular apical (masticatory) margin. Node of petiole much reduced, inconspicuous . . . . ***Xenomyrmex*** (part)

22 Antennal scrobes present (Figs. 260–263). Mandible with 0–4 teeth. Anterior clypeal margin concave medially **23**

— Antennal scrobes absent (Figs. 250–253). Mandible with 5 or 6 teeth. Anterior clypeal margin convex to indented medially **24**

23 Apical (masticatory) margin of mandible with 4 teeth (Fig. 260). Median impression of anterior clypeal margin broad and very shallow . . . . ***Protomognathus***

— Apical (masticatory) margin of mandible edentate (Fig. 262). Median impression of anterior clypeal margin narrow and deep . . . . ***Harpagoxenus***

24 Eyes with conspicuous, short, erect pubescence projecting between the ommatidia. [Xenobionts in nests of *Myrmica, Manica,* and *Formica*] . . . . ***Formicoxenus***

— Eyes without erect pubescence projecting between the ommatidia . . . . ***Leptothorax*** (part)

25 Palp formula 6,4 . . . . **26**

— Palp formula less than 6,4 (up to 5,3 maximum) **27**

26 Propodeum bidentate to bispinose (Fig. 309). Mandible with 6–10 teeth (Fig. 308). Metasternal process a closely approximated pair of raised flanges or plates, the ventral midline not visible between them . . . . ***Myrmica***

— Propodeum unarmed and rounded. Mandible with more than 12 teeth. Metasternal process a pair of low, eroded ridges or arched-convex, thickened lobes, the ventral midline clearly visible between them . . . . ***Manica***

27 Ventrolateral margin of head delineated by a sharp longitudinal carina on each side. The carina starts close to the inner-ventral mandibular base, runs the length of the head below the eye, and ascends the occipital surface posteriorly (Fig. 303). Petiole sessile, lacking a large ventral process; propodeum bispinose . . . . ***Myrmecina***

— Ventrolateral margin of head without a sharp longitudinal carina on each side (Figs. 311, 333, 335, 339, 341, 389, 417). Petiole usually pedunculate, but if sessile then either the petiole with a large ventral process, or the propodeum unarmed, or both **28**

28 Petiole sessile, without an anterior peduncle; the petiole equipped with a large, plate-like ventral process (Fig. 295) . . . . ***Vollenhovia***

— Petiole with an anterior peduncle; the petiole without a plate-like ventral process but usually with a small, anteroventral tooth on the peduncle (Figs. 251, 311, 339, 341, 379, 389, 417) **29**

29 Sting with an apicodorsal triangular lamellate appendage projecting from the shaft at an angle to its long axis (Fig. 417). Lateral portions of clypeus raised into a sharp-edged ridge or shield wall on each side, in front of the antennal insertions (Fig. 416). Spurs of middle and hind tibiae simple or absent . . . . ***Tetramorium*** (part)

— Sting without an apicodorsal triangular lamellate appendage projecting from the shaft at an angle to its long axis, though rarely the sting apex may be straight-spatulate. Lateral portions of clypeus usually not raised into a sharp-edged ridge in front of the antennal insertions, but if so then the spurs of the middle and hind legs are pectinate **30**

30 Metatibial spurs finely pectinate. Ventral alitrunk with an extensive V-shaped or narrowly U-shaped emargination running from the posterior border forward and terminating at or close to the metasternal pit, well in front of the hind coxal cavities. Metasternal process large, represented by 2 prominent triangles or teeth, at the apex of or on each side of the median emargination . . . . ***Pogonomyrmex***

— Metatibial spurs simple to absent. Ventral alitrunk at most with a shallow, U-shaped posterior emargination, which does not project forward beyond the hind coxal cavities and which ends far short of the metasternal pit. Metasternal process massive to absent; when present not usually represented by 2 triangles or teeth and always widely separated from the apex of the median emargination **31**

31 Median portion of clypeus abruptly raised and relatively narrow, the raised portion with a pair of fine longitudinal carinae which diverge anteriorly (Figs. 378, 386, 388); clypeal dorsum between the carinae concave to more or less flat . . **32**

Nearctic MYRMICINAE *(continued)*

— Median portion of clypeus broad, not abruptly raised, flat to transversely convex, and without a pair of fine longitudinal carinae medially; clypeus sometimes with a single median longitudinal carina (Figs. 250, 258, 332, 334, 338, 340) . **34**

32 Propodeum angular to rounded but unarmed, without spines or teeth (Fig. 379). Apical (masticatory) margin of mandible with 3 or 4 teeth or denticles in total (Fig. 378). With petiole in profile the spiracle situated at the node or on the peduncle immediately in front of the ascending face of the node . . . . ***Monomorium***

— Propodeum armed with a pair of teeth or spines (Figs. 387, 389). Apical (masticatory) margin of mandible with 5 or more teeth or denticles in total (Figs. 386, 388). With petiole in profile the spiracle situated on the peduncle close to the articulation with the alitrunk, far in front of the ascending face of the node **33**

33 Antennal club of 3 segments (Fig. 387). Propodeum armed with a pair of spines (Fig. 387). Propodeal spiracle relatively large and close to margin of declivity, the distance separating the margin of the declivity from the nearest point of the spiracle annulus usually no greater than the diameter of the spiracle . . . . ***Rogeria***

— Antennal club of 4 segments (Fig. 389). Propodeum armed with a pair of teeth or short, narrow spines (Fig. 389). Propodeal spiracle relatively small and removed from the margin of the declivity, the distance separating the margin of the declivity from the nearest point of the spiracle annulus usually much greater than the diameter of the spiracle . . . . . . ***Stenamma***

34 Apical (masticatory) margin of mandible with 0–6 teeth or denticles in total (Figs. 250, 258, 332), the margin never with a series of ill-defined crenulations or semi-effaced denticles near the basal angle, nor with teeth behind the second (counting from the apex) alternating in size along the margin **35**

— Apical (masticatory) margin of mandible with 7 or more teeth or denticles in total (Figs. 334, 338, 340), the margin often with ill-defined crenulations or denticles between the main teeth, or with a series of ill-defined crenulations or denticles near the basal angle. Sometimes teeth behind the second (counting from the apex) alternating in size along the margin **38**

35 Mandible massively constructed (as in Figs. 332, 340), edentate or with rounded vestiges of teeth, or with 2 teeth apically followed by a diastema and 1 or 2 basal teeth. With alitrunk in profile the pronotum or pronotum plus anterior mesonotum forming a high, dome-like or markedly convex arc (Figs. 333, 341). Behind this the mesonotum may or may not form a second eminence before sloping steeply to the metanotal groove. Dorsal surface of propodeum depressed far below the level of the promesonotum **36**

— Mandible delicately constructed, with 5 or 6 regular teeth which decrease in size from apex to base (Figs. 250, 258). With alitrunk in profile the dorsal outline simple, more or less flat to evenly shallowly convex from front to back (Figs. 251, 259), without breaks in the outline or at most with the metanotal groove present. Dorsal surface of propodeum on same level as promesonotum at least anteriorly, or only slightly lower **37**

36 With head in full-face view the eyes at or slightly behind the midlength of the sides (Fig. 340). Palp formula 4,3. Metasternal process large to massive . . . . ***Messor*** (part)

— With head in full-face view the eyes in front of the midlength of the sides (Fig. 332). Palp formula 2,2 or 3,2. Metasternal process vestigial to absent . . . . ***Pheidole*** (part)

37 Midpoint of anterior clypeal margin with an unpaired, long seta, which projects forward over the mandibles. Lateral portions of clypeus flattened and projecting over the mandibles (Fig. 258). Dorsum of head and body without standing setae (Fig. 259) . . . . ***Cardiocondyla***

— Midpoint of anterior clypeal margin without an unpaired, long seta; instead the midpoint usually straddled by a pair of setae. Lateral portions of clypeus not flattened nor projecting over the mandibles (Fig. 250). Dorsum of head and body with standing setae present (Fig. 251) . . . . ***Leptothorax***

38 Palp formula 2,2 or 3,2. Apical (masticatory) margin of mandible with the third tooth (counting from the apex) smaller than the fourth, or the reduced third tooth followed by a minute denticle before the larger fourth tooth (Fig. 334) [all minor workers]; or mandible with 2 large apical and 1 or 2 enlarged basal teeth, the margin between these teeth crenulate to low-dentate [most major workers] . . ***Pheidole*** (part)

— Palp formula 4,3 or 5,3. Apical (masticatory) margin of mandible with the third tooth (counting from the apex) larger than the fourth (Figs. 338, 340). Mandible never with dentition as described for major workers above **39**

39 Metasternal process large or very large. Mandibles short and powerful, their outer margins strongly curved toward the midline (Fig. 340). Head in full-face view blocky, square to rectangular. Mandible usually with 7 or 8 teeth, rarely with more . . . . ***Messor*** (part)

— Metasternal process vestigial to absent. Mandibles elongate-triangular, their outer margins not strongly curved toward the midline (Fig. 338). Head in full-face view elongate and narrow. Mandible usually with more than 8 teeth or denticles in total . . . . ***Aphaenogaster***

Key to Neotropical MYRMICINAE (Workers)

1 With head in profile antennal scrobes present which run longitudinally below the eyes (Figs. 194–205). Eyes usually distinct but sometimes may be minute and situated on the underside of the upper scrobe margin, apparently absent in full-face view **2**

— With head in profile either antennal scrobes completely absent (Figs. 178, 180, 218, 230, 250, 292, 311, 313, 320–323, 338, 374–383, 411), or present but running longitudinally above the eye (Figs. 207, 209, 213, 235, 245, 385, 392, 393, 416, 417), or the eye is at the posterior apex of the scrobe so that the scrobe runs forward from the anterior ocular margin (Figs. 177, 215, 217). Rarely eyes and scrobes may both be absent (e.g., Figs. 356, 357, 366, 367) **9**

2 Antenna with 12 segments **3**

— Antenna with 7–11 segments **4**

3 Dorsal surface of basal half of mandible with a deep, transverse-oblique groove running outward from the masticatory margin; apical part of mandible sharply down-curved . . . . ***Creightonidris***

— Dorsal surface of basal half of mandible without a transverse-oblique groove (Fig. 194); apical part of mandible moderately and evenly convex to the apex (Fig. 195) . . . . ***Basiceros***

4 Antenna with 11 segments. Petiole sessile, without an anterior peduncle (as Fig. 217); petiole in profile lacking a differentiated node. With head in full-face view the frontal lobes projecting farther forward than the anterior margin of the median portion of the clypeus (as Fig. 216). Lateral margins of alitrunk armed with spines, teeth, or foliaceous outgrowths . . . . ***Zacryptocerus*** (part)

— Antenna with 7–9 segments. Petiole with an elongate anterior peduncle (Figs. 197, 199, 201, 203, 205); petiole in profile with a node of some form. With head in full-face view the frontal lobes well behind the level of the anterior margin of the median portion of the clypeus (Figs. 196, 198, 200, 202, 204). Lateral margins of alitrunk without spines, teeth, or foliaceous outgrowths, though propodeal teeth may be present **5**

5 Mandibles triangular, their whole serially dentate apical (masticatory) margins engaging directly at full closure (Figs. 196, 198) **6**

— Mandibles linear, their insertions remote so that their apical (masticatory) margins cross or engage only near their apices (Figs. 200, 202, 204) **7**

6 Antenna with 8 segments . . . . ***Octostruma***

— Antenna with 7 segments . . . . ***Eurhopalothrix***

7 Antenna with 9 segments. Mandible very long and slender, up-curved along the length (Fig. 201). Basal third of each mandibular blade unarmed; at about one-third the length each blade with a long, slender, ventrally curved tooth which overlaps its opposite number. Distal to this tooth are 1 or 2 denticles, another long, spiniform tooth, and then a few denticles before the extremely long, curved, apical tooth (Fig. 200) . . . . ***Protalaridris***

— Antenna with 7 segments. Mandible narrow and with variable dentition, but never with the extremely specialized dentition described above (Figs. 202–205) **8**

8 Blade of mandible with a long, conspicuous spiniform tooth near the apex (Fig. 202) .... ***Rhopalothrix***

— Blade of mandible with a series of interlocking small teeth apically, without a long spiniform tooth (Fig. 204) .... ***Talaridris***

9 Postpetiole articulated on dorsal surface of first gastral segment (Fig. 219); the gaster in dorsal view roughly heart-shaped and capable of reflexion over the alitrunk ...... ***Crematogaster***

— Postpetiole articulated on anterior surface of first gastral segment (e.g., Figs. 187, 197, 211, 255, 267, 303, 327, 337, 371, 389, 407, 417); the gaster in dorsal view not heart-shaped, not capable of reflexion over the alitrunk **10**

10 Apical and preapical antennal segments distinctly much larger than preceding funicular segments and forming a conspicuously differentiated, 2-segmented club (e.g., Figs. 177, 203, 220, 240, 288, 319, 347, 351, 355, 357, 391, 393, 411) **11**

— Antenna never terminating in a conspicuously differentiated, 2-segmented club. Either the apical plus 2 preapical segments enlarged and forming a weak to conspicuous, 3-segmented club, or less commonly the club with more than 3 segments (e.g., Figs. 179, 257, 271, 272, 277, 285, 297, 309, 313, 329, 383, 389, 399, 407, 417). Sometimes the funiculus filiform or the segments gradually increasing in size apically, without a differentiated club **35**

11 Mandibles elongate and linear (Figs. 230, 234, 236), produced into narrow, projecting blades, which are much longer than broad and each of which terminates apically in a vertically arranged series of 2–4 teeth **12**

— Mandibles triangular to subtriangular (Figs. 176, 208, 242, 244, 258, 292, 350, 356, 392, 410), not produced into narrow blades and each not terminating apically in a vertically arranged series of 2–4 teeth **16**

12 Antenna with 11 segments. Palp formula 5,3. Antennal scrobes absent and eyes lateral (Figs. 230, 231) ... ***Acanthognathus***

— Antenna with 4–6 segments. Palp formula 1,1. Antennal scrobes present and eyes ventrolateral, usually on or near the lower scrobe margin (Figs. 234–237) **13**

13 Antenna with 4 segments. Dorsum of head with orbicular setae present .... ***Quadristruma***

— Antenna with 6 segments. Dorsum of head without orbicular setae **14**

14 Mandible broad at the base and rapidly evenly tapering to the apex. Mandible without an apical fork of 2 teeth, instead the mandible in apical (direct frontal) view with 4 long, spiniform teeth in a vertical series. Frontal lobes with outer margins straight in full-face view, confluent in a straight line with the upper scrobe margins .... ***Dorisidris***

— Mandible linear, not broad at the base and rapidly tapering to the apex (Figs. 234, 236). Mandible with an apical fork of 2 teeth arranged in a vertical series, sometimes also with intercalary denticles. In apical (direct frontal) view mandible without 4 long, spiniform teeth. Frontal lobes short and usually convex in full-face view, joining the upper scrobe margins through an angle or sinuous curve **15**

15 Mandible with a large horizontal basal lamella present. Inner margin of mandible proximal to apical fork serially denticulate. With 3–8 denticles on margin immediately behind the apical fork, these followed by a larger, submedian tooth; proximal to this tooth is a series of indistinct denticles after which the margin is unarmed to the basal lamella (Fig. 236) .... ***Neostruma***

— Mandible without a large basal lamella. Inner margin of mandible proximal to the apical fork usually with 0–3 teeth or denticles (Fig. 234), but if more present (up to 8 denticles) then they are evenly spaced on the distal two-thirds of the margin and do not have an enlarged submedian tooth .... ***Strumigenys***

16 Antenna with 4–7 segments **17**

— Antenna with 9–12 segments **25**

17 Antenna with 4 segments .... ***Codioxenus***

— Antenna with 6 or 7 segments **18**

18 Antenna with 7 segments. Eyes at extreme posterior ends of antennal scrobes (Fig. 177). Petiole sessile; postpetiole very broadly attached to gaster, the first gastral segment large and strongly down-curved. Petiole and postpetiole lacking spongiform or lamelliform appendages laterally or ventrally (Fig. 177) .... ***Tatuidris***

— Antenna with 6 segments. Eyes ventrolateral, at or close to lower margins of the antennal scrobes (Figs. 243, 245). Petiole pedunculate; postpetiole narrowly attached to gaster, the first gastral segment not down-curved. Either petiole, postpetiole, or both with spongiform or lamelliform appendages laterally or basally (Figs. 243, 245) **19**

19 Broad, thin, semitranslucent lamellae present bordering the upper scrobe margins, occipital margin, anterior margin of pronotum, and sides of propodeum; similar lamellae present on petiole and postpetiole .... ***Tingimyrmex***

— Broad, thin, semitranslucent lamellae usually absent from the upper scrobe margins; always absent from the occipital margin (Figs. 242, 244), anterior pronotal margin, and sides of propodeum. Petiole and postpetiole usually with spongiform appendages present (Figs. 243, 245) **20**

20 Closed mandibles with a very strongly defined basal margin, which is separated from the anterior clypeal margin by a conspicuous impression or gap. Basal lamella of mandible situated ventral to the basalmost tooth, in a plane almost at right angle to the anterior portion of the mandible, not visible in full-face view with the mandibles open .... ***Trichoscapa***

— Closed mandibles without a strongly defined basal margin, the basal region of the mandible contiguous with or overlapped by the anterior clypeal margin, the two not separated by an impression or gap. Basal lamella of mandible following basalmost tooth in the same plane, visible in full-face view with the mandibles open **21**

21 With the head in profile the mandible increasing in width from base to apex and the distal portion of the blade passing into a strongly down-curved arc so that part or most of the apical margin is at right angle to the long axis of the head (Fig. 244) **22**

— With the head in profile the mandible with upper and lower margins approximately parallel for most of their length or evenly tapering anteriorly. At most the extreme tip of the mandible down-curved or truncated (Fig. 243) **24**

22 With head in full-face view the anterior clypeal margin convex and prominent medially **23**

— With head in full-face view the anterior clypeal margin transverse or more commonly concave medially ***Glamyromyrmex***

23 Mandible with a series of 3 enlarged teeth basally, distal to which is a series of denticles that rapidly decrease in size toward the apical tooth . . . . ***Chelystruma***

— Mandible with a series of 7 or 8 spiniform teeth basally, distal to which is a series of denticles before the apical tooth . . . . ***Codiomyrmex***

24 Upper margins of scrobes bounded by broad, laterally projecting cuticular flanges throughout their lengths; the flanges may be semitranslucent or have translucent fenestrae in places . . . . ***Gymnomyrmex***

— Upper margins of scrobes bounded by narrow, sinuate cuticular rims throughout their lengths; never with semitranslucent areas nor with translucent fenestrae . . . . ***Smithistruma***

25 Frontal carinae and antennal scrobes present (Figs. 206–209, 392, 393). Antenna always with 11 segments **26**

— Frontal carinae and antennal scrobes entirely absent (Figs. 258, 259, 292, 293, 350, 351, 356, 357, 376, 377, 410, 411). Antenna with 9–12 segments **29**

26 Mandible with 10 teeth in total, the teeth alternating large and small from base to apex; teeth on basal half of mandible larger than those on apical half. With head in full-face view the frontal lobes massively expanded laterally, overhanging the anterolateral angles of the head so that the latter are invisible . . . . ***Phalacromyrmex***

— Mandible with 5 teeth or denticles in total, which decrease in size from the apical; the apical tooth the largest. Frontal lobes moderate to large but not massively expanded laterally; with the head in full-face view the anterolateral angles of the head visible, not overhung and concealed by the frontal lobes **27**

27 Median portion of clypeus longitudinally bicarinate centrally (Fig. 392). Propodeum in profile with the spiracle very close to the margin of the declivity (Fig. 393). Promesonotum in profile swollen, forming an evenly dome-like convexity . . . . ***Lachnomyrmex***

— Median portion of clypeus not longitudinally bicarinate centrally (Figs. 206, 208), though the lateral edges may be sharp. Propodeum in profile with the spiracle well in front of the margin of the declivity (Figs. 207, 209). Promesonotum in profile shallowly to quite strongly convex, but not forming an evenly dome-like convexity **28**

28 Upper scrobe margins almost parallel in full-face view (Fig. 206); the occipital corners extended backwards as a pair of triangular processes. Frontal lobes large, projecting forward almost as far as anterior margin of the median clypeal lobe. Lateral portions of clypeus not raised into a ridge or wall in front of the antennal insertions . . . . ***Blepharidatta***

— Upper scrobe margins conspicuously sinuate in full-face view (Fig. 208); the occipital corners not extended backwards as a pair of triangular processes. Frontal lobes small and inconspicuous, not approaching the anterior margin of the median clypeal lobe. Lateral portions of clypeus raised into a sharp ridge or wall in front of the antennal insertions . . . . ***Wasmannia***

29 Petiole sessile, in profile without an anterior peduncle (Fig. 293). Petiole in profile roughly cylindrical, with dorsal margin weakly convex but without a differentiated node . . . . ***Xenomyrmex*** (part)

— Petiole in profile with a conspicuous anterior peduncle (Figs. 259, 351, 353, 357, 377, 411). Petiole in profile not cylindrical, with a developed node **30**

30 In full-face view the midpoint of the anterior clypeal margin with a long, unpaired median seta, which projects forward over the mandibles (Figs. 258, 376, 410) **31**

— In full-face view the midpoint of the anterior clypeal margin without a long, unpaired median seta; instead the midpoint of the margin straddled by a pair of setae (Figs. 350, 352, 356) **34**

31 Antenna with 10 segments. Maxillary palp geniculate. Propodeum unarmed (Fig. 377) **32**

— Antenna with 12 segments. Maxillary palp not geniculate. Propodeum with a pair of teeth or spines (Figs. 259, 411) **33**

32 Median portion of clypeus abruptly raised and flat-topped, the raised platform thus formed bounded by the parallel paired clypeal carinae, which then turn mesad anteriorly and meet anteromedially to form the anterior margin of the raised portion . . . . ***Carebarella***

— Median portion of clypeus without a raised, flat-topped section; instead the 2 clypeal carinae divergent anteriorly and running to the margin, where they often project as a pair of teeth or denticles (Fig. 376). Clypeal carinae never turning mesad anteriorly nor meeting anteromedially . . . . ***Solenopsis***

33 Frontal lobes closely approximated; median portion of clypeus, where it is inserted between the lobes, narrower than either of the frontal lobes (Fig. 410). Basal margin of mandible with a tooth or tumulus close to or behind its midlength. Palp formula 2,2 or less . . . . ***Adelomyrmex***

— Frontal lobes widely separated; median portion of clypeus, where it is inserted between the lobes, much broader than either of the frontal lobes (Fig. 258). Basal margin of mandible unarmed. Palp formula 5,3 ...... ***Cardiocondyla*** (part)

34 Propodeum bispinose, bidentate (Figs. 351, 353), or sharply angulate in profile. Mandible with 4 teeth. Worker caste dimorphic, without intermediates .... ***Oligomyrmex***

— Propodeum unarmed and rounded in profile (Fig. 357). Mandible with 5 or 6 teeth. Worker caste monomorphic .... ***Carebara***

35 Antenna with 7–11 segments **36**

— Antenna with 12 segments **66**

36 Mandibles produced into a pair of elongate blades, which are edentate on their inner margins and terminate apically in a fork of 2 teeth arranged in a vertical series. Antennal scapes, when laid back from their insertions, passing below the eye (Figs. 228, 229) .... ***Daceton***

— Mandibles triangular to subtriangular, edentate to multidentate, but never terminating in an apical fork of 2 teeth in a vertical series. Antennal scapes, when laid back from their insertions, running either to the eye or above the eye, not passing below the eye **37**

37 Horizontal frontal lobes absent, the antennal sockets fully exposed in full-face view (as Fig. 300). Apical (masticatory) margin of mandible with 3 teeth; basal margin of mandible with a tooth at about its midlength. Anterior clypeal margin denticulate. Antenna 9-segmented .... ***Perissomyrmex***

— Horizontal frontal lobes present, partially to entirely covering the sockets in full-face view (Figs. 178, 182, 184, 190, 208, 216, 250, 322, 332, 334, 374, 416). Apical (masticatory) margin of mandible with more than 3 teeth or denticles in total; basal margin of mandible unarmed. Anterior clypeal margin not denticulate. Antenna with 7–11 segments, but only very rarely with 9 .... **38**

38 Frontal lobes enormously expanded laterally and anteriorly (Figs. 184, 214, 216), overhanging and concealing the lateral portions of the clypeus so that the anterolateral corners of the head capsule cannot be seen in full-face view. Frontal lobes projecting forward so far that their anteriormost points are level with, or project beyond, the anterior margin of the median portion of the clypeus **39**

— Frontal lobes small to moderate (Figs. 178, 182, 190, 208, 250, 322, 332, 334, 374, 416), never overhanging and concealing the anterolateral corners of the head capsule, the latter always visible in full-face view. Anteriormost points of frontal lobes usually well behind the level of the anterior margin of the median clypeus; when otherwise (rarely), the anterolateral corners of the head capsule are clearly exposed **42**

39 With the head in profile the eye situated below the antennal scrobe, the scrobe extending posteriorly beyond the level of the eye **40**

— With the head in profile the eye situated at the posterior apex of the antennal scrobe, the scrobe not extending posteriorly above and beyond the eye (Figs. 215, 217) **41**

Neotropical MYRMICINAE *(continued)*

40 With head in full-face view the eyes at or very close to the posterolateral corners. Outer margins of frontal lobes continuous with upper scrobe margins, the two not separated by a constriction, indentation, or change of direction .... ***Cephalotes***

— With head in full-face view the eyes at or just in front of the midlength of the sides. Outer margins of frontal lobes differentiated from upper scrobe margins by a constriction, indentation, or change of direction (Fig. 184) .... ***Cyphomyrmex***

41 Eyes subglobose and strongly prominent. Sides of petiole and postpetiole unarmed (Fig. 215). Occipital angles bispinose (Fig. 214) .... ***Eucryptocerus***

— Eyes gently convex to flat. Sides of petiole and postpetiole always with projecting teeth, spines, or lobes (Fig. 217). Occipital angles of head either rounded, laminate, dentate, or armed with a single spine (Fig. 216) .... ***Zacryptocerus***

42 Apical (masticatory) margin of mandible with 3–6 teeth or denticles in total (Figs. 184, 190, 208, 250, 322, 332, 362, 374) **43**

— Apical (masticatory) margin of mandible with 7 or more teeth or denticles in total (Figs. 178, 180, 182, 186, 188, 192, 334, 416) **56**

43 Frontal carinae present, strongly and conspicuously developed; frontal carinae running backwards from the posteriormost points of the frontal lobes as a pair of sharp ridges or flanges, which reach, or almost reach, the occipital margin (Figs. 184, 185, 208, 209, 212, 213) **44**

— Frontal carinae entirely absent or vestigially present; if the latter the frontal carinae are represented only by a pair of faint lines, which fade out posteriorly (Figs. 188, 190, 250, 292, 320, 322, 332, 362, 374) **47**

44 With alitrunk in profile the promesonotal dorsum with one to several pairs of vertically directed conical prominences or tumuli (Fig. 185). Propodeal spiracle set far back on the side, very close to the margin of the declivity (Fig. 185) **45**

— With alitrunk in profile the promesonotal dorsum without vertical conical prominences or tumuli (Figs. 209, 213). Propodeal spiracle set farther forward on the side, at or close to the midlength of the sclerite (Figs. 209, 213) **46**

45 Well-defined, impressed antennal scrobes absent. Frontal lobes narrow, not entirely concealing the antennal sockets. Frontal carinae close together, subparallel, and terminating near the occipital midpoint, on each side of a median impression. Propodeum bispinose .... ***Mycetarotes***

— Well-defined, impressed antennal scrobes present. Frontal lobes broad, concealing the antennal sockets. Frontal carinae divergent posteriorly and terminating at the occipital corners. Propodeum unarmed to bidentate .... ***Cyphomyrmex*** (part)

46 Petiole sessile, without an elongate anterior peduncle (Fig. 213). Frontal carinae in full-face view widely separated, divergent posteriorly, not sinuate; frontal carinae overhanging and usually concealing lateral margins of head behind level of eyes

(Fig. 212). Antennal scrobes narrow and deep (Fig. 213). Midpoint of anterior clypeal margin concave and lateral portions of clypeus not raised into a narrow wall or ridge in front of the antennal insertions . . . . ***Procryptocerus***

— Petiole with an elongate anterior peduncle (Fig. 209). Frontal carinae in full-face view not widely separated, sinuate along their length; frontal carinae not overhanging nor concealing lateral margins of head behind level of eyes (Fig. 208). Antennal scrobes broad and shallow (Fig. 209). Midpoint of anterior clypeal margin convex and lateral portions of clypeus raised into a narrow wall or ridge in front of the antennal insertions . . . . ***Wasmannia*** (part)

47 Lateral portions of clypeus dorsoventrally flattened and raised into a sharp transverse ridge or shield wall in front of the antennal insertions . . . . ***Mycetophylax***

— Lateral portions of clypeus not dorsoventrally flattened, not raised into a sharp transverse ridge or shield wall in front of the antennal insertions **48**

48 With head in full-face view the frontal lobes closely approximated (Figs. 188, 190, 374); median portion of clypeus posteriorly, where it passes between the frontal lobes, very narrow, its width equal to or less than the width of one of the frontal lobes **49**

— With head in full-face view the frontal lobes separated (Figs. 250, 292, 320, 322, 332, 362); median portion of clypeus posteriorly, where it passes between the frontal lobes, broad, its width distinctly much greater than the width of one of the frontal lobes **51**

49 With alitrunk in profile the promesonotal dorsum adorned with tubercles, spines, conical processes, elevated plateaux, raised cuticular ridges, or a combination of 2 or more of these (Figs. 189, 191). Median portion of clypeus not longitudinally bicarinate and eyes behind midlength of sides of head (Figs. 188, 190) **50**

— With alitrunk in profile the promesonotal dorsum evenly convex (Fig. 375). Median portion of clypeus sharply longitudinally bicarinate and eyes in front of midlength of sides of head (Fig. 374) . . . . ***Oxyepoecus***

50 Curved, squamiform setae abundant on head and body (Figs. 188, 189). Pronotal dorsum not armed with 10–12 teeth or spines (Fig. 189) . . . . ***Myrmicocrypta*** (part)

— Curved squamiform setae absent from head and body (Figs. 190, 191). Pronotal dorsum armed with 10–12 teeth or spines (Fig. 191) . . . . ***Mycocepurus***

51 Midpoint of anterior clypeal margin with an unpaired median seta (Fig. 362). Proximal section of each antennal club segment constricted into a distinct neck. Antenna usually with 7–10 segments, only rarely with 11 . . . . ***Allomerus***

— Midpoint of anterior clypeal margin without an unpaired median seta, instead usually with a pair of setae that straddle the midpoint (Figs. 250, 292, 322). Proximal section of each antennal club segment not constricted into a neck. Antenna usually with 11 segments, only rarely with 10 . . . . **52**

52 Petiole in profile sessile (Fig. 293), subcylindrical, with a weakly convex area dorsally but lacking a differentiated node. Palp formula 4,2 . . . . ***Xenomyrmex***

— Petiole in profile with a conspicuous anterior peduncle (Figs. 251, 321, 323, 333), not subcylindrical, with a distinctly differentiated node. Palp formula variable but never 4,2 . . . **53**

53 Mandibular dentition irregular. Either with 2 large teeth apically, followed by a long diastema (which may be minutely crenulate) and 1–3 smaller teeth basally (Fig. 332); or with serial dentition but the third tooth (or rarely third and fourth teeth) disproportionately small. Reduced tooth or teeth followed by a larger tooth and at least 1 small tooth or denticle at the basal angle; basalmost tooth not enlarged . . . . ***Pheidole*** (part)

— Mandibular dentition regular. Apical (masticatory) margin with 4–6 teeth which decrease in size from apex to base (Figs. 250, 320, 322), or with the basalmost tooth enlarged. Without diastemata, without a disproportionately reduced tooth behind the second **54**

54 Palp formula 5,3 . . . . ***Leptothorax*** (part)

— Palp formula usually 3,2, rarely 4,3 . . . . **55**

55 Vestigial frontal carinae present which run from the apices of the frontal lobes approximately to the level of the midlengths of the eyes (Fig. 322). Median portion of clypeus with a trace of a median longitudinal carina. Anterior border of pronotum marginate. Eyes relatively large, with 10 or more ommatidia in the longest row (Fig. 323) . . . . ***Ochetomyrmex***

— Frontal carinae entirely absent behind the frontal lobes (Fig. 320). Median portion of clypeus entirely lacking a median longitudinal carina. Anterior border of pronotum not marginate. Eyes small to absent, when present with less than 10 ommatidia in the longest row (Fig. 321) . . . . . . ***Tranopelta***

56 Sting with a conspicuous apical or apicodorsal lamellate appendage, which projects at an angle from the shaft (Fig 417). Lateral portions of clypeus raised into a sharp transverse ridge or shield wall in front of the antennal insertions (Fig. 416). Mandible with 3 larger teeth apically followed by a row of 4 (rarely more) smaller teeth or denticles . . . . ***Tetramorium*** (part)

— Sting lacking a lamellate appendage projecting at an angle apically or apicodorsally from the shaft; sting frequently vestigial. Lateral portions of clypeus not raised into a sharp transverse ridge or shield wall in front of the antennal insertions (Figs. 178, 180, 182, 184, 186, 188, 192, 332, 334). Mandibular dentition not of 3 larger teeth apically followed by a row of 4 smaller teeth or denticles **57**

57 Apical (masticatory) margin of mandible with 2 large teeth apically followed by 1 or sometimes 2 reduced teeth or denticles; these followed by a distinctly larger tooth, behind which the margin with irregular teeth or denticles to the basal angle (Fig. 334). Dorsal alitrunk without dorsally directed spines, teeth, tumuli, or conical processes (Fig. 335) . . . . ***Pheidole*** (part)

— Apical (masticatory) margin of mandible either with all teeth approximately the same size or the teeth gradually reducing in size from apex to base (Figs. 178, 180, 182, 186, 188, 192); sometimes the basalmost tooth enlarged. Dorsal alitrunk usually, but not always, with dorsally directed spines, teeth, tumuli, or conical processes (Figs. 179, 181, 183, 185, 187, 189, 193) **58**

58 Frontal carinae either entirely absent or the frontal lobes followed by short, feeble ridges which rapidly peter out, not forming the upper margin of a scrobe nor approaching the occipital margin (Figs. 180, 188, 192) **59**

— Frontal carinae present and usually very conspicuous, running back to, or almost to, the occipital margin. In strength of development the frontal carinae varying from massive flanges, to ridges which border the dorsal margin of a conspicuous scrobe, to a pair of narrow, posteriorly divergent, ridge-like carinae which may or may not delimit a scrobal boundary (Figs. 182, 184, 186) **62**

59 With head in full-face view the frontal lobes closely approximated, with only a narrow strip of cuticle separating them (Figs. 188, 192) **60**

— With head in full-face view the frontal lobes widely separated, with the broad arch of the posterior clypeal border between them (Figs. 178, 180) **61**

60 Squamiform setae present on head and body (Figs. 188, 189). Propodeum bidentate to bispinose. Promesonotum with dorsum raised into a number of peaks, tubercles or spiniform processes. Junction of postpetiole and gaster dorsoventrally compressed, extremely narrow in profile (Fig. 189) . . . . ***Myrmicocrypta*** (part)

— Fine, acute setae present on head and body, usually densely so (Figs. 192, 193). Propodeum rounded to angulate. Promesonotum with dorsum rounded or with a pair of longitudinal ridges. Junction of postpetiole and gaster not markedly compressed, not extremely narrow in profile (Fig. 193) . . . . ***Apterostigma***

61 Promesonotal dorsum with 3 or more pairs of spines or teeth. First gastral tergite tuberculate, usually distinctly so (Fig. 179) . . . . ***Acromyrmex*** (part)

— Promesonotal dorsum with 2 pairs of spines or teeth. First gastral tergite smooth, sometimes sculptured but never tuberculate (Fig. 181) . . . . ***Atta***

62 First gastral tergite with numerous to abundant tubercles, which may be irregular, subconical, or conical but which are always distinct and widely distributed on the sclerite (Figs. 179, 183) **63**

— First gastral tergite without tubercles; the sclerite usually unadorned except for fine surface sculpture but sometimes with parallel longitudinal ridges (Figs. 185, 187) **64**

63 Monomorphic species. Antennal scrobes present, shallow and broad. Superocular carina sometimes running the length of the scrobe and forming its lower margin, but sometimes the carina abruptly curving mesad and petering out immediately behind the level of the eye, in which case the scrobes are conspicuous and the pronotal dorsum is multituberculate but lacks 3 pairs of acute teeth or spines (Figs. 182, 183) . . . . ***Trachymyrmex***

— Polymorphic species. Antennal scrobes usually absent but sometimes shallowly present. Whether scrobes absent or present the superocular carina always curves abruptly mesad and peters out immediately behind the level of the eye, never forming a lower margin to the scrobe behind the eye. Pronotal dorsum with 3 isolated pairs of acute teeth or spines, not multituberculate (Figs. 178, 179) . . . . . . ***Acromyrmex*** (part)

64 Dorsal surfaces of head and body with numerous to abundant simple setae (Figs. 186, 187) **65**

— Dorsal surfaces of head and body without setae of any description (Figs. 184, 185) . . . . ***Cyphomyrmex*** (part)

65 Antennal scrobes without sharply defined ventral margins behind the level of the eyes (Fig. 187) . . . . ***Sericomyrmex***

— Antennal scrobes with sharply defined ventral margins behind the level of the eyes . . . . ***Mycetosoritis***

66 Apical (masticatory) margin of mandible with only 3–6 teeth or denticles in total, the dentition usually sharply defined and the teeth decreasing in size from the apical tooth (Figs. 250, 258, 310, 332, 378, 380, 382, 386) **67**

— Apical (masticatory) margin of mandible with 7 or more teeth or denticles in total (Figs. 334, 338, 384, 388, 416), the dentition sometimes decreasing in size from the apical tooth but sometimes the masticatory margin with ill-defined crenulations or denticles between the main teeth, or with a series of ill-defined crenulations or denticles near the basal angle **77**

67 Tibial spurs of hind legs finely pectinate. Ventral alitrunk with an extensive V-shaped or narrowly U-shaped emargination running from the posterior border forward and terminating at or close to the metasternal pit, well in front of the hind coxal cavities. Metasternal process large, represented by 2 prominent triangles or teeth, at the apex on each side of the median emargination **68**

— Tibial spurs of hind legs simple to absent. Ventral alitrunk at most with a shallow, U-shaped, posterior emargination, which does not project forward beyond the hind coxal cavities and which ends far short of the metasternal pit. Metasternal process moderate to absent, when present always widely separated from the apex of the median emargination . . . **69**

68 Propodeal lobes biaculeate, the upper and lower points of the lobe both sharp (Fig. 313). Apical (masticatory) margin of mandible markedly oblique in full-face view (Fig. 312) . . . . ***Hylomyrma***

— Propodeal lobes rounded, triangular, or spiniform; when prominent always with a single point, never biaculeate (Fig. 311). Apical (masticatory) margin of mandible perpendicular or nearly so in full-face view (Fig. 310) . . . . ***Pogonomyrmex*** (part)

69 Midpoint of anterior clypeal margin with a long, unpaired median seta, which projects forward over the mandibles (Figs. 258, 372, 378, 382), or occasionally the anterior clypeal margin with an undifferentiated, dense row of long, strong setae **70**

— Midpoint of anterior clypeal margin without an unpaired, long median seta, instead with a pair of long setae that straddle the midpoint (Figs. 250, 332, 386, 388); anterior clypeal margin never with an undifferentiated, dense row of long, strong setae **74**

70 Maxillary palp with 5 segments. Lateral portions of clypeus dorsoventrally flattened and thin (Fig. 258), strongly prominent over the mandibles, and sometimes projecting farther forward than the median portion of the clypeus . . . . ***Cardiocondyla*** (part)

— Maxillary palp with 1–4 segments. Lateral portions of clypeus not dorsoventrally flattened nor projecting over the mandibles (Figs. 372, 378, 380, 382); never projecting as far forward as the median portion of the clypeus **71**

71 Maxillary palp with 3 or 4 segments **72**

— Maxillary palp with 1 or 2 segments **73**

72 Propodeal declivity with a transversely arched rim or carina running across the declivity between the uppermost points of the propodeal lobes. Propodeal spiracle not preceded on side of alitrunk by a thin-walled vestibule that is conspicuous through the cuticle (Fig. 373) . . . . ***Megalomyrmex*** (part)

— Propodeal declivity without a transversely arched rim or carina running across the declivity between the propodeal lobes. Propodeal spiracle preceded on side of alitrunk by a thin-walled vestibule that is conspicuous through the cuticle (Fig. 381) . . . . ***Nothidris***

73 Mandible with 5 teeth (Fig. 382). Anterior tentorial pit midway between antennal sockets and lateral margin of clypeus. Propodeum bidentate to bispinose (Fig. 383) . . . . ***Antichthonidris***

— Mandible with 4 teeth (Fig. 378). Anterior tentorial pit close to antennal socket. Propodeum usually unarmed and rounded (Fig. 379) but rarely may be angulate or minutely bidenticulate . . . . ***Monomorium***

74 Apical (masticatory) margin of mandible with 2 teeth apically, followed by a long diastema and 1–3 teeth basally (Fig. 332) (major workers) . . . . ***Pheidole*** (part)

— Apical (masticatory) margin of mandible with 5 or 6 teeth which decrease in size from the apical to the basal, the dentition lacking a long diastema between the apical and basal groups of teeth (Figs. 250, 386) **75**

75 Median portion of clypeus longitudinally bicarinate, the carinae running on each side of, and close to, the midline (Figs. 386, 388); the carinae diverging anteriorly. Maxillary palps with fewer than 5 segments, usually 2 or 3, rarely 4 . . . . **76**

— Median portion of clypeus not longitudinally bicarinate (Fig. 250); either clypeus without carinae or a median carina present on part or all of the midline. Maxillary palps with 5 segments . . . . ***Leptothorax*** (part)

76 Antennal club of 3 segments (Fig. 387). With head in lateroventral view the posteroventral corner with a short, narrow groove running forward from the corner . . . . ***Rogeria*** (part)

— Antennal club of 4 segments (Fig. 389). With head in lateroventral view the posteroventral corner entire, without a groove running forward from the corner . . . . ***Stenamma*** (part)

77 Frontal lobes enormously expanded laterally and anteriorly, projecting farther forward than the anterior clypeal margin in full-face view (Fig. 384). In profile the median portion of the clypeus vertical and some distance back from the anteriormost points of the frontal lobes. Long and extremely deep antennal scrobes present (Fig. 385) . . . . ***Stegomyrmex***

— Frontal lobes small to moderately expanded, never projecting farther forward than the anterior clypeal margin in full-face view. In profile the median portion of the clypeus not vertical, not behind the anteriormost points of the frontal lobes. Antennal scrobes usually absent but if present then only weakly developed **78**

78 Sting with an apicodorsal triangular to pennant-shaped lamellate appendage, which projects from the sting shaft at an angle to its long axis (Fig. 417) . . . . ***Tetramorium*** (part)

— Sting simple, without an apicodorsal triangular to pennant-shaped lamellate appendage projecting from the sting shaft at an angle to its long axis **79**

79 Tibial spurs of hind legs finely pectinate. Ventral alitrunk with an extensive V-shaped or narrowly U-shaped emargination running from the posterior border forward and terminating at or close to the metasternal pit, well in front of the hind coxal cavities. Metasternal process large, represented by 2 prominent triangles or teeth, at the apex on each side of the median emargination . . . . ***Pogonomyrmex*** (part)

— Tibial spurs of hind legs simple to absent. Ventral alitrunk at most with a shallow, U-shaped posterior emargination, which does not project forward beyond the hind coxal cavities and which ends far short of the metasternal pit. Metasternal process small to absent; when present always widely separated from the apex of the median emargination **80**

80 Median portion of clypeus longitudinally bicarinate, the carinae located one on each side of the midline (Figs. 372, 386, 388) **81**

— Median portion of clypeus not longitudinally bicarinate (Figs. 334, 338); either the clypeus without carinae or a median carina present on part or all of the midline **83**

81 With petiole in profile the spiracle situated at the node or on the peduncle about level with the ascending anterior face of the node (Fig. 373); the spiracle always behind the midlength of the peduncle, never far forward . . . ***Megalomyrmex*** (part)

— With petiole in profile the spiracle situated far anteriorly on the peduncle, at or very close to the articulation of the peduncle

Neotropical MYRMICINAE *(continued)*

with the alitrunk (Figs. 387, 389); spiracle always in front of the midlength of the peduncle, never close to the node **82**

82 Antennal club of 3 segments (Fig. 387). With head in lateroventral view the posteroventral corner with a short, narrow groove running forward from the corner . . . . ***Rogeria*** (part)

— Antennal club of 4 segments (Fig. 389). With head in lateroventral view the posteroventral corner entire, without a groove running forward from the corner . . . . ***Stenamma*** (part)

83 Apical (masticatory) margin of mandible with the third tooth (counting from the apex) smaller than the fourth, or the reduced third tooth followed by a minute denticle before the larger fourth tooth [all minor workers] (Fig. 334); or the mandible with 2 large apical and 1 or 2 enlarged basal teeth, the margin between these groups of teeth irregularly crenulate to bluntly dentate [major workers of some species]. Palp formula usually 2,2 or 3,2, rarely more . . . . ***Pheidole*** (part)

— Apical (masticatory) margin of mandible with the third tooth (counting from the apex) larger than the fourth (Fig. 338). Mandible never with dentition as described for major workers above. Palp formula 4,3 or 5,3 . . . . ***Aphaenogaster***

Synoptic Classification

A name prefixed by * indicates an extinct taxon. The placement of many genera, including most fossil forms, is speculative.

Subfamily **MYRMICINAE.**

Tribe **Agroecomyrmecini.** Genera: ****Agroecomyrmex, *Eulithomyrmex*** (= **Lithomyrmex*), ***Tatuidris*** (Figs. 176, 177).

Tribe **Attini.** Genera: ***Acromyrmex*** (Figs. 178, 179) [subgenera: nominal plus *Moellerius*], ***Apterostigma*** (Figs. 192, 193), ***Atta*** (Figs. 180, 181) (= *Archeatta,* = *Epiatta,* = *Myrmegis* (nomen nudum), = *Neoatta,* = *Oecodoma,* = *Palaeatta*), **Attaichnus* (ichnotaxon, unavailable name), ***Cyphomyrmex*** (Figs. 184, 185) (= *Cyphomannia*), ***Mycetarotes, Mycetophylax*** (= *Paramycetophylax*), ***Mycetosoritis, Mycocepurus*** (Figs. 190, 191) (= *Descolemyrma,* = *Mycetopurus* (misspelling)), ***Myrmicocrypta*** (Figs. 188, 189) (= *Glyptomyrmex*), ***Pseudoatta*** (workerless), ***Sericomyrmex*** (Figs. 186, 187), ***Trachymyrmex*** (Figs. 182, 183).

Tribe **Basicerotini.** Genera: ***Basiceros*** (Figs. 194, 195) (= *Aspididris,* = *Ceratobasis* (homonym)), ***Creightonidris, Eurhopalothrix*** (Figs. 196, 197), ***Octostruma*** (Figs. 198, 199), ***Protalaridris*** (Figs. 200, 201), ***Rhopalothrix*** (Figs. 202, 203) (= *Acanthidris,* = *Heptastruma*), ***Talaridris*** (Figs. 204, 205).

Tribe ***Blepharidattini.*** Genera: ***Blepharidatta*** (Figs. 206, 207), ***Wasmannia*** (Figs. 208, 209) (= *Hercynia*).

Tribe ***Cataulacini.*** Genus: ***Cataulacus*** (Figs. 210, 211) (= *Otomyrmex*).

Tribe **Cephalotini** (= Cryptoceridae). Genera: ***Cephalotes*** (= *Cryptocerus*), ***Eucryptocerus*** (Figs. 214, 215), ***Procryptocerus*** (Figs. 212, 213), ***Zacryptocerus*** (Figs. 216, 217) (= *Cyathocephalus* (homonym), = *Cyathomyrmex,* = *Harnedia,* = *Hypocryptocerus,* = *Paracryptocerus*).

Tribe **Crematogastrini.** Genus: ***Crematogaster*** (Figs. 218, 219) (= *Acrocoelia,* = *Cremastogaster* (misspelling), = *Tranopeltoides*) [subgenera: nominal plus *Apterocrema, Atopogyne, Colobocrema, Decacrema* (= *Decracrema* (misspelling)), *Eucrema, Mesocrema, Nematocrema, Neocrema, Orthocrema, Oxygyne, Paracrema, Physocrema, Rhachiocrema, Sphaerocrema, Xiphocrema*].

Tribe **Dacetonini.** Genera: ***Acanthognathus*** (Figs. 230, 231), ***Asketogenys, Chelystruma, Cladarogenys, Codiomyrmex, Codioxenus, Colobostruma*** (Figs. 220, 221) (= *Alistruma,* = *Clarkistruma*), ***Daceton*** (Figs. 228, 229) (= *Dacetum*), ***Dorisidris, Dysedrognathus*** (Figs. 238, 239), ***Epitritus*** (Figs. 240, 241), ***Epopostruma*** (Figs. 224, 225) (= *Hexadaceton*), ***Glamyromyrmex*** (Figs. 244, 245) (= *Borgmeierita*), ***Gymnomyrmex, Kyidris*** (Figs. 248, 249) (= *Polyhomoa*), ***Mesostruma*** (Figs. 222, 223), ***Microdaceton*** (Figs. 226, 227), ***Neostruma*** (Figs. 236, 237), ***Orectognathus*** (Figs. 232, 233) (= *Arnoldidris*), ***Pentastruma, Quadristruma, Serrastruma*** (Figs. 246, 247), ***Smithistruma*** (Figs. 242, 243) (= *Cephaloxys* (homonym), = *Miccostruma,* = *Platystruma,* = *Weberistruma,* = *Wessonistruma*), ***Strumigenys*** (Figs. 234, 235) (= *Eneria,* = *Labidogenys,* = *Proscopomyrmex,* = *Pyramica*), ***Tingimyrmex, Trichoscapa.***

Tribe ***Formicoxenini*** (= Cardiocondylini, = Leptothoracini, = Podomyrmini, = Stereomyrmicini). Genera: ***Ankylomyrma*** (Figs. 284, 285), ***Atopomyrmex*** (Figs. 272, 273), ***Cardiocondyla*** (Figs. 258, 259) (= *Dyclona,* = *Emeryia,* = *Loncyda,* = *Prosopidris,* = *Xenometra*), ***Chalepoxenus*** (Figs. 256, 257) (= *Leonomyrma*), ***Dilobocondyla*** (Figs. 276, 277) (= *Mesomyrma*), ***Doronomyrmex*** (workerless), ***Epimyrma*** (Figs. 254, 255) (= *Myrmetaerus,* = *Myrmoxenus*) [subgenera: nominal plus *Gonepimyrma*], ***Formicoxenus*** (Figs. 252, 253) (= *Symmyrmica*), ***Harpagoxenus*** (Figs. 262, 263) (= *Tomognathus* (homonym)), ***Ireneopone*** (Figs. 278, 279), ***Leptothorax*** (Figs. 250, 251) (= *Antillaemyrmex,* = *Caulomyrma,* = *Croesomyrmex,* = *Dichothorax,* = *Goniothorax* (homonym), = *Icothorax,* = *Limnomyrmex,* = *Macromischa,* = *Meia,* = *Mychothorax,* = *Myrmammophilus,* = *Myrafant,* = *Nesomyrmex,* = *Temnothorax,* = *Tetramyrma*), *Leptoxenus* (nomen nudum), ***Paratopula*** (Figs. 280, 281), ***Peronomyrmex, Podomyrma*** (Figs. 270, 271) (= **Acrostigma,* = *Dacryon,* = *Pseudopodomyrma*), ***Poecilomyrma, Protomognathus*** (Figs. 260, 261), ***Romblonella*** (Figs. 264, 265), ***Rotastruma*** (Figs. 268, 269), ***Stereomyrmex*** (Figs. 282, 283), ****Stigmomyrmex, Terataner*** (Figs. 274, 275) (= *Tranetera*), ***Tricytarus*** (male only, dubiously placed here), ***Vombisidris*** (Figs. 266, 267), ***Willowsiella.***

Tribe **Melissotarsini.** Genera: ***Melissotarsus*** (Figs. 288, 289), ***Rhopalomastix*** (Figs. 286, 287).

Tribe **Meranoplini.** Genera: ***Meranoplus*** (Figs. 304, 305) (= *Cryptocephalus*), ****Parameranoplus.***

Tribe **Metaponini.** Genera: ***Liomyrmex*** (Figs. 290, 291) (= *Laparomyrmex,* = *Promyrma*), ***Metapone, Vollenhovia*** (Figs. 294, 295) (= *Acalama,* = *Aratromyrmex,* = *Dorothea,* = *Dyomorium,* = *Gauromyrmex,* = *Heteromyrmex,* = **Propodomyrma,* = *Solenomyrma,* = *Vollenhovenia*), ***Xenomyrmex*** (Figs. 292, 293) (= *Myrmecinella*) (dubiously placed here).

Tribe **Myrmecinini** (= Archaeomyrmicini). Genera: ***Acanthomyrmex*** (Figs. 296–299), ****Enneamerus, Myrmecina*** (Figs. 302, 303) (= *Archaeomyrmex*), ***Perissomyrmex, Pristomyrmex*** (Figs. 300, 301) (= *Dodous,* = *Hylidris,* = *Odontomyrmex*), ****Stiphromyrmex.***

Tribe **Myrmicariini.** Genus: ***Myrmicaria*** (Figs. 306, 307) (= *Heptacondylus,* = *Physatta*).

Tribe **Myrmicini.** Genera: ***Eutetramorium*** (Figs. 314, 315), ***Huberia*** (Figs. 316, 317), ***Hylomyrma*** (Figs. 312, 313) (= *Lundella*), ***Manica*** (= *Neomyrma,* = *Oreomyrma*), ***Myrmica*** (Figs. 308, 309) (= *Dodecamyrmica,* = *Paramyrmica,* = *Sifolinia,* = *Sommimyrma,* = *Symbiomyrma*), **Nothomyrmica, Pogonomyrmex*** (Figs. 310, 311) (= *Ephebomyrmex,* = *Forelomyrmex,* = *Janetia* (homonym)).

Tribe **Ochetomyrmecini.** Genera: ***Ochetomyrmex*** (Figs. 322, 323) (= *Brownidris*), ***Tranopelta*** (Figs. 320, 321).

Tribe ***Phalacromyrmecini.*** Genera: ***Ishakidris*** (Figs. 318, 319), ***Phalacromyrmex, Pilotrochus.***

Tribe **Pheidolini** (= Anergatidini, = Aphaenogastrini, = Ocymyrmicini). Genera: ***Aphaenogaster*** (Figs. 338, 339) (= *Attomyrma,* = *Brunella,* = *Deromyrma,* = *Novomessor,* = *Nystalomyrma,* = *Planimyrma*), ***Chimaeridris*** (Figs. 324, 325), ***Goniomma*** (Figs. 326, 327), *Hypopheidole* (nomen nudum), ***Kartidris*** (Figs. 336, 337), ****Lonchomyrmex, Messor*** (Figs. 340, 341) (= *Cratomyrmex,* = *Lobognathus,* = *Sphaeromessor,* = *Veromessor*), ***Ocymyrmex*** (Figs. 330, 331), ***Oxyopomyrmex*** (Figs. 328, 329), ****Paraphaenogaster*** (= **Paraphaeogaster* (misspelling)), ***Pheidole*** (Figs. 332–335) (= *Allopheidole,* = *Anergatides,* = *Bruchomyrma,* = *Cardiopheidole,* = *Cephalomorium,* = *Ceratopheidole,* = *Conothoracoides,* = *Conothorax* (homonym), = *Decapheidole,* = *Elasmopheidole,* = *Electropheidole,* = *Epipheidole,* = *Eriopheidole,* = *Gallardomyrma,* = *Hendecapheidole,* = *Ischnomyrmex,* = *Isopheidole,* = *Leptomyrma,* = *Macropheidole,* = *Oecophthora,* = *Parapheidole,* = *Pheidolacanthinus,* = *Phidola,* = *Phidole,* = *Scrobopheidole,* = *Stegopheidole,* = *Sympheidole,* = *Trachypheidole,* = *Xenoaphaenogaster*).

Tribe **Pheidologetonini** (= Lophomyrmicini). Genera: ***Adlerzia*** (= *Stenothorax*),***Afroxyidris, Anisopheidole*** (Figs. 358, 359), ***Carebara*** (Figs. 356, 357), ****Hypopomyrmex, Lophomyrmex*** (Figs. 348, 349), ***Machomyrma*** (Figs. 360, 361), ***Oligomyrmex*** (Figs. 350–353) (= *Aeromyrma,* = *Aneleus,* = *Crateropsis,* = *Erebomyrma,* = *Hendecatella,* = *Lecanomyrma,* = *Nimbamyrma,* = *Solenops,* = *Spelaeomyrmex,* = *Sporocleptes*), ****Oxyidris, Paedalgus*** (Figs. 354, 355), ***Pheidologeton*** (Figs. 344–347) (= *Amauromyrmex,* = *Idrisella,* = *Phidologeton*), ***Recurvidris*** (Figs. 342, 343) (= *Trigonogaster* (homonym)).

Tribe **Solenopsidini** (= Megalomyrmecini, = Monomoriini). Genera: ***Allomerus*** (Figs. 362, 363), ***Anillomyrma*** (Figs. 366, 367), ***Antichthonidris*** (Figs. 382, 383), ***Bondroitia*** (Figs. 368, 369), ***Carebarella*** (= *Carebarelloides*), ***Diplomorium*** (Figs. 364, 365), ***Epelysidris*** (Figs. 370, 371), ***Megalomyrmex*** (Figs. 372, 373) (= *Cepobroticus,* = *Wheelerimyrmex*), ***Monomorium*** (Figs. 378, 379) (= *Chelaner,* = *Corynomyrmex,* = *Epixenus,* = *Epoecus,* = *Equessimessor* (misspelling), = *Equestrimessor,* = *Holcomyrmex,* = *Ireneidris,* = *Isholcomyrmex* (misspelling), = *Isolcomyrmex,* = **Lampromyrmex,* = *Mitara,* = *Notomyrmex,* = *Paraholcomyrmex* (misspelling), = *Paraphacota,* = *Parholcomyrmex,* = *Pharaophanes* (nomen nudum), = *Protholcomyrmex,* = *Schizopelta,* = *Syllophopsis,* = *Syllopsis* (misspelling), = *Trichomyrmex,* = *Wheeleria* (homonym), = *Wheeleriella,* = *Xenhyboma,* = *Xeromyrmex*), ***Nothidris*** (Figs. 380, 381), ***Oxyepoecus*** (Figs. 374, 375) (= *Forelifidis,* = *Martia* (homonym)), ***Phacota, Solenopsis*** (Figs. 376, 377) (= *Bisolenopsis,* = *Diagyne,* = *Diplorhoptrum,* = *Disolenopsis* (misspelling), = *Euophthalma,* = *Granisolenopsis,* = *Labauchena,* = *Lilidris,* = *Octella,* = *Oedaleocerus,* = *Paranamyrma,* = *Synsolenopsis*).

Tribe **Stegomyrmecini.** Genus: ***Stegomyrmex*** (Figs. 384, 385).

Tribe **Stenammini** (= Calyptomyrmecini, = Proattini). Genera: ***Ancyridris, Bariamyrma, Calyptomyrmex*** (Figs. 404, 405) (= *Weberidris*), ***Cyphoidris*** (Figs. 394, 395), ***Dacatria, Dacetinops*** (Figs. 398, 399), ***Dicroaspis*** (Figs. 406, 407) (= *Geognomicus*), ****Ilemomyrmex, Indomyrma*** (Figs. 400, 401), ***Lachnomyrmex*** (Figs. 392, 393), ***Lordomyrma*** (Figs. 396, 397) (= *Prodicroaspis,* = *Promeranoplus*), ***Proatta*** (Figs. 402, 403), ***Rogeria*** (Figs. 386, 387) (= *Irogera*), ***Rostromyrmex, Stenamma*** (Figs. 388, 389) (= *Asemorhoptrum,* = *Theryella*), ***Tetheamyrma*** (Figs. 390, 391). Dubiously also included here: ***Mayriella*** (Figs. 408, 409). Very dubiously included here: ***Adelomyrmex*** (Figs. 410, 411) (= *Apsychomyrmex,* = *Arctomyrmex*), ***Baracidris*** (Figs. 412, 413).

Tribe **Tetramoriini** (= Anergatini, = Teleutomyrmini). Genera: ***Anergates*** (workerless), ***Decamorium*** (Figs. 414, 415), ***Rhoptromyrmex*** (Figs. 418, 419) (= *Acidomyrmex,* = *Hagioxenus,* = *Ireneella*), ***Secostruma*** (Figs. 422, 423), ***Strongylognathus*** (Figs. 420, 421) (= *Myrmus* (homonym)), ***Tetramorium*** (Figs. 416, 417) (= *Atopula,* = *Lobomyrmex,* = *Macromichoides* (misspelling), = *Macromischoides,* = *Sulcomyrmex* (unavailable), = *Tetrogmus,* = *Triglyphothrix,* = *Xiphomyrmex*), ***Teleutomyrmex*** (workerless).

Fossil myrmicine genera unplaced to tribe: ****Archimyrmex,*** ****Attopsis,*** ****Cephalomyrmex,*** ****Electromyrmex,*** ****Eocenidris,*** ****Eoformica,*** ****Eomyrmex,*** **Myrmicites* (unavailable), ****Myrmicium*** (= **Myrmecium* (misspelling)), ****Promyrmicium*** (= **Myrmicium* (homonym)).

[Material of the unavailable names Eumyrmicinae, Mycetomyrmicinae and Rhagiomyrmicinae is referable to Myrmicinae; that of the unavailable name Promyrmicinae is referable to Metaponini.]

Distribution

The subfamily Myrmicinae is found in all zoogeographical regions, as shown in the table given in the introduction. The total number of myrmicine genera shared by 2 or more regions is as follows, where PAL = Palaearctic, AFR = Afrotropical, MAL = Malagasy, ORI = Oriental, INA = Indo-Australian, AUS = Australasian, NEA = Nearctic, NEO = Neotropical. The table excludes endemic genera and those accidentally introduced by human activities.

	PAL	AFR	MAL	ORI	INA	AUS	NEA
AFR	12						
MAL	11	17					
ORI	18	22	16				
INA	15	19	15	37			
AUS	11	15	13	19	27		
NEA	20	11	11	14	15	11	
NEO	11	11	10	12	16	13	21

A number of these genera have extremely wide distributions and are shared by 5 or more zoogeographical regions. There are 11 such genera *(Aphaenogaster, Cardiocondyla, Crematogaster, Leptothorax, Monomorium, Oligomyrmex, Pheidole, Smithistruma, Solenopsis, Strumigenys, Tetramorium),* and if they are subtracted from the above table its form changes as follows.

AFR	2						
MAL	—	6					
ORI	7	12	5				
INA	5	10	5	28			
AUS	2	7	4	10	17		
NEA	9	1	—	4	4	3	
NEO	1	2	—	2	6	4	11
	PAL	AFR	MAL	ORI	INA	AUS	NEA

The most striking feature of this table is that the Malagasy region shares only these world-distributed genera with the Palaearctic, Nearctic, and Neotropical regions.

Taxonomic References

Identification of extant species

Some older references have a suffixed comment "[out of date]." These references are included as they contain the only identification keys ever attempted for the taxon in question. They should be used with great caution as, for the most part, they contain numerous infraspecific and infrasubspecific taxa that are no longer recognized. Older references that have been superseded, or those rendered useless by the volume of later descriptions and synonymies, are omitted.

Acanthognathus: Brown and Kempf (1969). ***Acanthomyrmex:*** Moffett (1986). ***Acromyrmex:*** Gonçalves (1961) [Brazil]; Fowler (1988) [subgenus *Moellerius*]. ***Afroxyidris:*** Belshaw and Bolton (1994). ***Ankylomyrma:*** Bolton (1981b). ***Anillomyrma:*** Bolton (1987). ***Antichthonidris:*** Snelling (1975). ***Aphaenogaster:*** M. R. Smith (1961) [New Guinea]; Arnol'di (1976b) [former U.S.S.R.]. ***Asketogenys:*** Brown (1972). ***Atopomyrmex:*** Bolton (1981b). ***Atta:*** Borgmeier (1959). ***Baracidris:*** Bolton (1981b). ***Bariamyrma:*** Lattke (1990a). ***Basiceros:*** Brown and Kempf (1960); Brown (1974a). ***Blepharidatta:*** Kempf (1967a). ***Bondroitia:*** Bolton (1987). ***Calyptomyrmex:*** Baroni Urbani (1975a) [Oriental]; Bolton (1981a) [Afrotropical]. ***Cardiocondyla:*** Bernard (1956b) [Palaearctic]; Bolton (1982) [Afrotropical]; J. Kugler (1984) [males]. ***Carebara:*** W. M. Wheeler (1922) [Afrotopical, out of date]. ***Cataulacus:*** Bolton (1974a) [world]; Bolton (1982) [Afrotropical]. ***Cephalotes:*** Kempf (1951). ***Chalepoxenus:*** Kutter (1973a); Buschinger et al. (1988); Radchenko (1989a) [former U.S.S.R.]. ***Chelystruma:*** Kempf (1959c). ***Chimaeridris:*** Wilson (1989). ***Cladarogenys:*** Brown (1976b); Bolton (1983). ***Creightonidris:*** Brown and Kempf (1960). ***Crematogaster:*** Buren (1968b) [U.S.A,]; Johnson (1988) [eastern U.S.A.]. ***Cyphoidris:*** Bolton (1981b). ***Cyphomyrmex:*** Kempf (1964b) [*strigatus*-group]; Kempf (1966), Snelling and Longino (1992) [*rimosus*-group]. ***Dacetinops:*** Taylor (1985). ***Decamorium:*** Bolton (1976). ***Dicroaspis:*** Bolton (1981a). ***Dilobocondyla:*** W. M. Wheeler (1924) [out of date]. ***Diplomorium:*** Bolton (1987). ***Dorisidris:*** Brown (1948). ***Dysedrognathus:*** Taylor (1968c). ***Epelysidris:*** Bolton (1987). ***Epimyrma:*** Menozzi (1931) [out of date]; Kutter (1973b); Buschinger et al. (1987); Buschinger (1989). ***Epitritus:*** Bolton (1972) [world]; Bolton (1983) [Afrotropical]. ***Eucryptocerus:*** Kempf (1951). ***Eurhopalothrix:*** Brown and Kempf (1960) [world]; Taylor (1968a, 1980b, 1990) [Indo-Australian, Australasian]. ***Formicoxenus:*** Francoeur, Loiselle, and Buschinger (1985). ***Glamyromyrmex:*** Kempf (1960d) [Neotropical]; Bolton (1983) [Afrotropical]. ***Goniomma:*** Santschi (1929b) [out of date]. ***Gymnomyrmex:*** Kempf (1960d); Perrault (1986). ***Huberia:*** Brown (1958c). ***Hylomyrma:*** Kempf (1964c, 1973a). ***Indomyrma:*** Brown (1986). ***Ishakidris:*** Bolton (1984). ***Kartidris:*** Bolton (1991). ***Kyidris:*** Wilson and Brown (1956). ***Lachnomyrmex:*** Weber (1950c). ***Leptothorax:*** Bernard (1956a) [western Europe]; Kempf (1959b) ["subgenus" *Nesomyrmex,* Neotropical]. Baroni Urbani (1978a) ["subgenus" *Macromischa,* Neotropical]; Bolton (1982) [Afrotropical]; Dlussky and Soyunov (1988) ["subgenus" *Temnothorax,* former U.S.S.R.]. ***Lordomyrma:*** Donisthorpe (1941) [out of date]. ***Manica:*** G. C. Wheeler and J. Wheeler (1986) [Nearctic]. ***Mayriella:*** W. M. Wheeler (1935) [out of date]. ***Megalomyrmex:*** Brandao (1990). ***Melissotarsus:*** Bolton (1982). ***Meranoplus:*** Bolton (1981a) [Afrotropical]. ***Mesostruma:*** Brown (1952b); Taylor (1973). ***Messor:*** Bernard (1955) [*structor*-group, Mediterranean]; Bernard (1980) [*barbarus*-group]; Arnol'di (1977) [former U.S.S.R.]; G. Tohmé and H. Tohmé (1981) [Syria]; Bolton (1982) [Afrotropical]; M. R. Smith (1956a) [Nearctic]. ***Metapone:*** W. M. Wheeler (1919) [out of date]. ***Microdaceton:*** Bolton (1983). ***Monomorium:*** DuBois (1986) [Nearctic]; Bolton (1987) [Afrotropical]. ***Mycetarotes:*** Kempf (1960c). ***Mycetophylax:*** Santschi (1922) [out of date]. ***Mycocepurus:*** Kempf (1963b). ***Myrmecina:*** Brown (1967) [North America]. ***Myrmica:*** Menozzi (1939) [Himalaya, Tibet, out of date]; Arnol'di (1970) [former European U.S.S.R.]; Arnol'di (1976a) [former central U.S.S.R.]; Seifert (1988b) [west Palaearctic]; Kupyanskaya (1986) [*lobicornis*-group of far eastern Russia]; Weber (1947, 1948, 1950b) [Nearctic, synopsis of Palaearctic, out of date]. ***Myrmicaria:*** Santschi (1925) [Afrotropical, out of date]. ***Neostruma:*** Brown (1959). ***Nothidris:*** Snelling (1975); Bolton (1987). ***Octostruma:*** Brown and Kempf (1960). ***Ocymyrmex:*** Bolton (1981b); Bolton and Marsh (1989). ***Oligomyrmex:*** Weber (1950a, 1952) [partial, Afrotropical]. ***Orectognathus:*** Brown (1953c, 1958a); Taylor (1977, 1978b, 1980a). ***Oxyepoecus:*** Kempf (1974). ***Paedalgus:*** Bolton and Belshaw (1993). ***Paratopula:*** Bolton (1988b). ***Pentastruma:*** Brown and Boisvert (1979). ***Peronomyrmex:*** Taylor (1970). ***Phacota:*** Bolton (1987). ***Phalacromyrmex:*** Kempf (1960b); Bolton (1984). ***Pheidole:*** Kusnezov (1952a) [Argentina]; Gregg (1959) [Nearctic]; Naves (1985) [Florida]; Ogata (1982) [Japan]; Wilson and Brown (in preparation) [Neotropical]. ***Pilotrochus:*** Brown (1978a); Bolton (1984). ***Pogonomyrmex:*** Kusnezov (1951) [Argentina]; Cole (1968) [Nearctic]; Snelling (1982a) [partial, Nearctic]; Shattuck (1987) [partial, Nearctic]; MacKay et al. (1985) [Mexico]. Snelling and Hunt (1976) [Chile]. ***Pristomyrmex:*** Taylor (1965a, 1968b) [Australasian]; Bolton (1981b) [Afrotropical]. ***Procryptocerus:*** Kempf (1951). ***Protalaridris:*** Brown (1980a). ***Quadristruma:*** Bolton (1983). ***Recurvidris:*** Bolton (1992). ***Rhopalothrix:*** Brown and Kempf (1960) [world]; Taylor (1990) [Indo-Australian, Australasian]. ***Rhoptromyr-***

mex: Brown (1964b); Bolton (1976, 1986). ***Rogeria:*** Kempf (1963a, 1964c) [Neotropical]; C. Kugler (in preparation) [world]. ***Romblonella:*** M. R. Smith (1953, 1956b); Bolton (1976); Taylor (1991). ***Rotastruma:*** Bolton (1991). ***Secostruma:*** Bolton (1988c). ***Serrastruma:*** Brown (1952a); Bolton (1983). ***Smithistruma:*** Brown (1953a, 1964a) [world]; Bolton (1983) [Afrotropical]; Ward (1988) [west Nearctic]. ***Solenopsis:*** Creighton (1930) [New World, out of date]; Bernard (1950, 1978) [France]; Thompson and Johnson (1989) [Florida]; Snelling and Hunt (1976) [Chile]; Trager (1991) [*geminata*-group, world]. ***Stegomyrmex:*** Diniz (1990). ***Stenamma:*** M. R. Smith (1957), Snelling (1973) [Nearctic]; M. R. Smith (1962) [Mesoamerica]; Yasumatsu and Murakami (1960) [Japan]; Arnol'di (1975) [former U.S.S.R.]; DuBois (in preparation) [world]. ***Strongylognathus:*** Baroni Urbani (1969b) [*huberi*-group, Palaearctic]; Radchenko (1985, 1991) [former U.S.S.R.]. ***Strumigenys:*** Brown (1962), Kempf (1976) [Neotropical]; Brown (1954b), Bolton (1983) [Afrotropical]. ***Talaridris:*** Brown and Kempf (1960). ***Tatuidris:*** Brown and Kempf (1968). ***Terataner:*** Bolton (1981b). ***Tetheamyrma:*** Bolton (1991). ***Tetramorium:*** Bolton (1976) [partial, former *Triglyphothrix*, world]; Bolton (1977) [Oriental, Indo-Australian, Australasian]; Bolton (1979) [Malagasy, New World]; Bolton (1980) [Afrotropical]; Radchenko and Arakelian (1990) [*ferox*-complex, Caucasus]; Wang, Xiao, and Wu (1988) [China]; Radchenko (1992) [[former U.S.S.R.]. ***Trichoscapa:*** Bolton (1983). ***Vombisidris:*** Taylor (1989) [Australia]; Bolton (1991). ***Willowsiella:*** Taylor (1991). ***Xenomyrmex:*** Creighton (1957). ***Zacryptocerus:*** Kempf (1951, 1952, 1958a, 1973b).

Other taxonomic references

Anergates: Ettershank (1966). ***Aphaenogaster:*** Bolton (1982). **Basicerotini:** Brown and Kempf (1960). **Cephalotini:** Kempf (1951, 1952, 1958a, 1973b). **Dacetonini:** Bolton (1983) [Afrotropical]; Terayama and Kubota (1989) [Taiwan]. ***Liomyrmex:*** Ettershank (1966). ***Myrmica:*** Bolton (1988a). **Myrmicinae:** Brown (1954a, 1973); Snelling (1981); Dlussky and Fedoseeva (1988); Hölldobler and Wilson (1990); Baroni Urbani, Bolton, and Ward (1992); Ogata (1991b) [genera of Japan]. **Phalacromyrmecini:** Bolton (1984). **Pheidologetonini:** Ettershank (1966). **Solenopsidini:** Ettershank (1966); Bolton (1987). **Tetramoriini:** Bolton (1976).

See also References to Faunistic Studies.

Figures 176–423 MYRMICINAE workers, heads in full-face view and body profiles:

176–177, **Agroecomyrmecini,** *Tatuidris*

178–193, **Attini:** 178–179, *Acromyrmex;* 180–181, *Atta;* 182–183, *Trachymyrmex;* 184–185, *Cyphomyrmex;* 186–187, *Sericomyrmex;* 188–189, *Myrmicocrypta;* 190–191, *Mycocepurus;* 192–193, *Apterostigma*

194–205, **Basicerotini:** 194–195, *Basiceros;* 196–197, *Eurhopalothrix;* 198–199, *Octostruma;* 200–201, *Protalaridris;* 202–203, *Rhopalothrix;* 204–205, *Talaridris*

206–209, **Blepharidattini:** 206–207, *Blepharidatta;* 208–209, *Wasmannia*

210–211, **Cataulacini,** *Cataulacus*

212–217, **Cephalotini:** 212–213, *Procryptocerus;* 214–215, *Eucryptocerus;* 216–217, *Zacryptocerus*

218–219, **Crematogastrini,** *Crematogaster*

220–249, **Dacetonini:** 220–221, *Colobostruma;* 222–223, *Mesostruma;* 224–225, *Epopostruma;* 226–227, *Microdaceton;* 228–229, *Daceton;* 230–231, *Acanthognathus;* 232–233, *Orectognathus;* 234–235, *Strumigenys;* 236–237, *Neostruma;* 238–239, *Dysedrognathus;* 240–241, *Epitritus;* 242–243, *Smithistruma;* 244–245, *Glamyromyrmex;* 246–247, *Serrastruma;* 248–249, *Kyidris*

250–285, **Formicoxenini:** 250–251, *Leptothorax;* 252–253, *Formicoxenus;* 254–255, *Epimyrma;* 256–257, *Chalepoxenus;* 258–259, *Cardiocondyla;* 260–261, *Protomognathus;* 262–263, *Harpagoxenus;* 264–265, *Romblonella;* 266–267, *Vombisidris;* 268–269, *Rotastruma;* 270–271, *Podomyrma;* 272–273, *Atopomyrmex;* 274–275, *Terataner;* 276–277, *Dilobocondyla;* 278–279, *Ireneopone;* 280–281, *Paratopula;* 282–283, *Stereomyrmex;* 284–285, *Ankylomyrma*

286–289, **Melissotarsini:** 286–287, *Rhopalomastix;* 288–289, *Melissotarsus*

290–295, **Metaponini:** 290–291, *Liomyrmex;* 292–293, *Xenomyrmex;* 294–295, *Vollenhovia*

296–303, **Myrmecinini:** 296–297, *Acanthomyrmex* major worker; 298–299, *Acanthomyrmex* minor worker; 300–301, *Pristomyrmex;* 302–303, *Myrmecina*

304–305, **Meranoplini,** *Meranoplus*

306–307, **Myrmicariini,** *Myrmicaria*

308–317, **Myrmicini:** 308–309, *Myrmica;* 310–311, *Pogonomyrmex;* 312–313, *Hylomyrma;* 314–315, *Eutetramorium;* 316–317, *Huberia*

318–319, **Phalacromyrmecini,** *Ishakidris*

320–323, **Ochetomyrmecini:** 320–321, *Tranopelta;* 322–323, *Ochetomyrmex*

324–341, **Pheidolini:** 324–325, *Chimaeridris;* 326–327, *Goniomma;* 328–329, *Oxyopomyrmex;* 330–331, *Ocymyrmex;* 332–333, *Pheidole* major worker; 334–335, *Pheidole* minor worker; 336–337, *Kartidris;* 338–339, *Aphaenogaster;* 340–341, *Messor*

342–361, **Pheidologetonini:** 342–343, *Recurvidris;* 344–345, *Pheidologeton* major worker; 346–347, *Pheidologeton* minor worker; 348–349, *Lophomyrmex;* 350–351, *Oligomyrmex* major worker; 352–353, *Oligomyrmex* minor worker; 354–355, *Paedalgus;* 356–357, *Carebara;* 358–359, *Anisopheidole;* 360–361, *Machomyrma*

362–383, **Solenopsidini:** 362–363, *Allomerus;* 364–365, *Diplomorium;* 366–367, *Anillomyrma;* 368–369, *Bondroitia;* 370–371, *Epelysidris;* 372–373, *Megalomyrmex;* 374–375, *Oxyepoecus;* 376–377, *Solenopsis;* 378–379, *Monomorium;* 380–381, *Nothidris;* 382–383, *Antichthonidris*

384–385, **Stegomyrmecini,** *Stegomyrmex*

386–413, **Stenammini:** 386–387, *Rogeria;* 388–389, *Stenamma;* 390–391, *Tetheamyrma;* 392–393, *Lachnomyrmex,* 394–395, *Cyphoidris;* 396–397, *Lordomyrma;* 398–399, *Dacetinops;* 400–401, *Indomyrma;* 402–403, *Proatta;* 404–405, *Calyptomyrmex;* 406–407, *Dicroaspis;* 408–409, *Mayriella;* 410–411, *Adelomyrmex;* 412–413, *Baracidris*

414–423, **Tetramoriini:** 414–415, *Decamorium;* 416–417, *Tetramorium;* 418–419, *Rhoptromyrmex;* 420–421, *Strongylognathus;* 422–423, *Secostruma.*

176

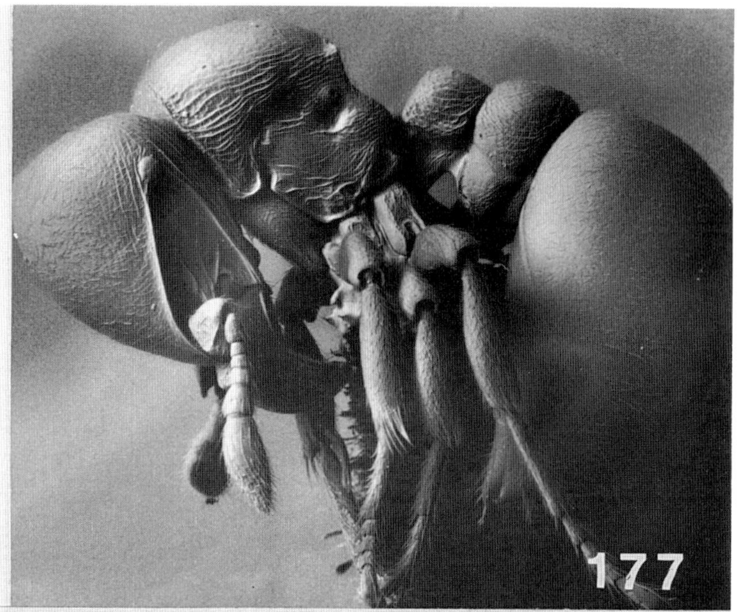

177

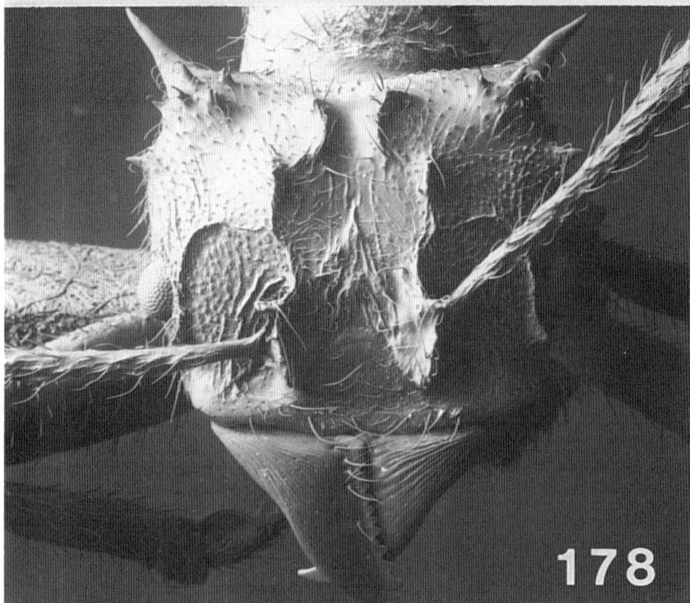

178

179

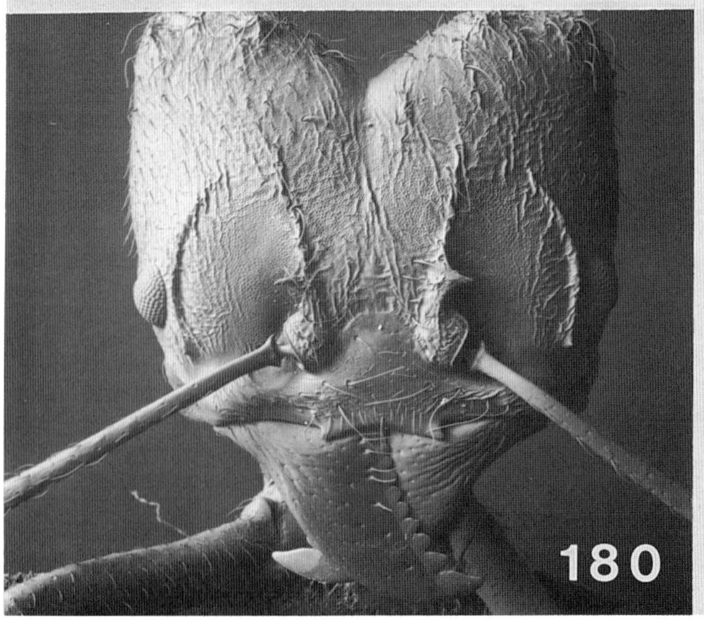

180

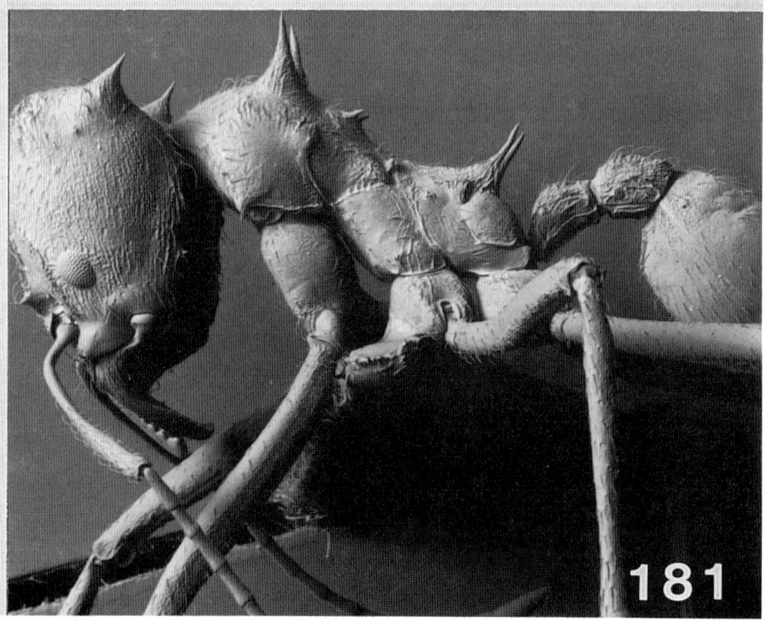

181

182

183

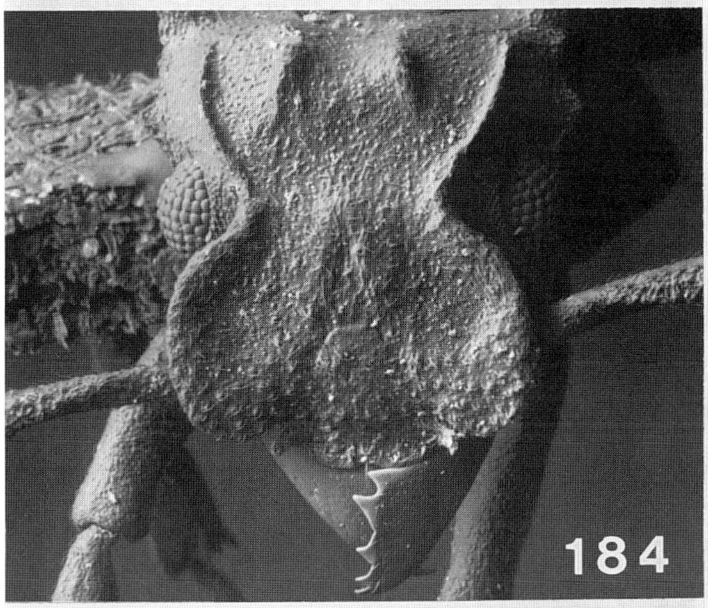

184

185

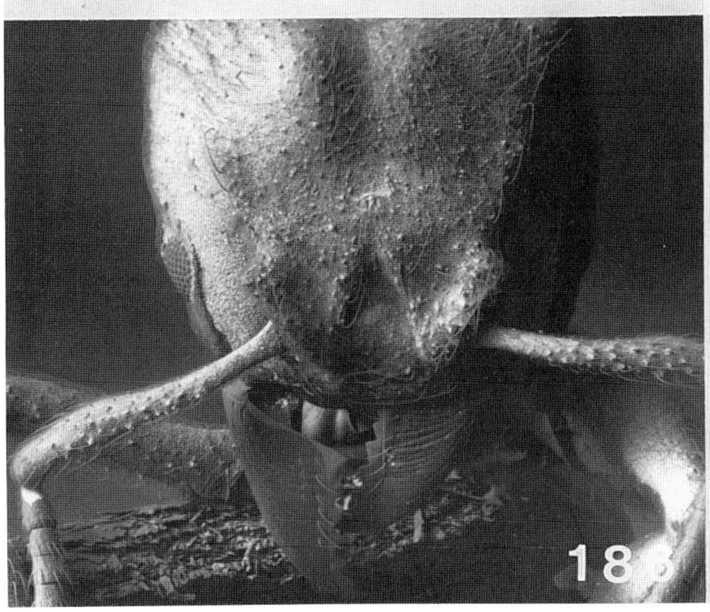

186

187

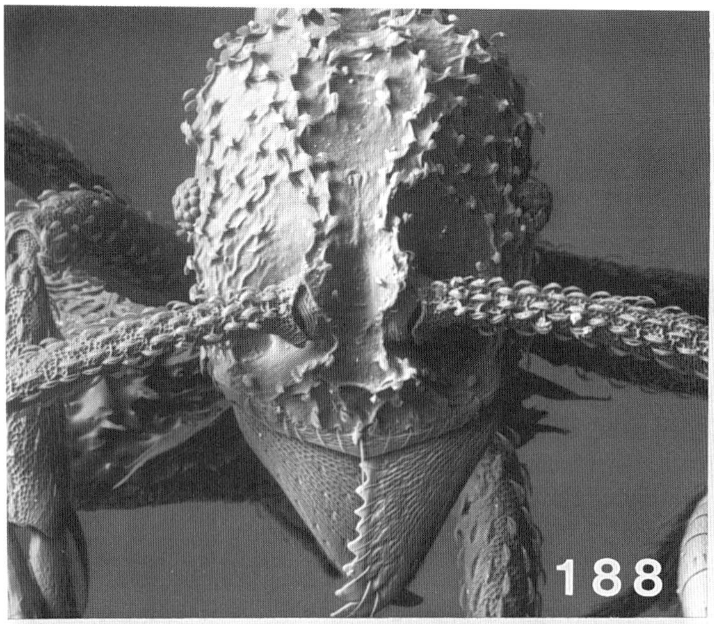
188

189

190

191

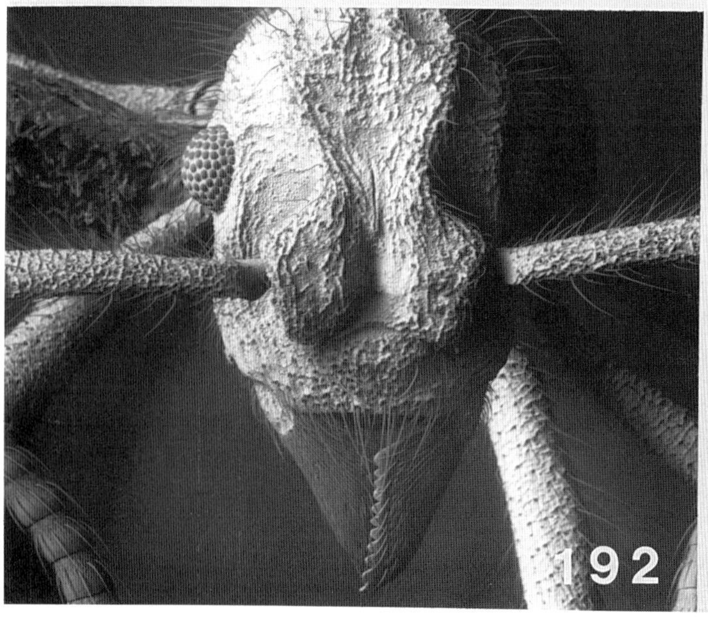
192

193

194

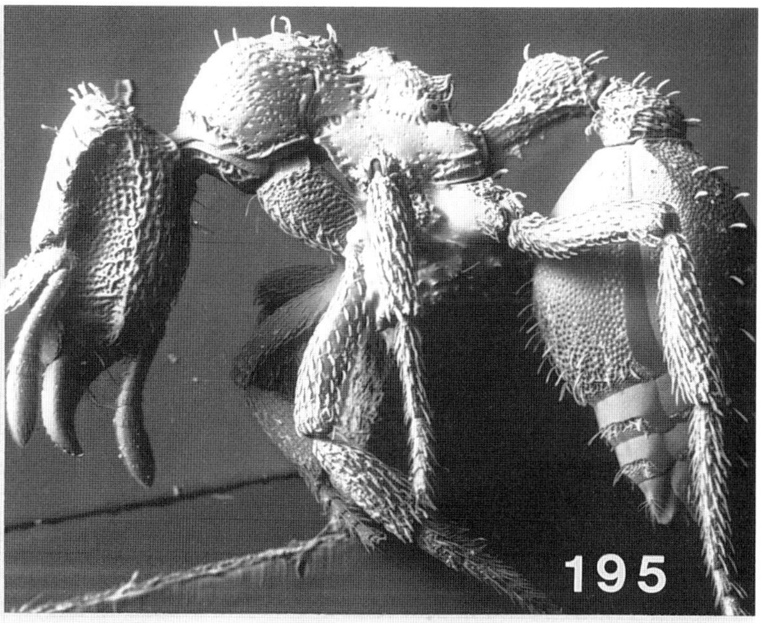

195

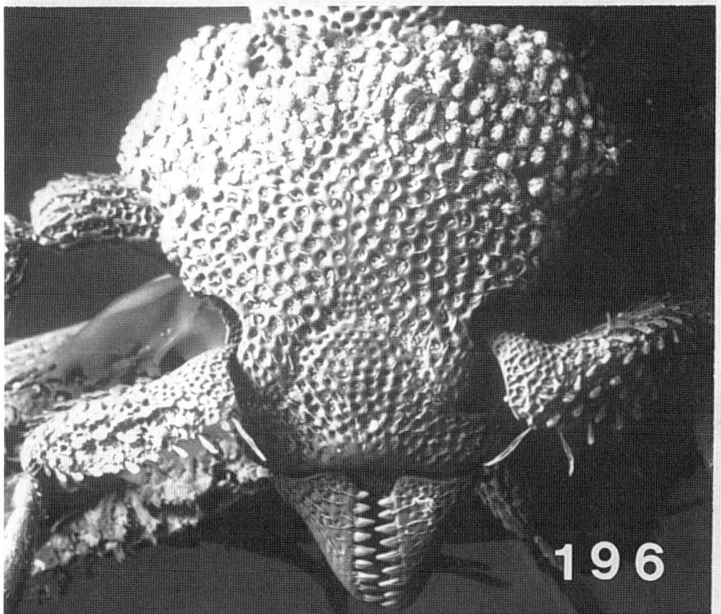

196

197

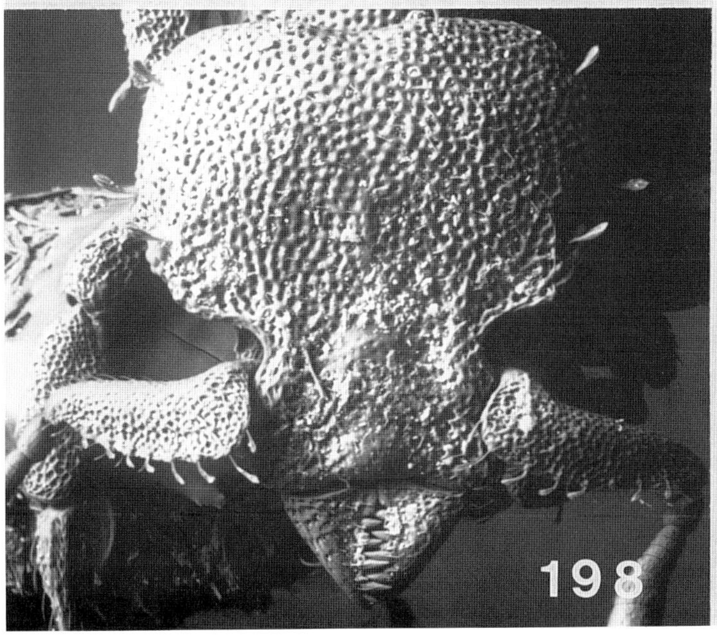

198

199

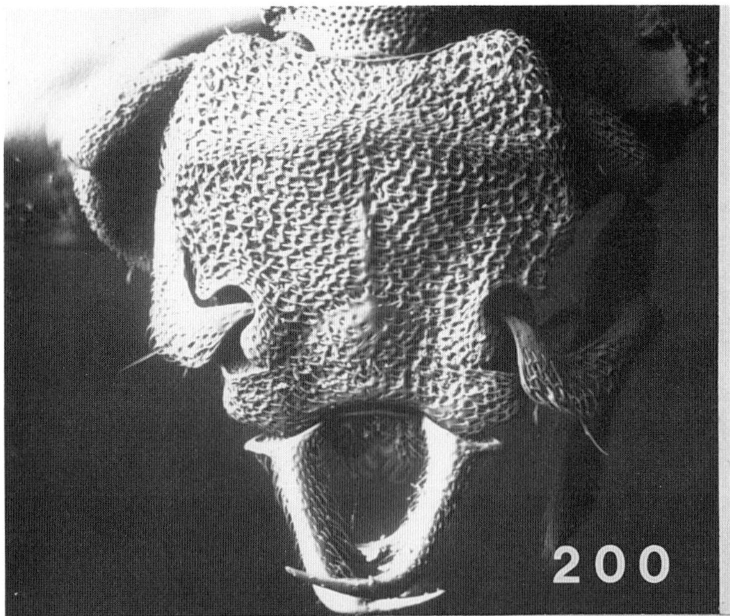
200

201

202

203

204

205

206

207

208

209

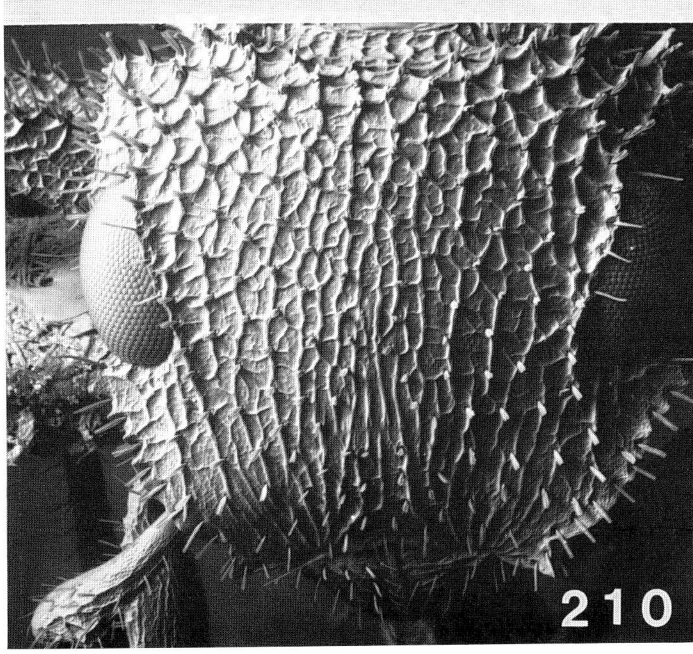

210

211

212

213
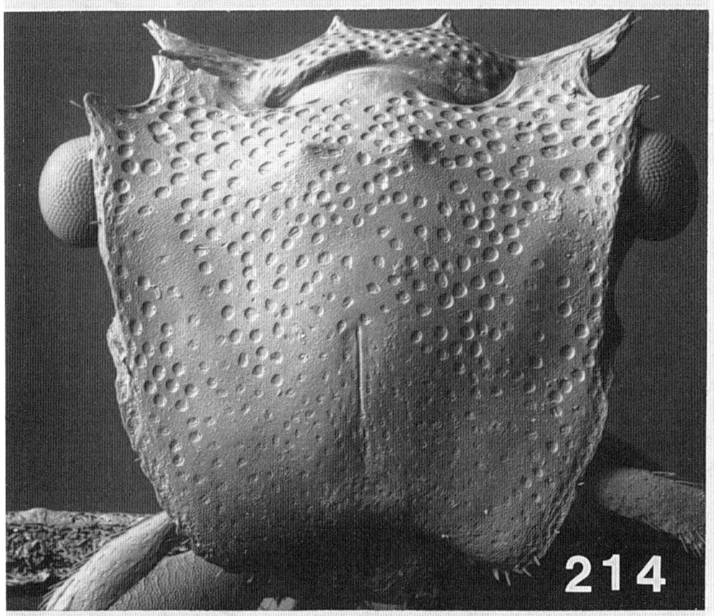
214

215
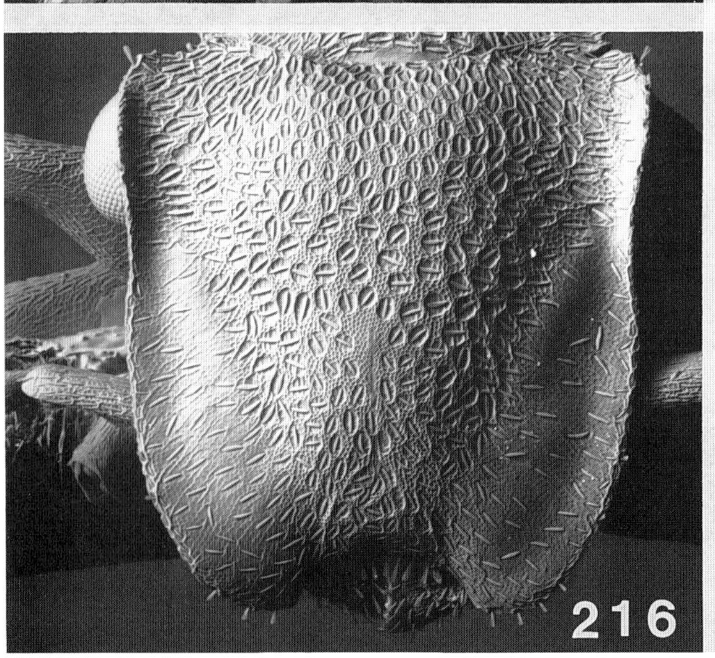
216

217

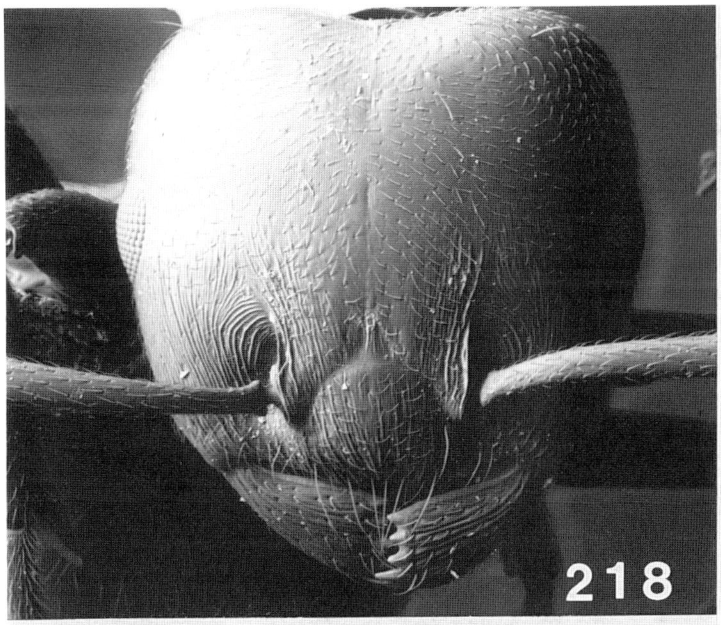
218

219

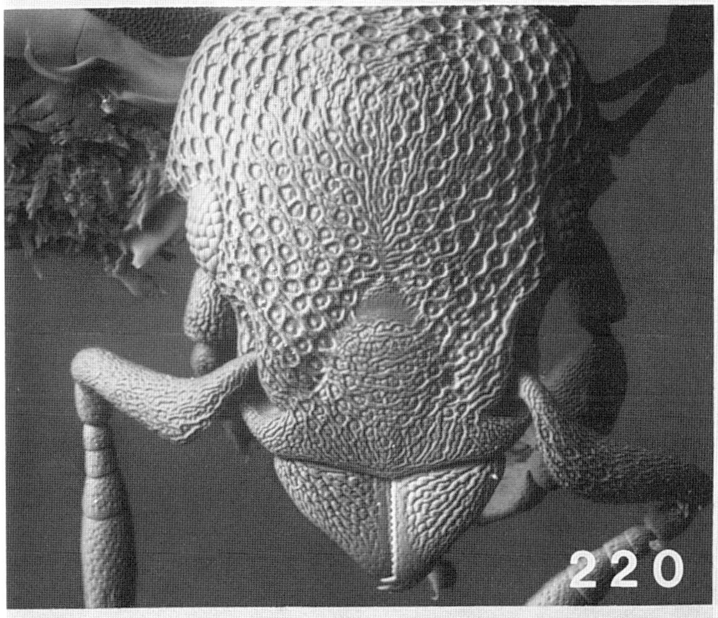
220

221

222

223

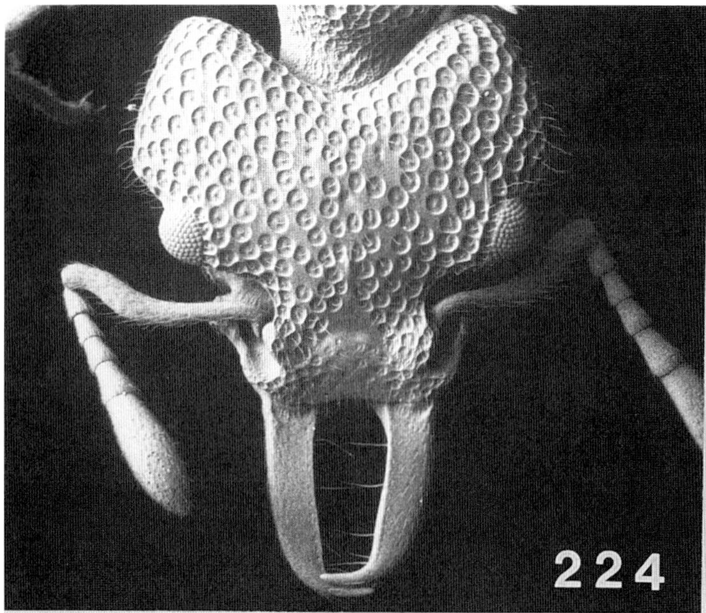

224

225

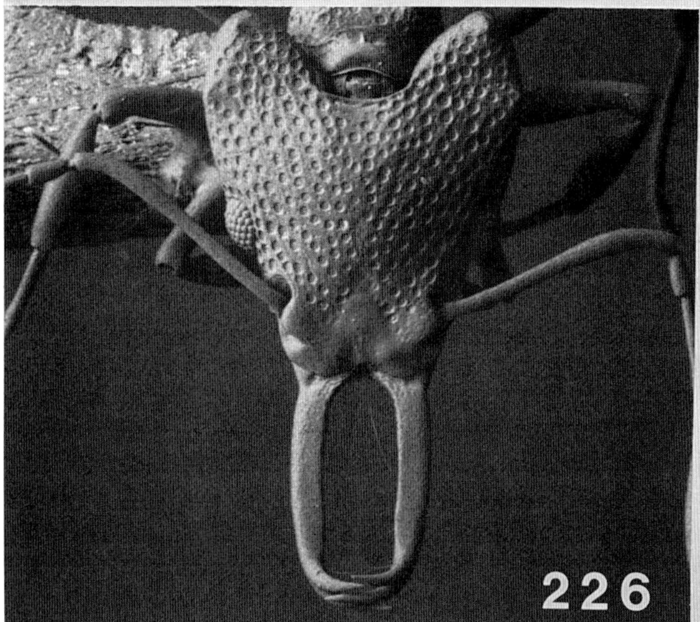

226

227

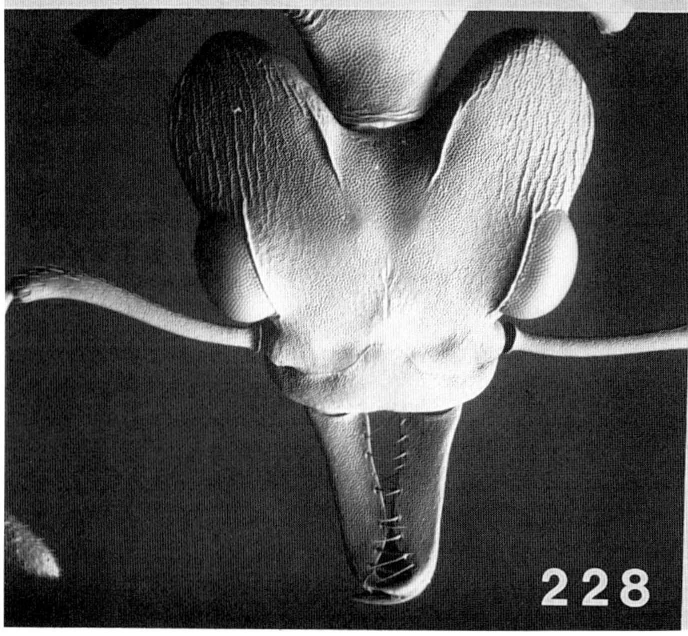

228

229

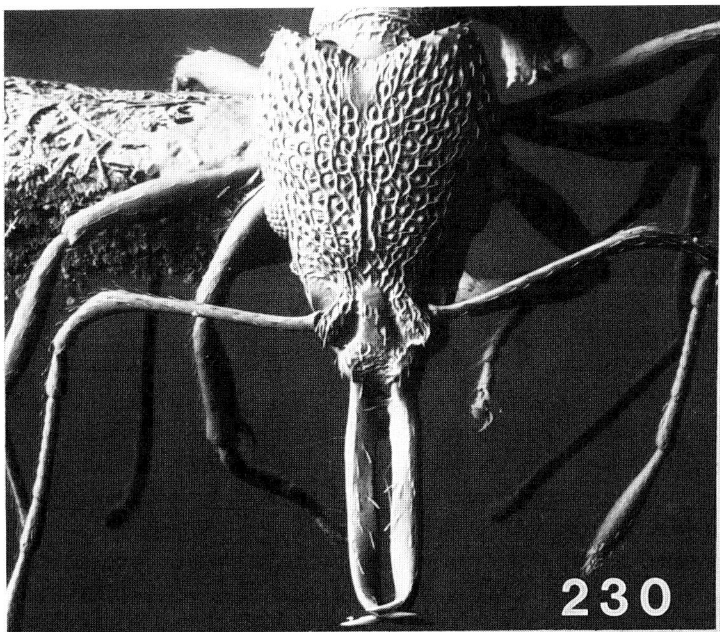
230

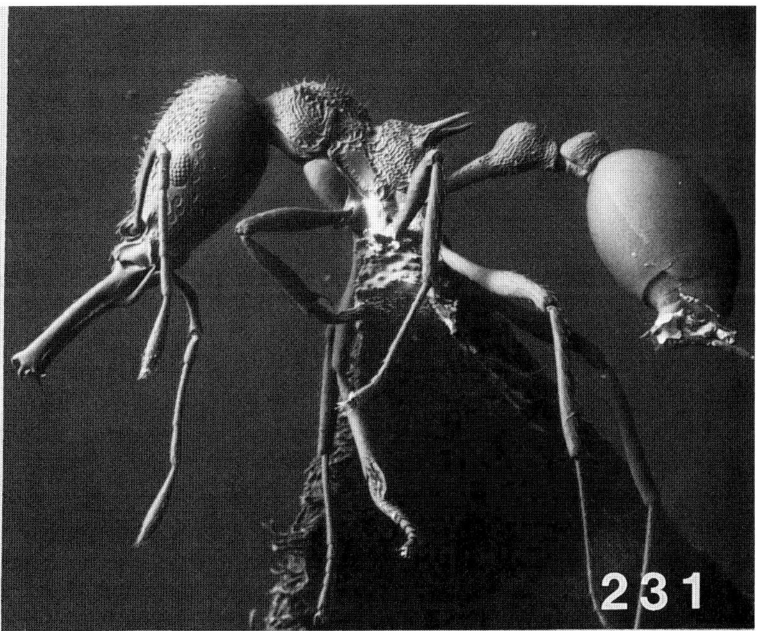
231

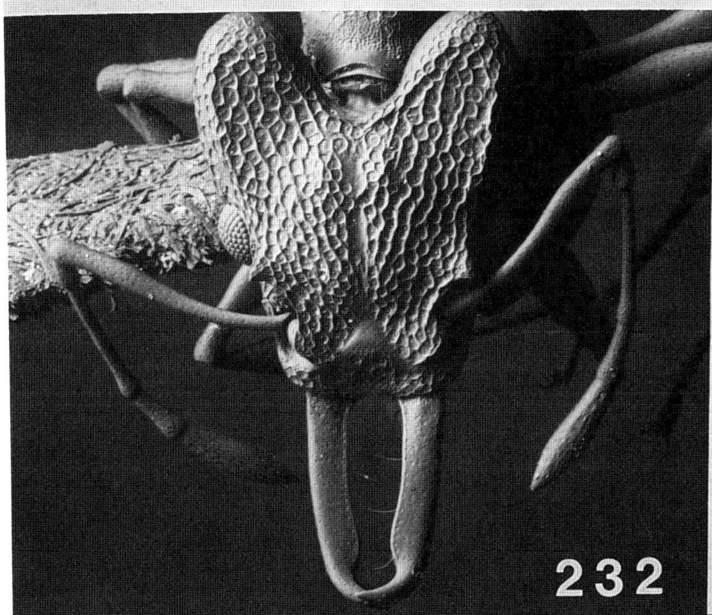
232

233

234

235

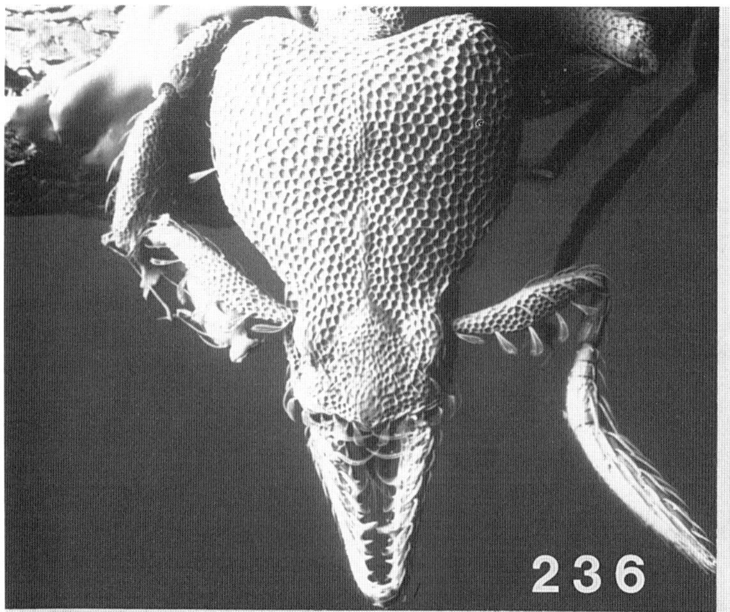
236

237

238

239

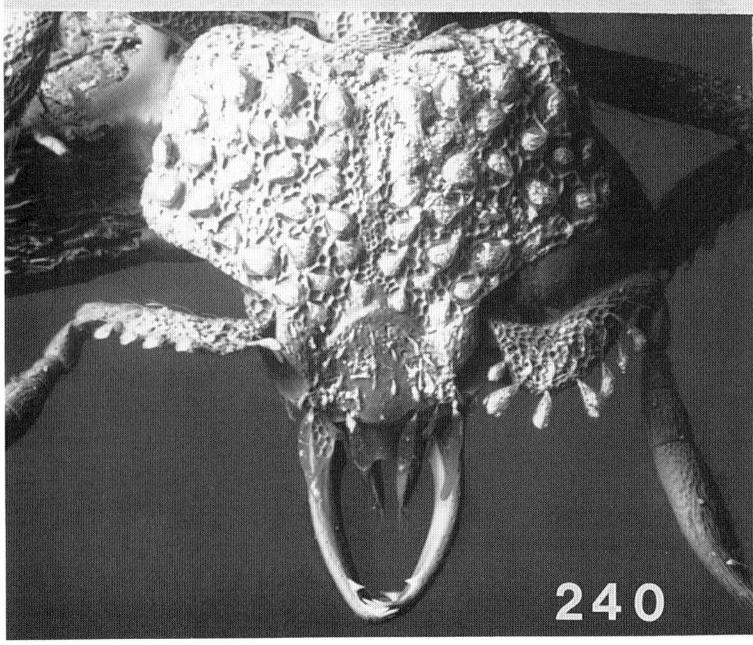
240

241

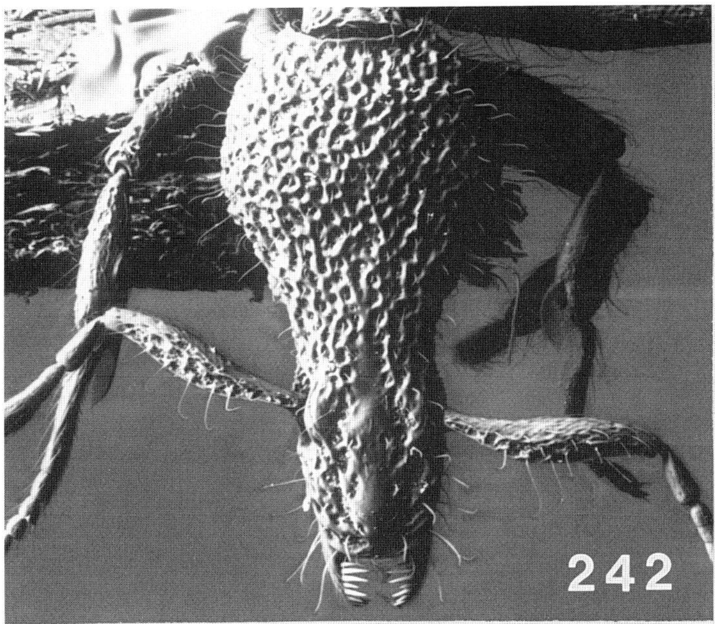
242

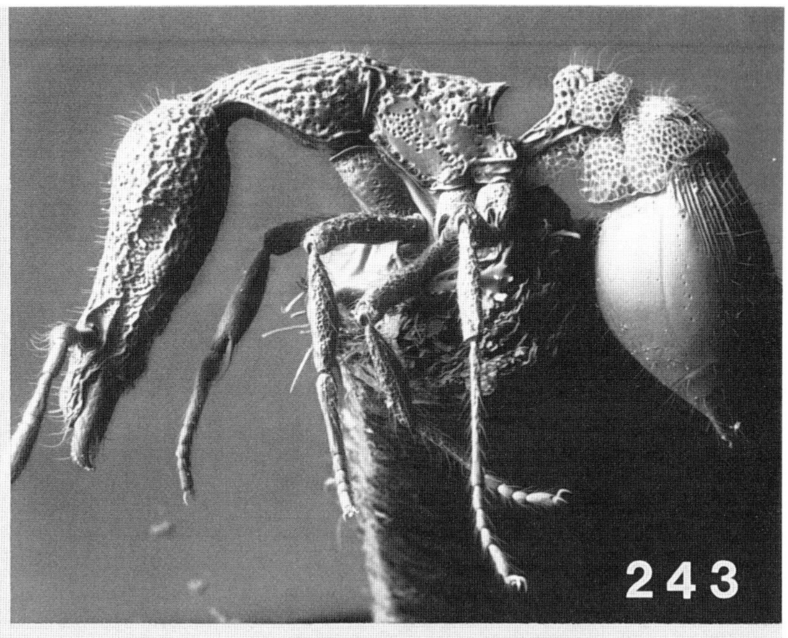
243

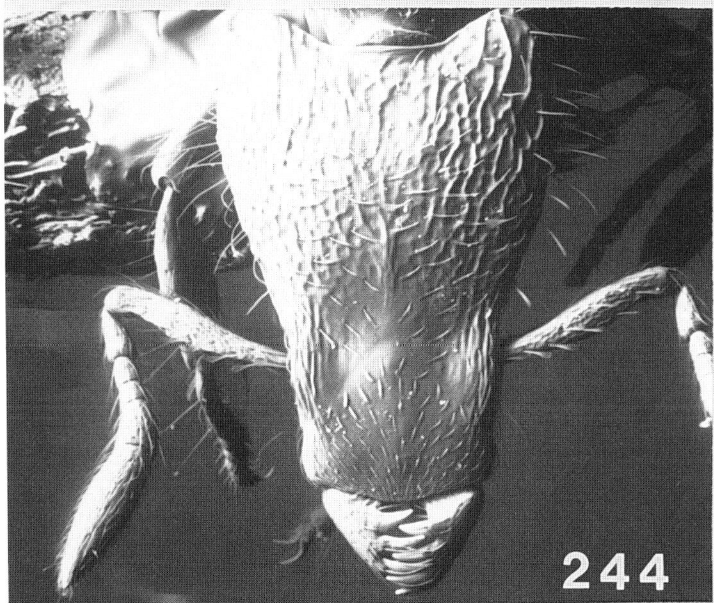
244

245

246

247

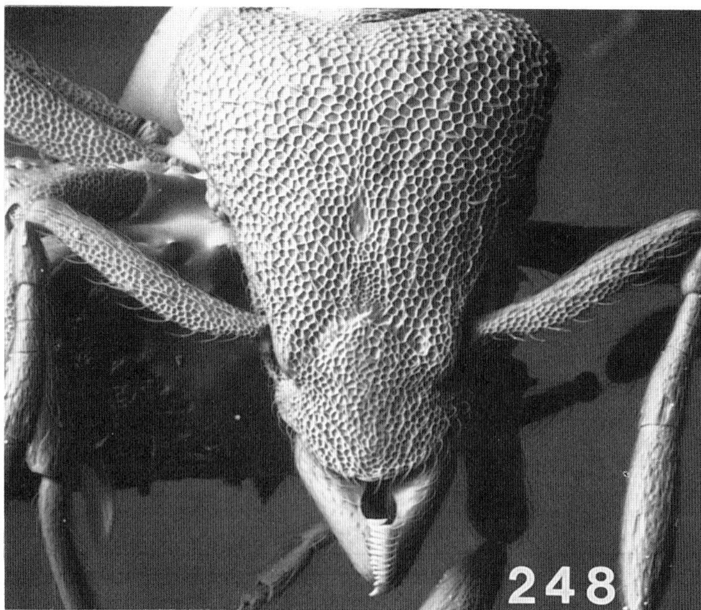

248

249

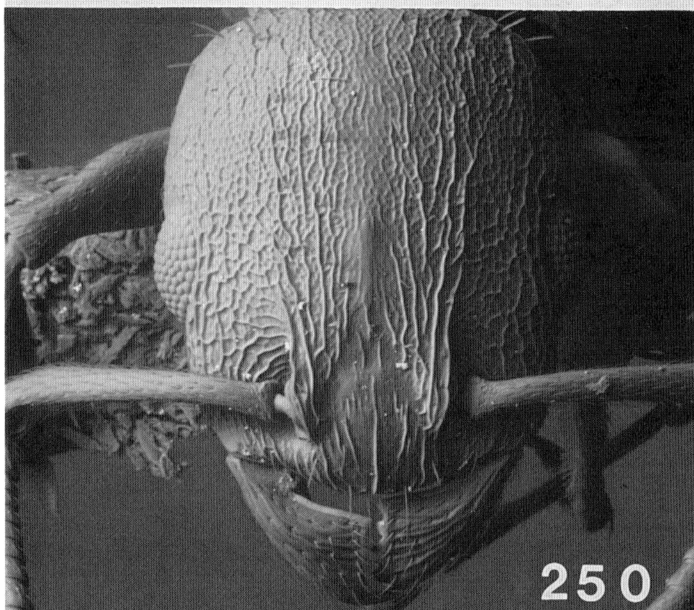

250

251

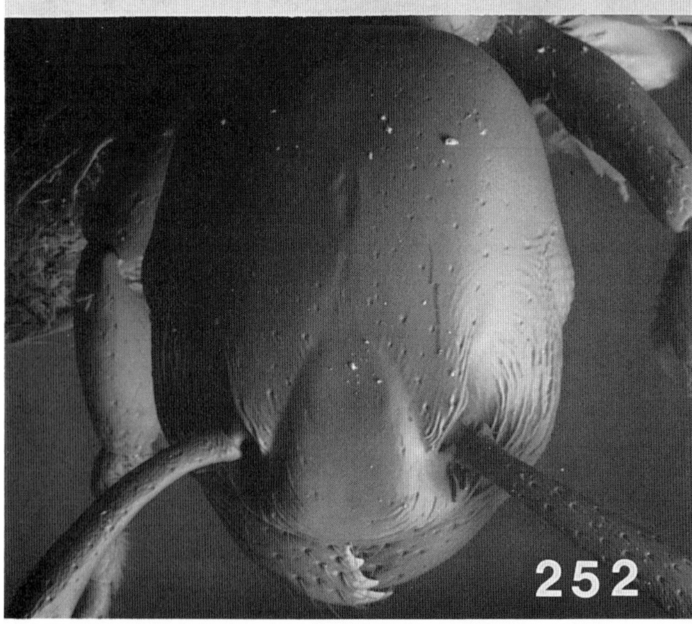

252

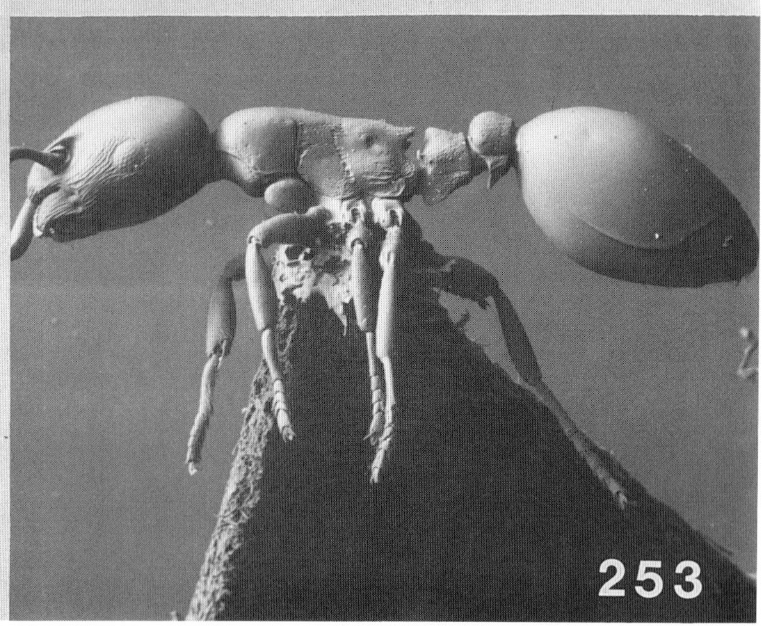

253

254

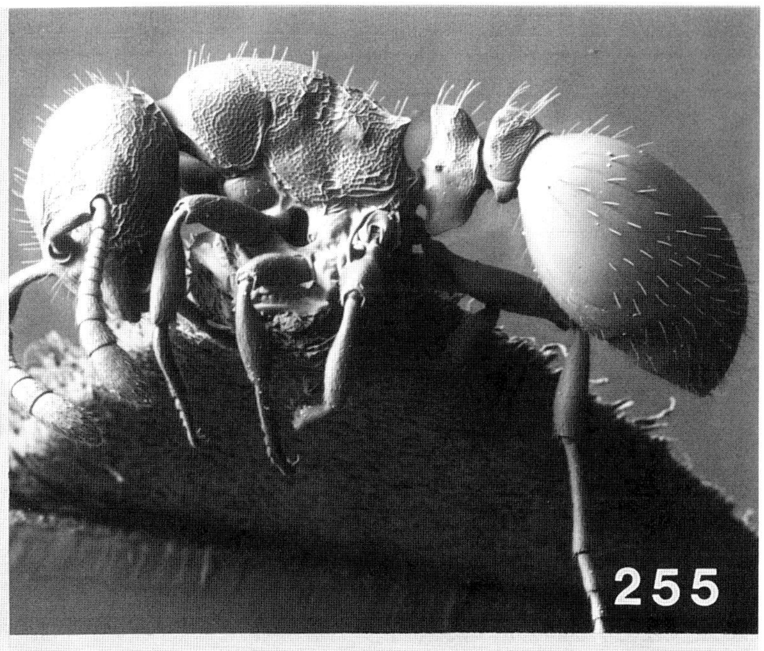

255

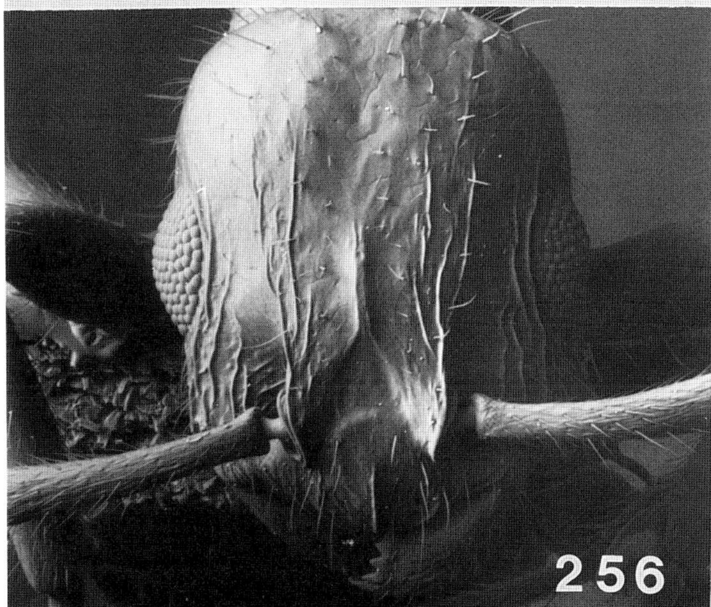

256

257

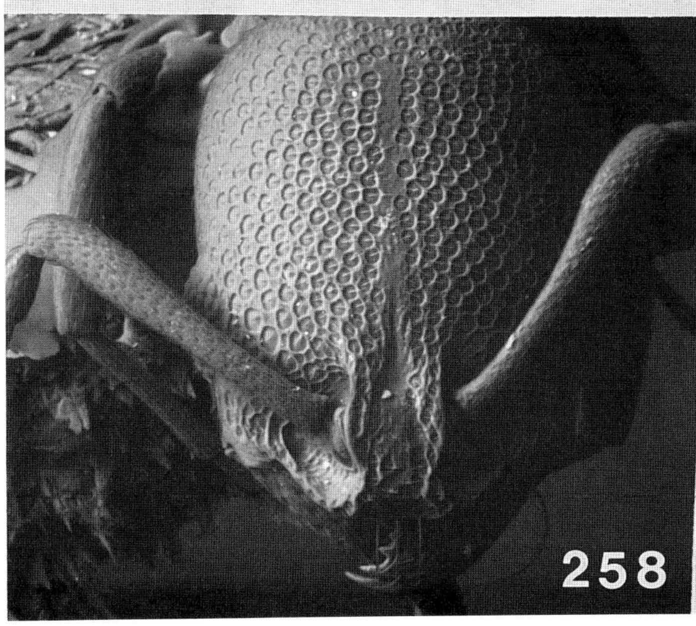

258

259

260

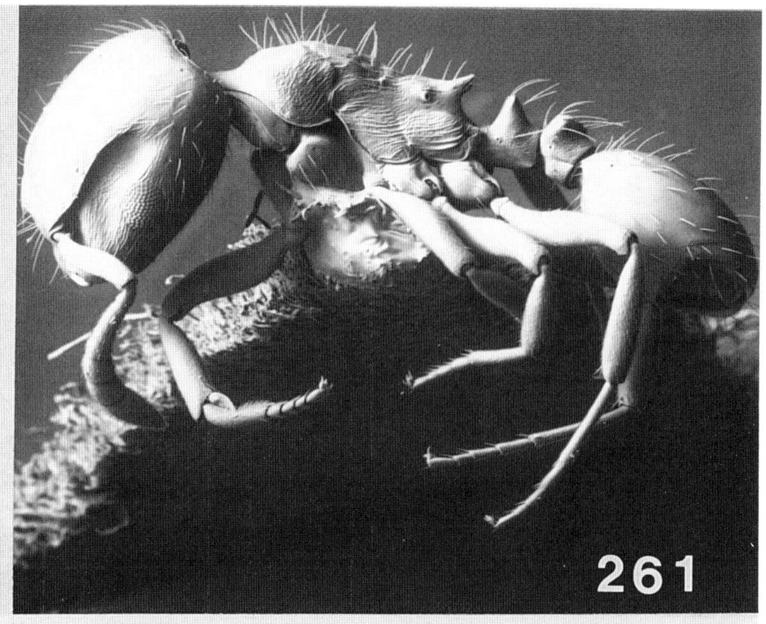

261

262

263

264

265

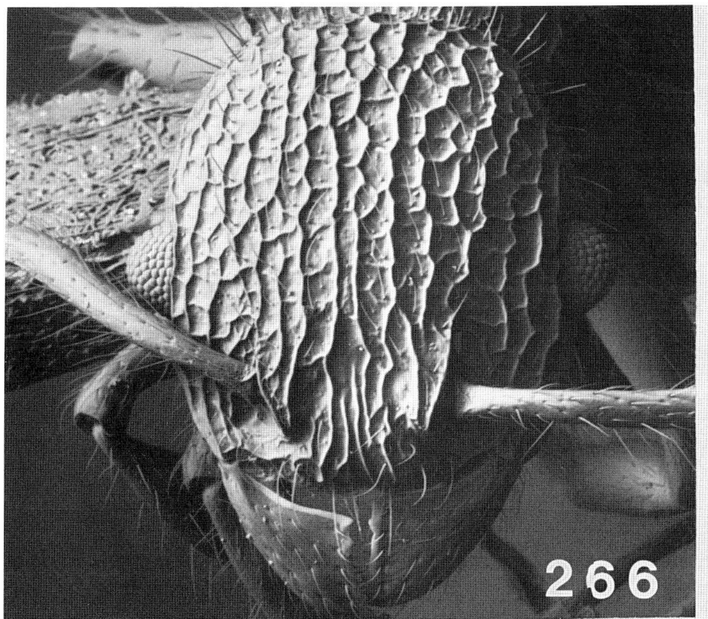
266

267

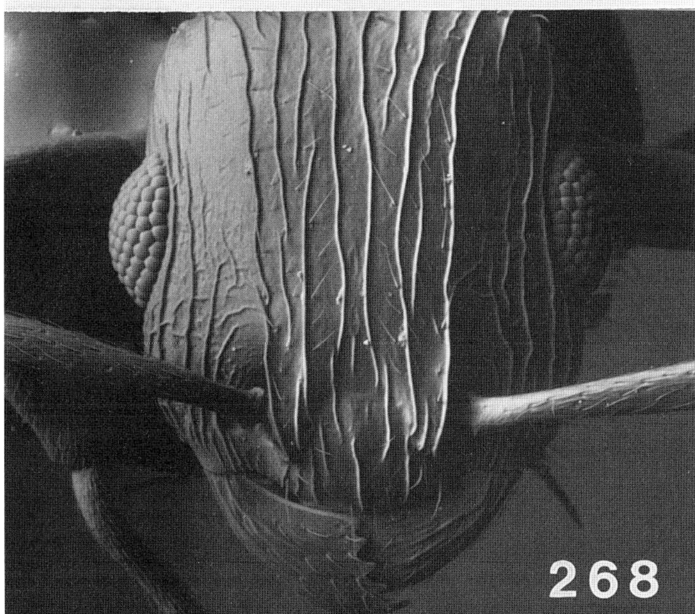
268

269

270

271

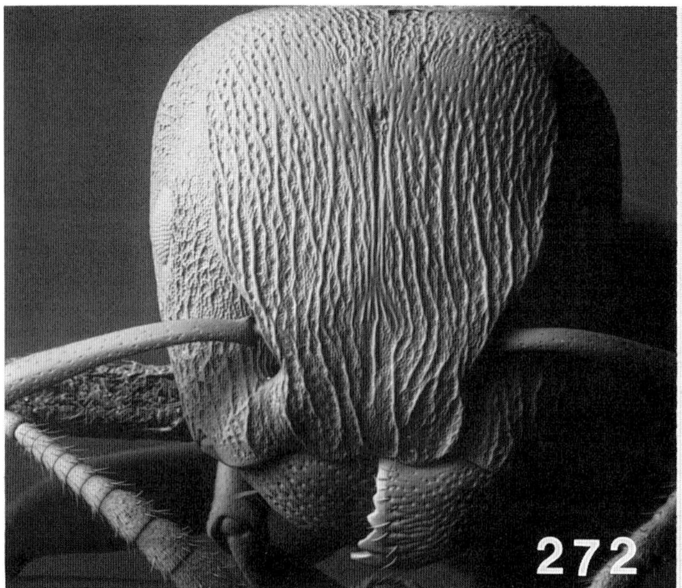
272

273

274

275

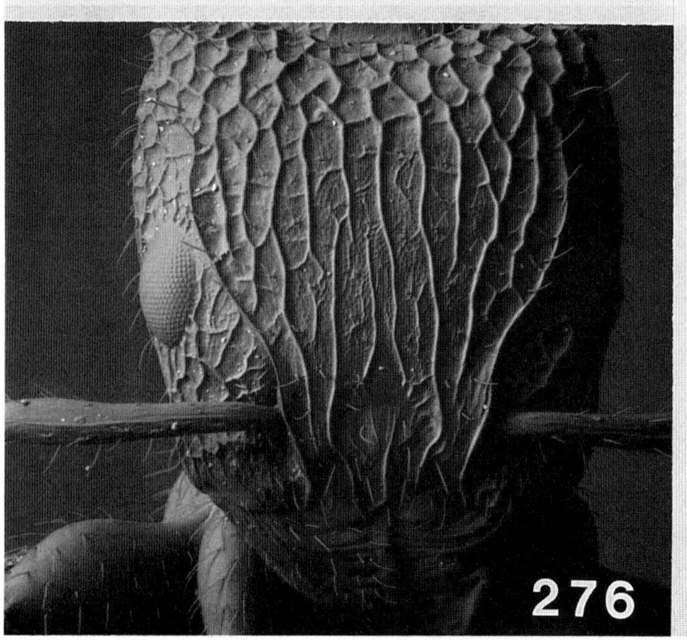
276

277

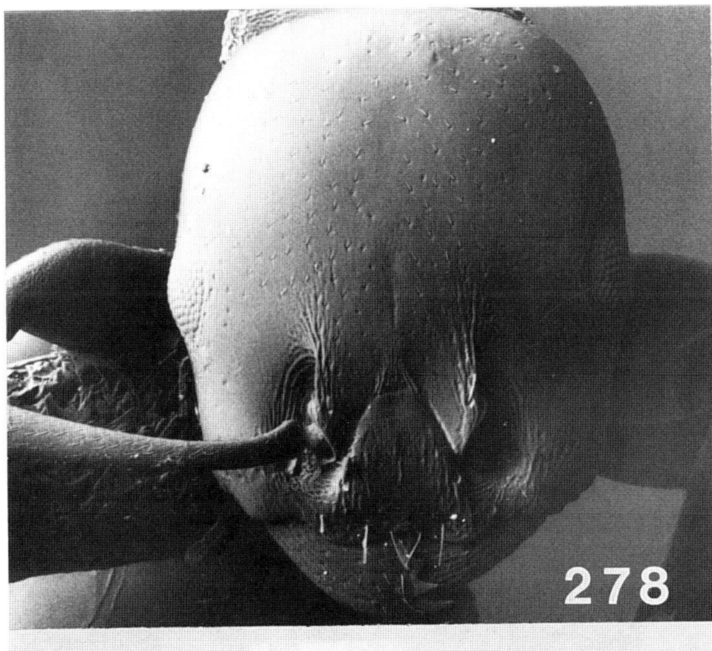

278

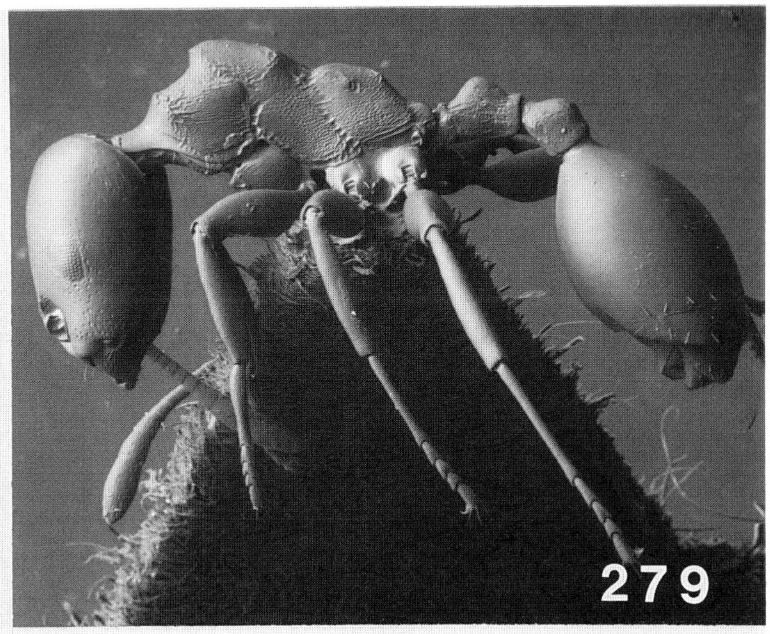

279

280

281

282

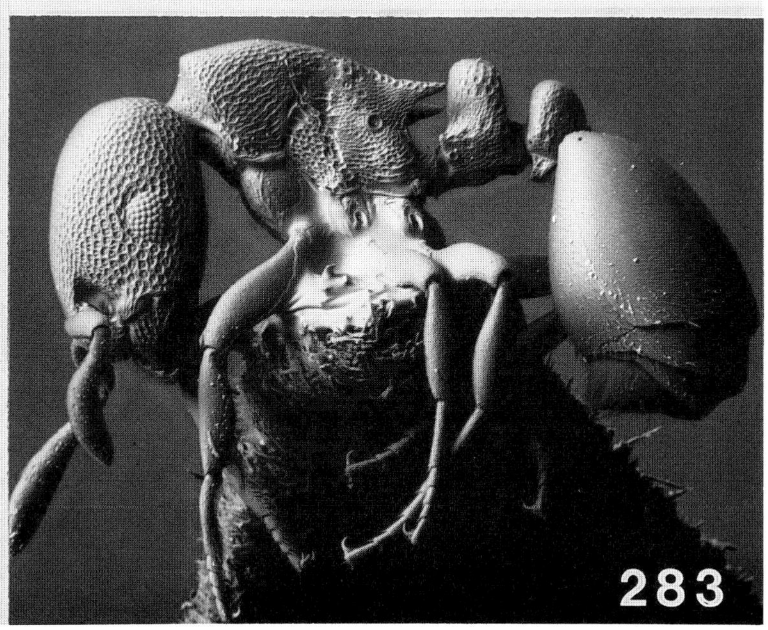

283

284

285

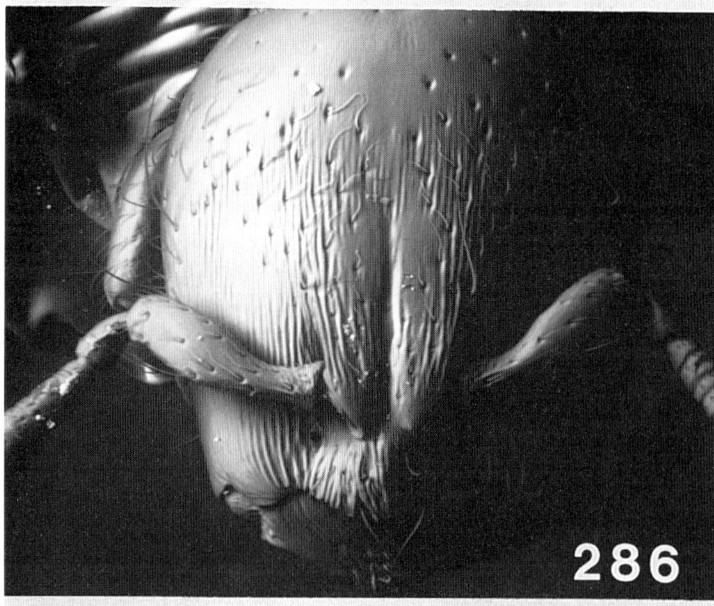
286

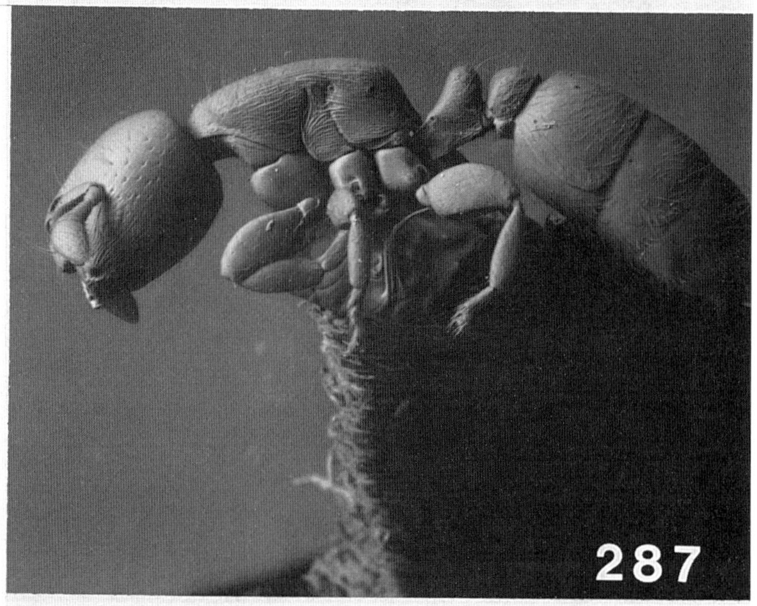
287

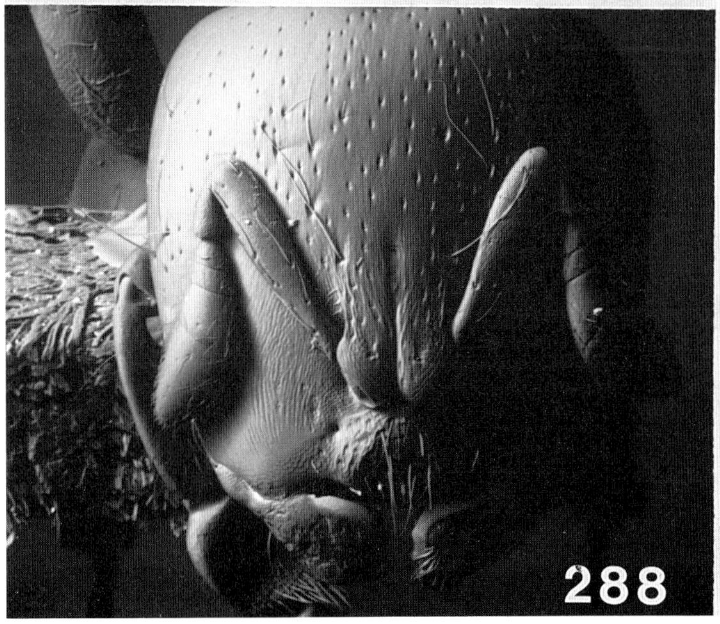
288

289

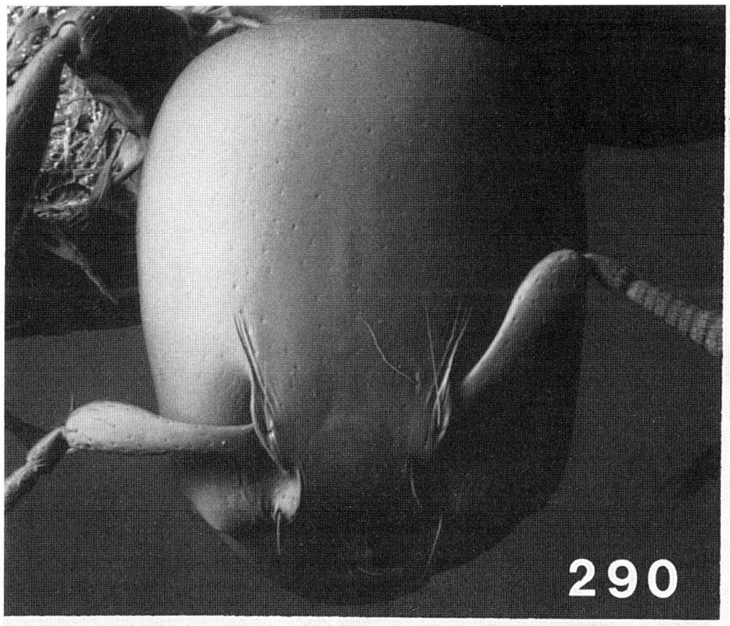
290

291

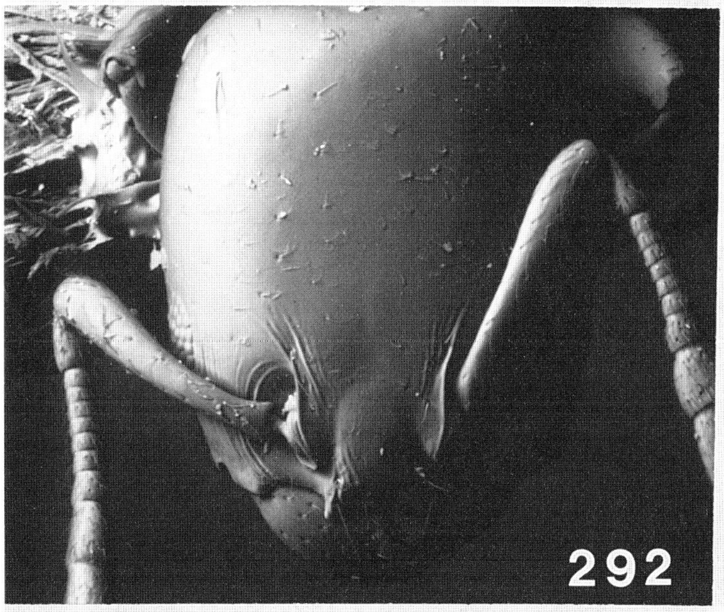
292

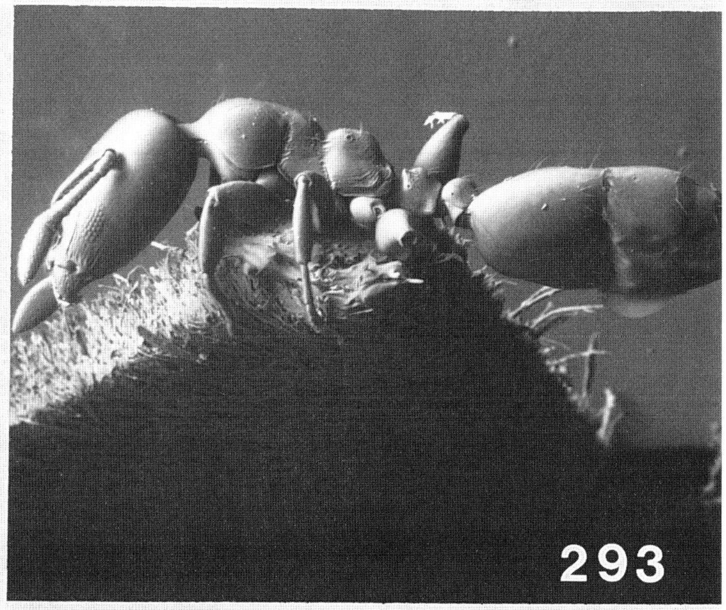
293

294

295

296
297
298
299
300
301

302

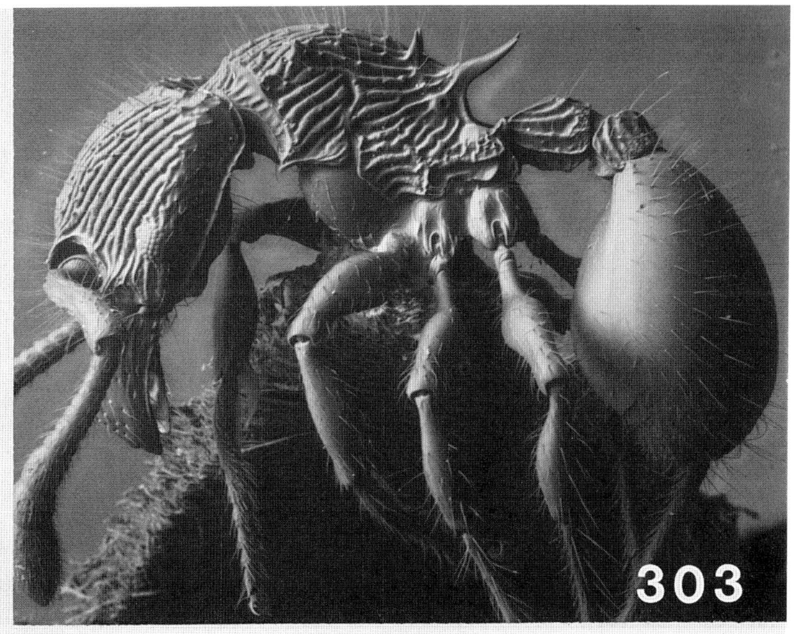
303

304

305

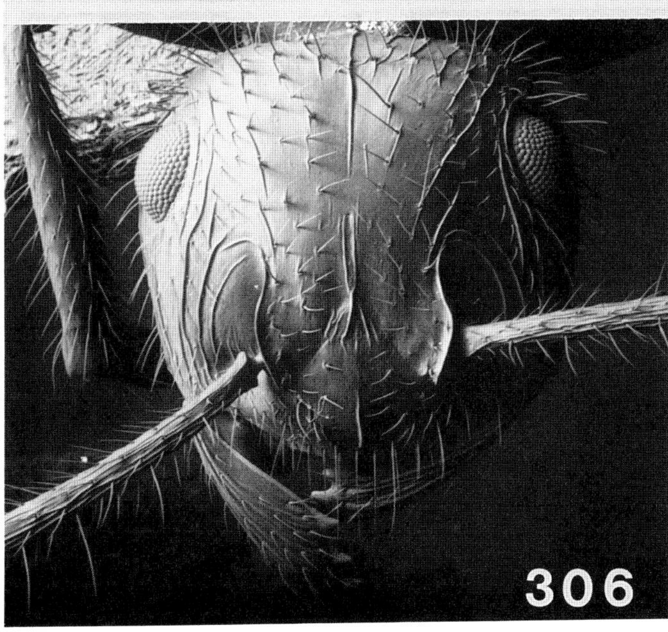
306

307

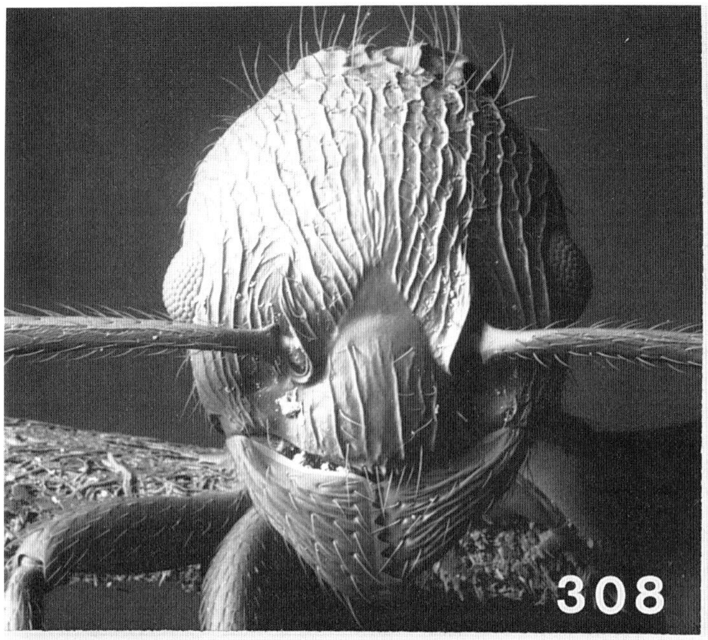
308

309
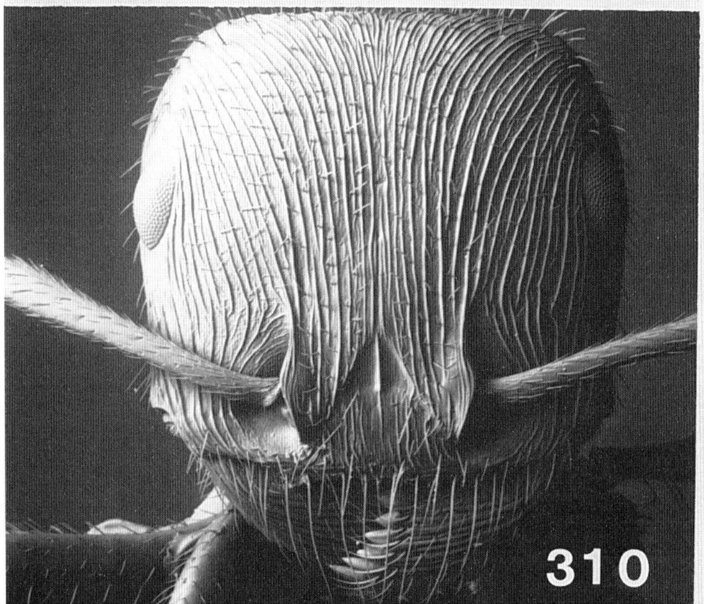
310

311
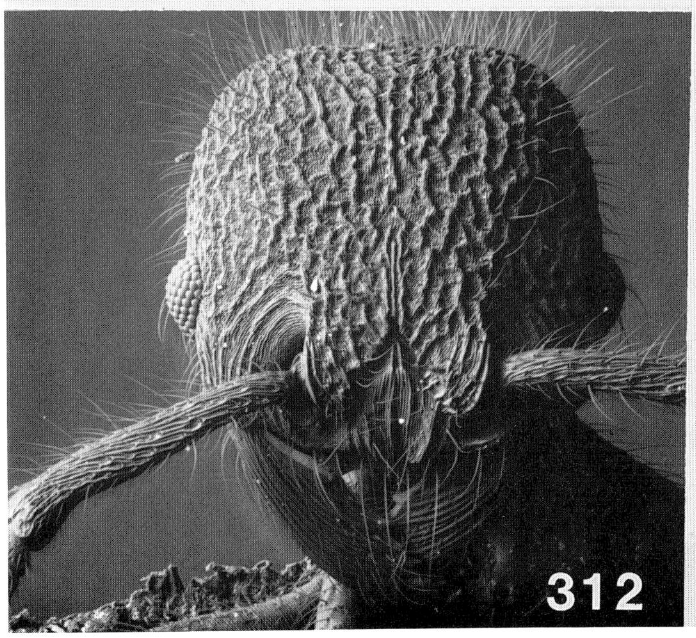
312

313

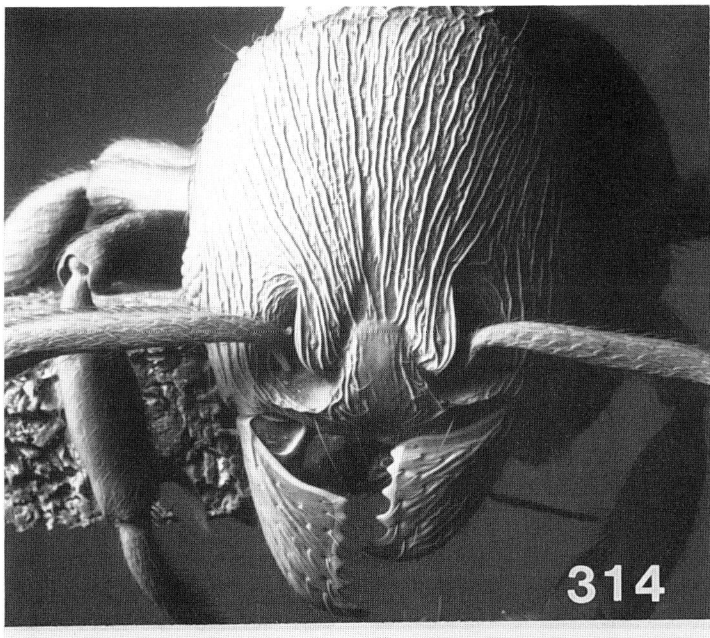
314

315

316

317

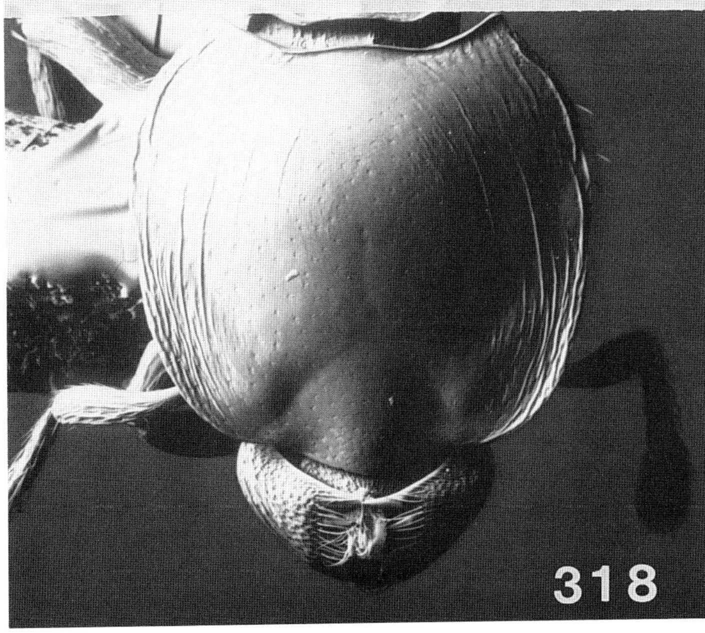
318

319

320

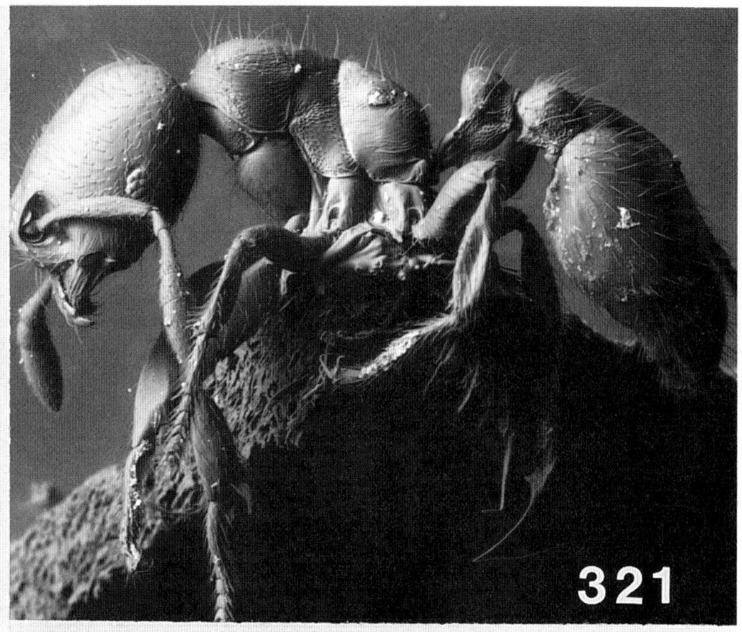
321

322

323

324

325

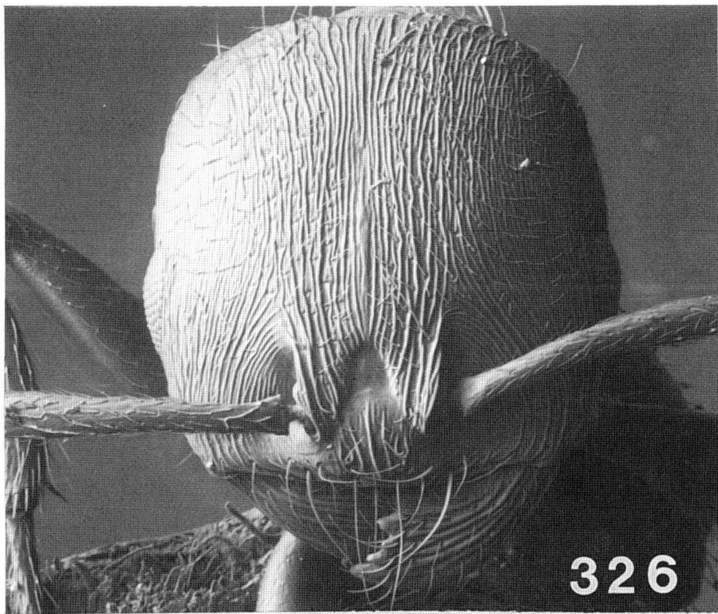
326

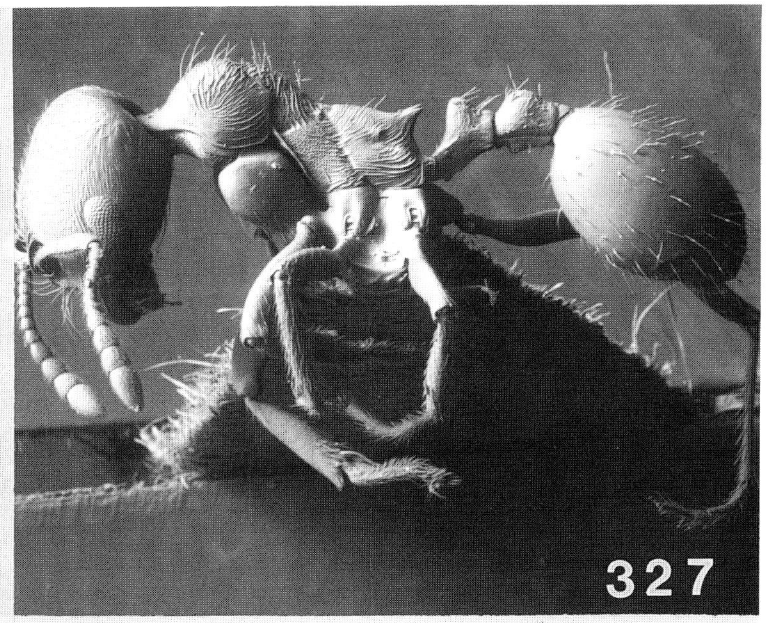
327

328

329

330

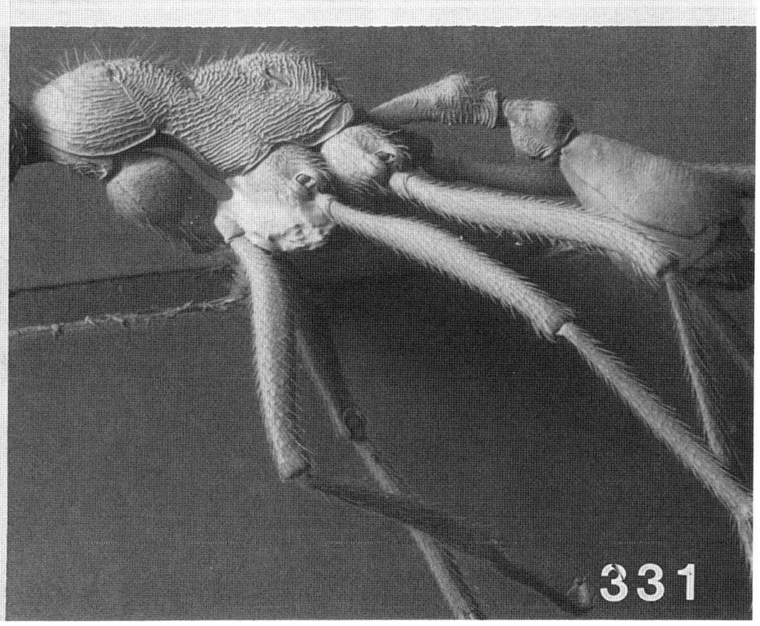
331

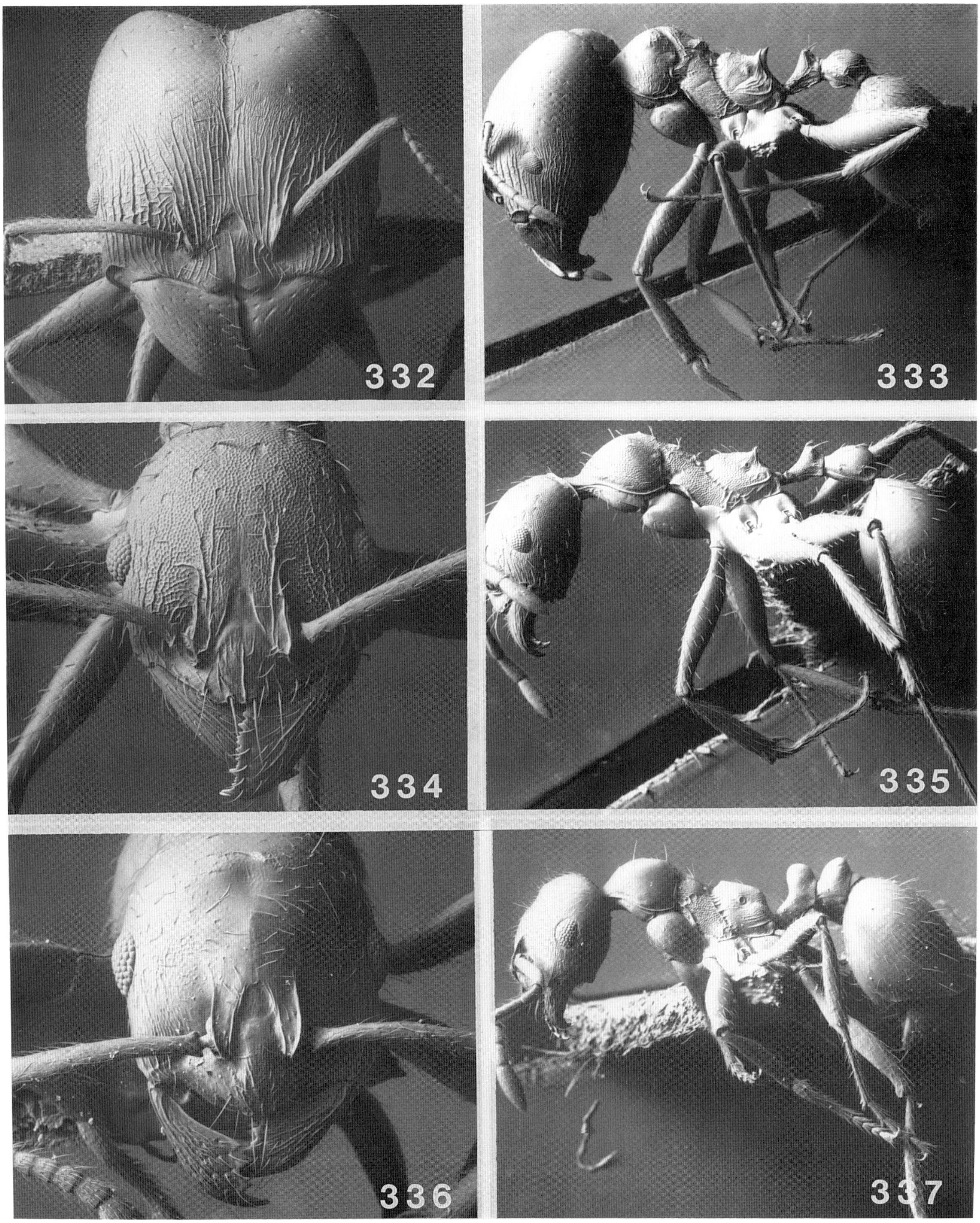
332
333
334
335
336
337

338

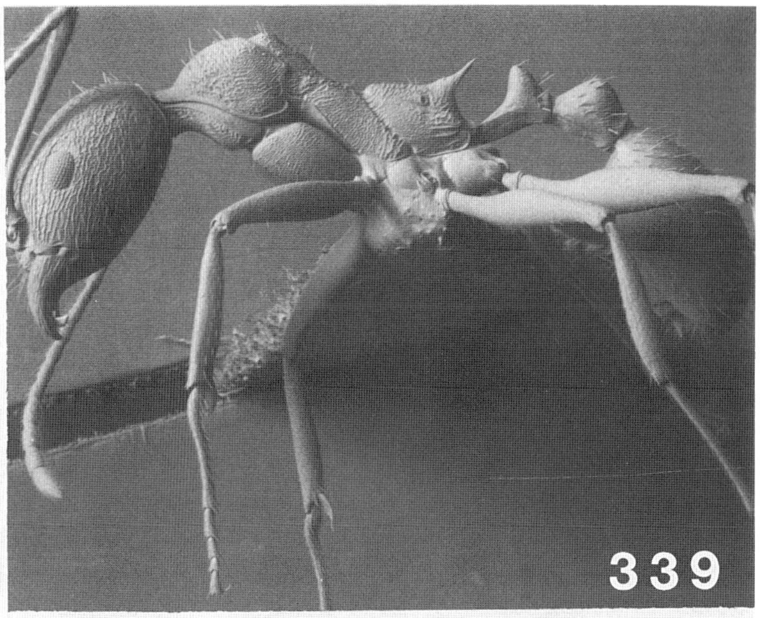

339

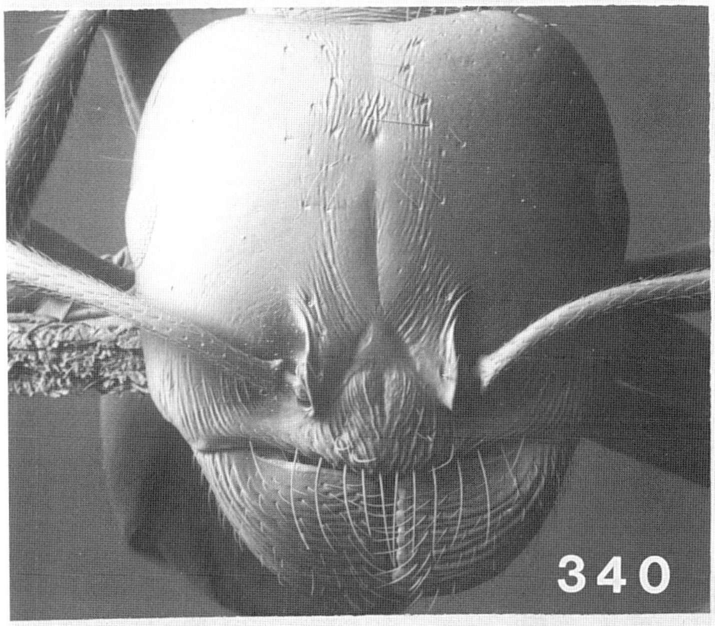

340

341

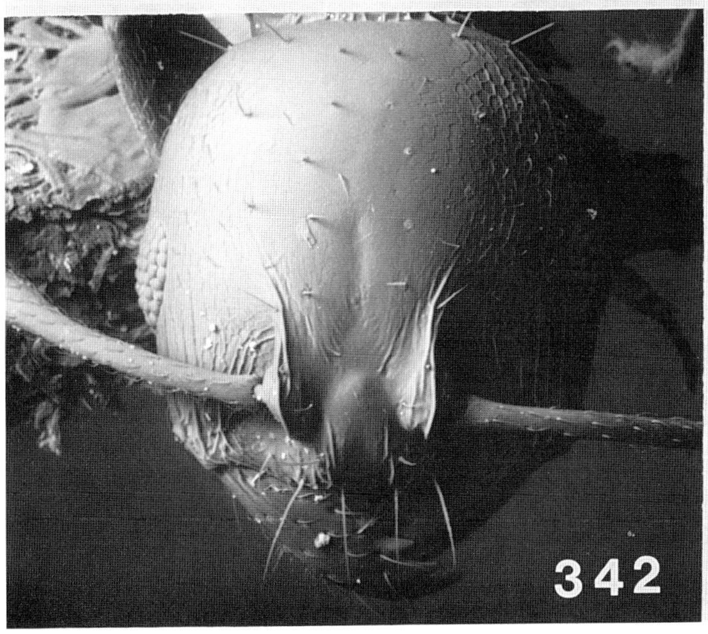

342

343

344

345

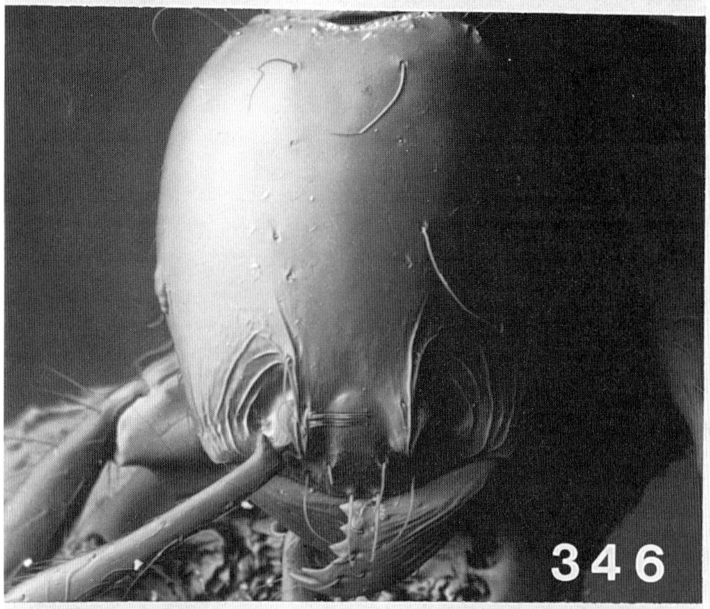

346

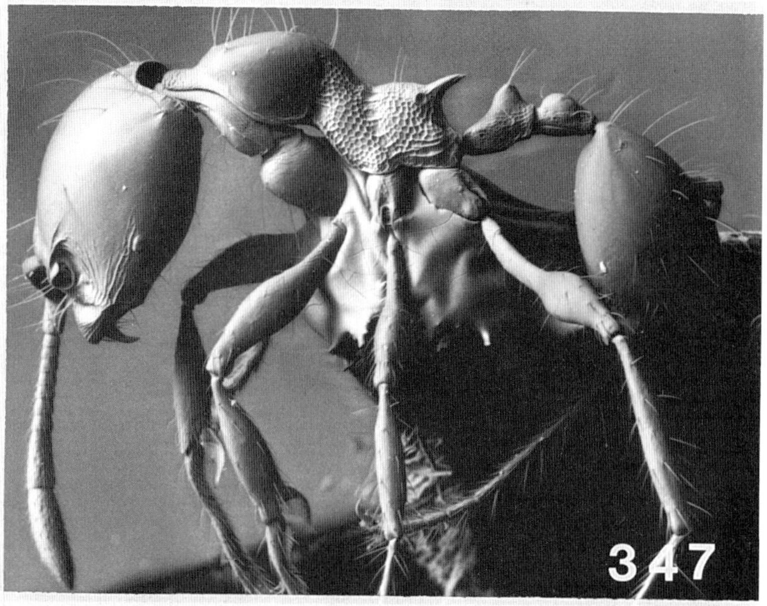

347

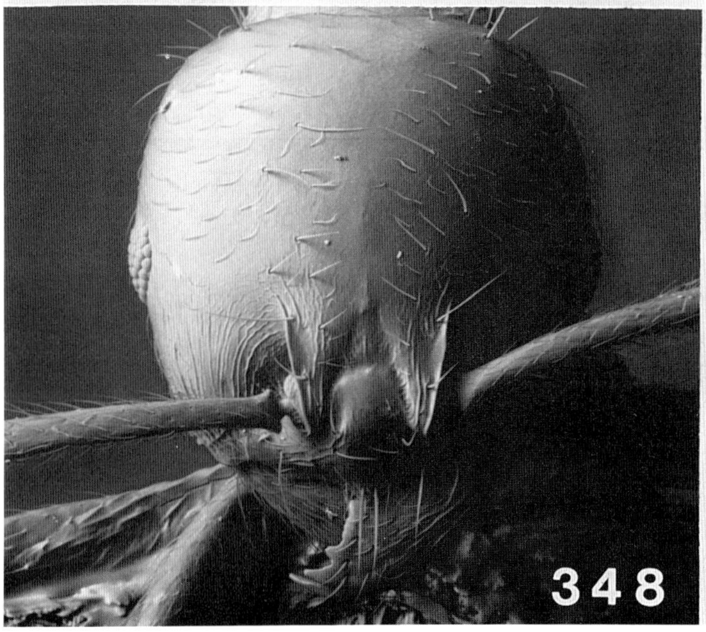

348

349

350

351

352

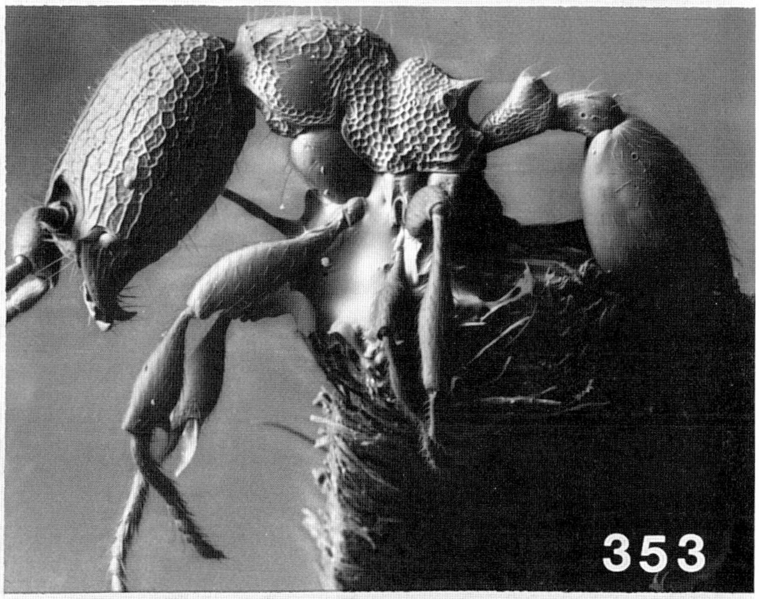
353

354

355

356

357

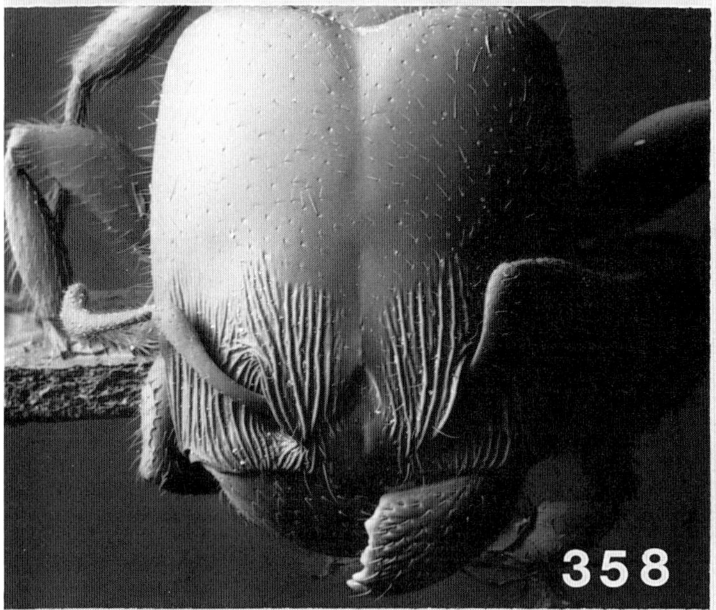
358

359

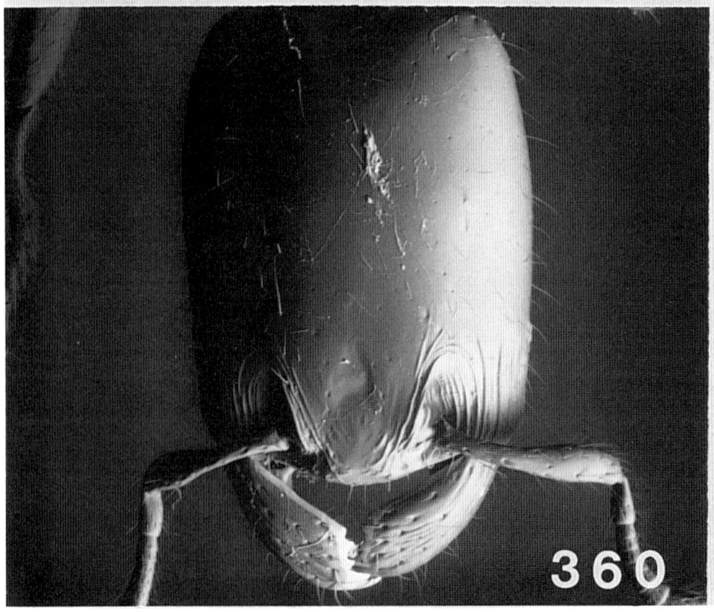
360

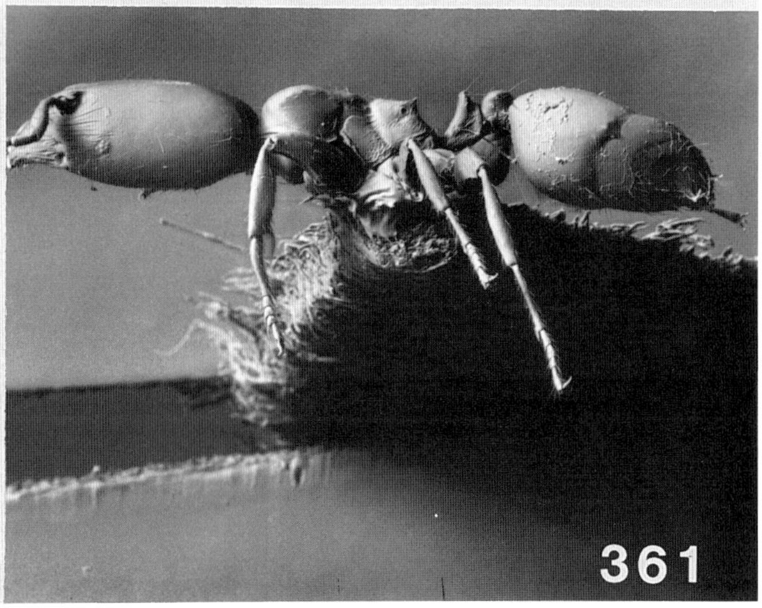
361

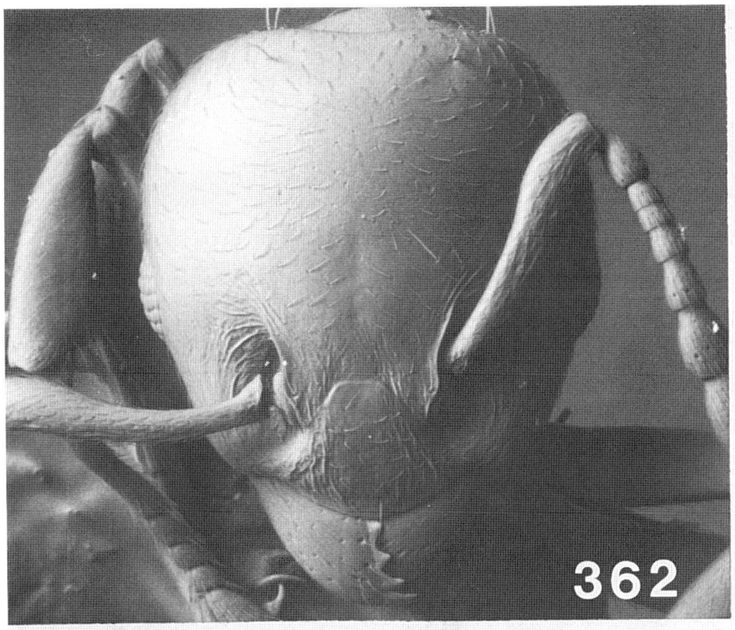
362

363

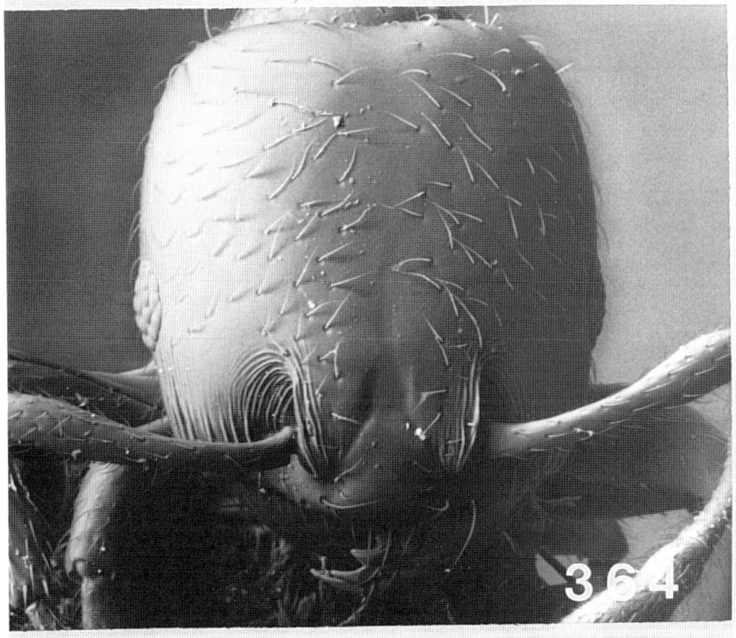
364

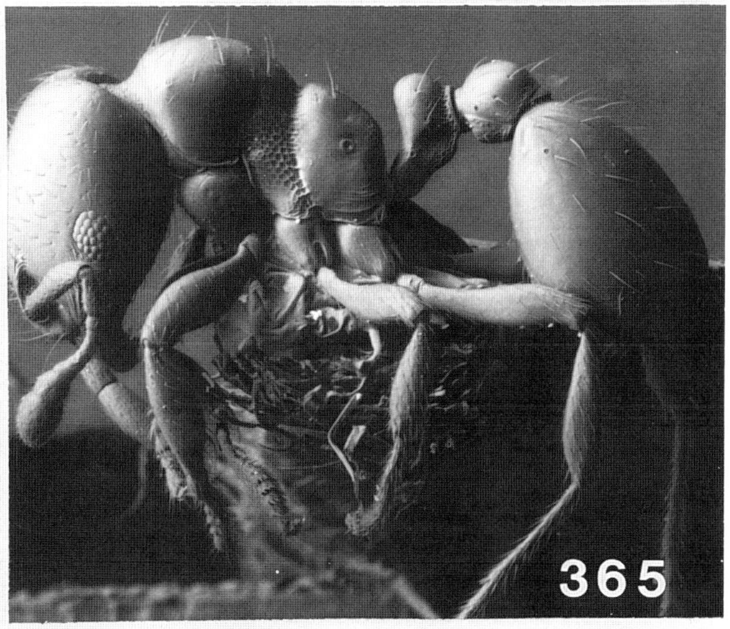
365

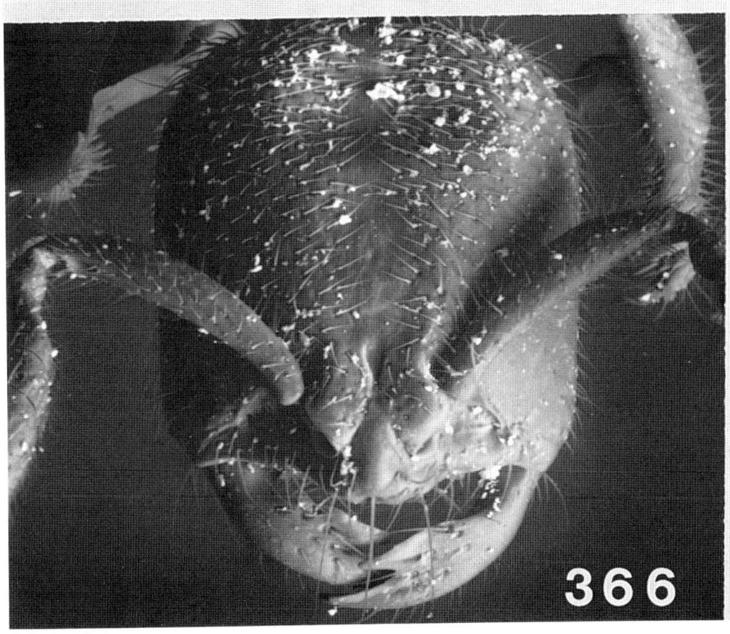
366

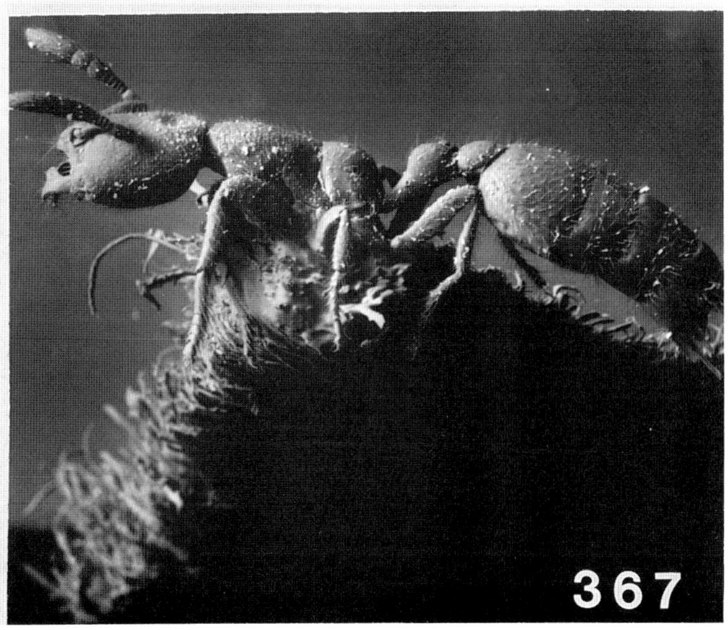
367

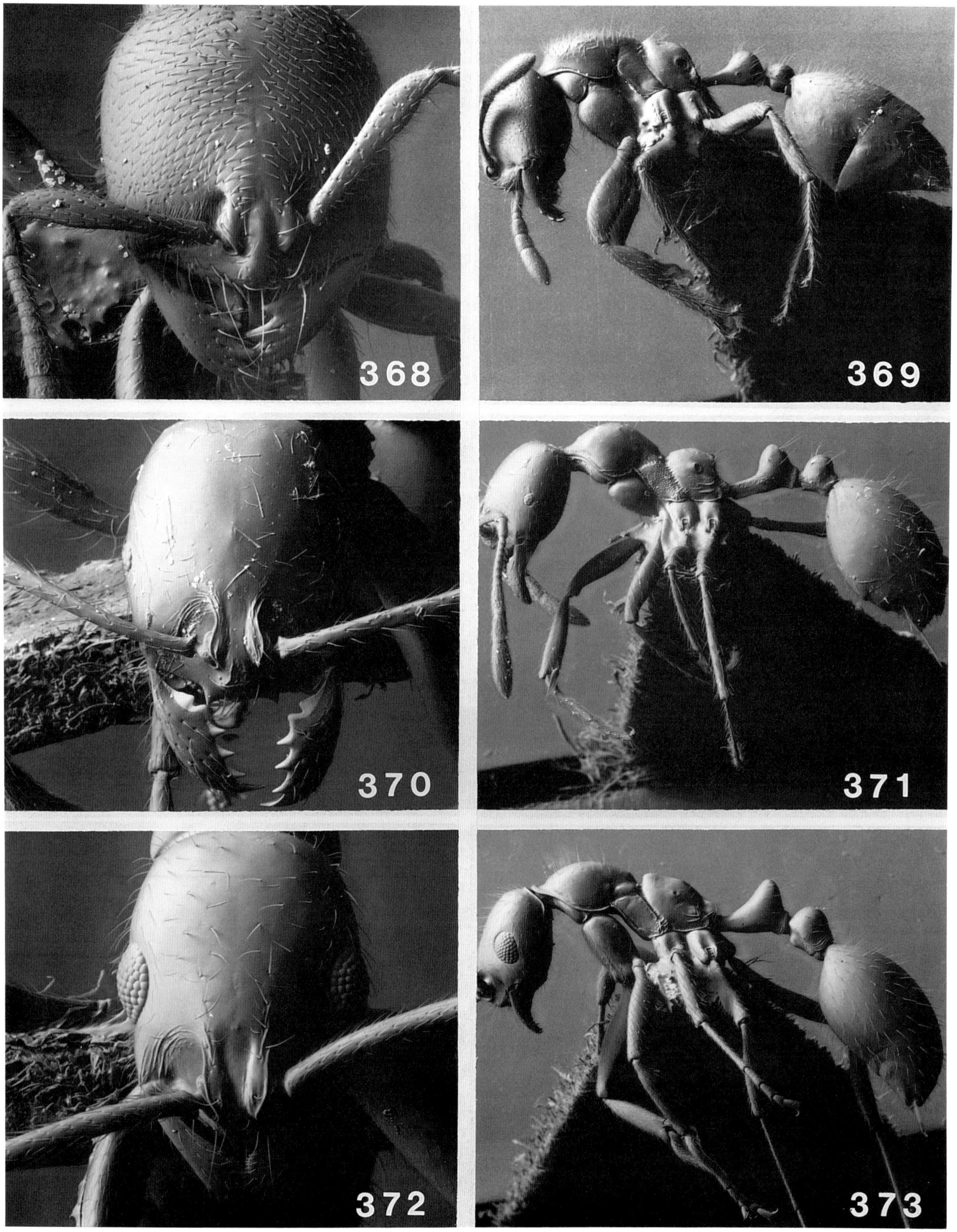
368
369
370
371
372
373

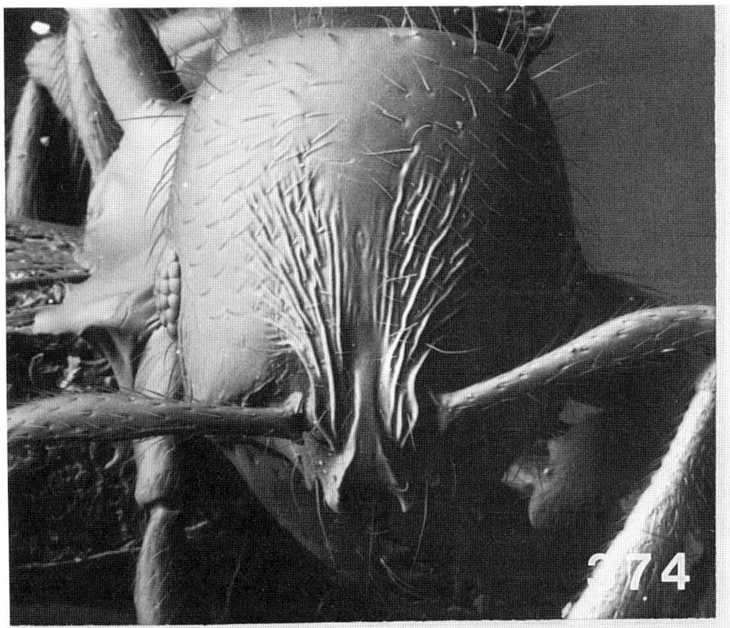
374

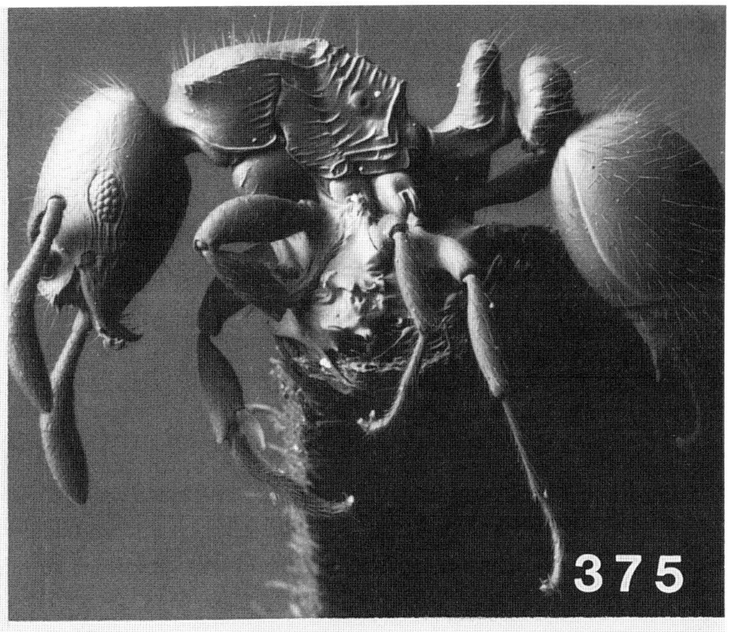
375

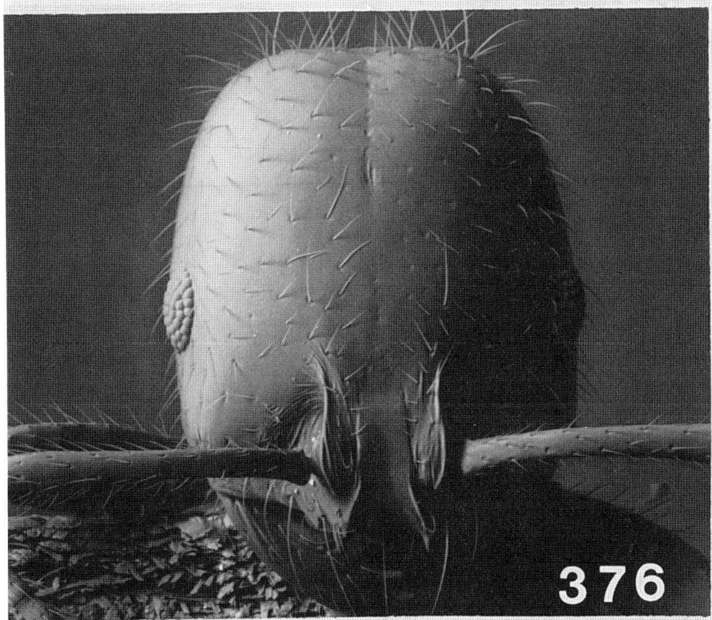
376

377

378

379

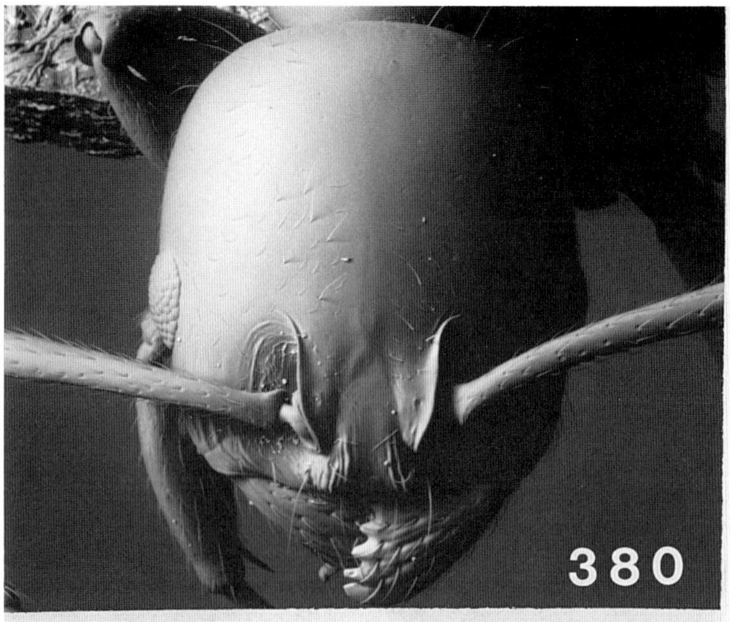

380

381

382

383

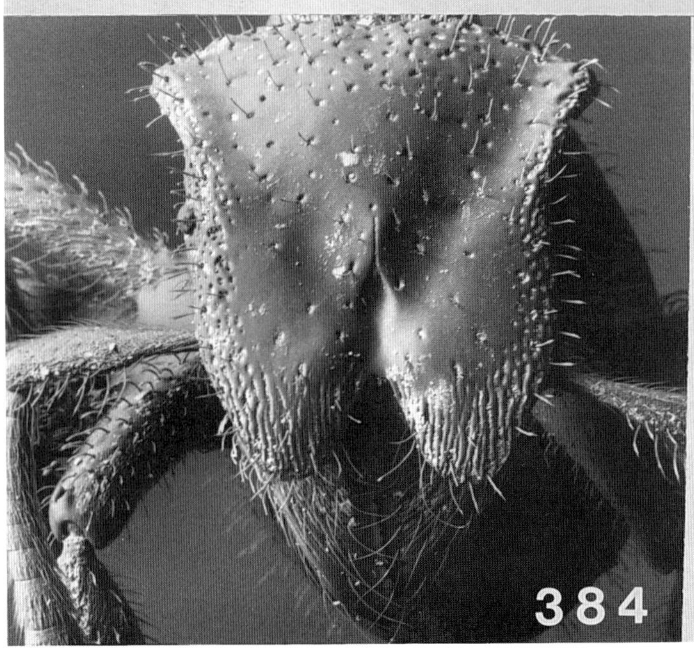

384

385

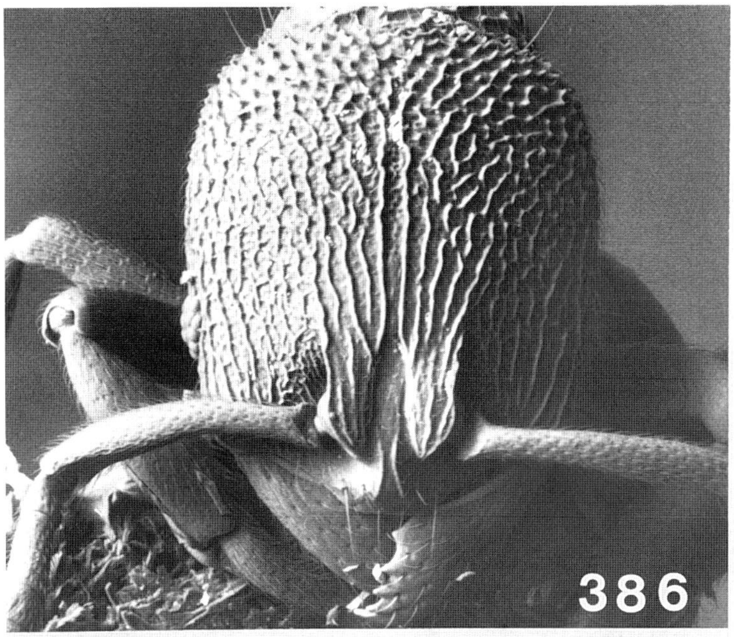
386
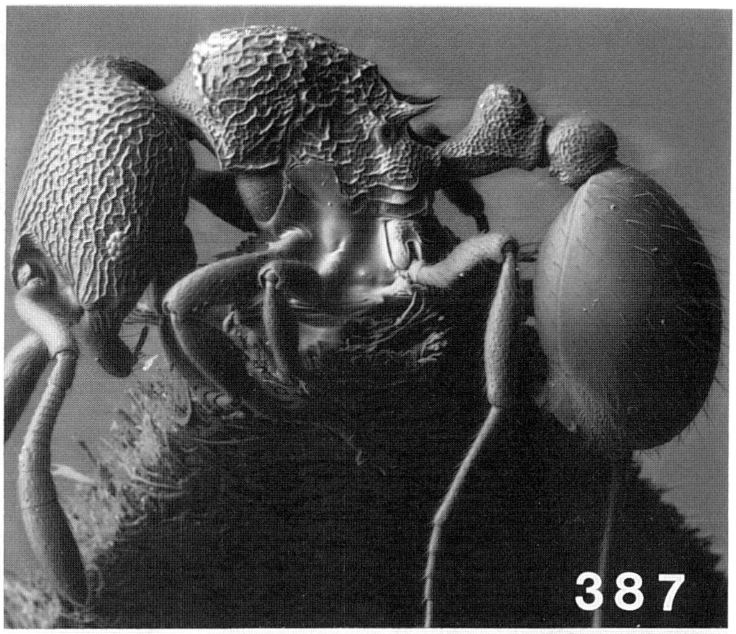
387

388
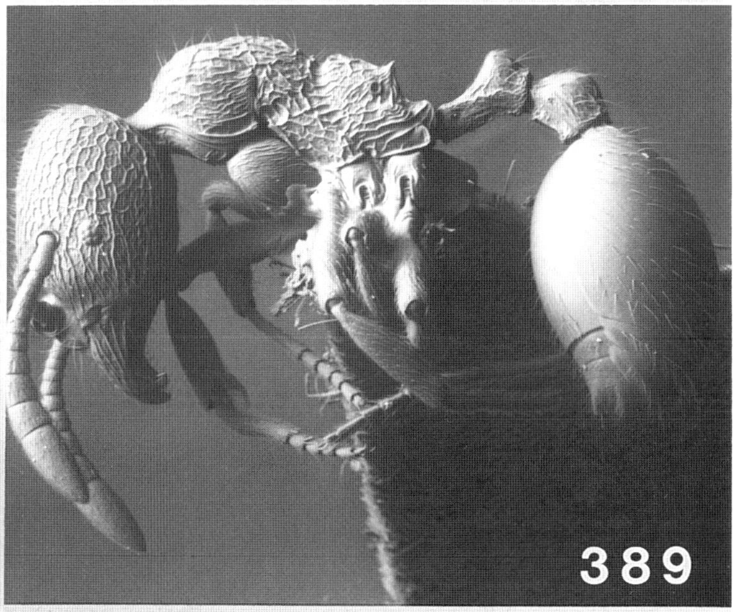
389

390
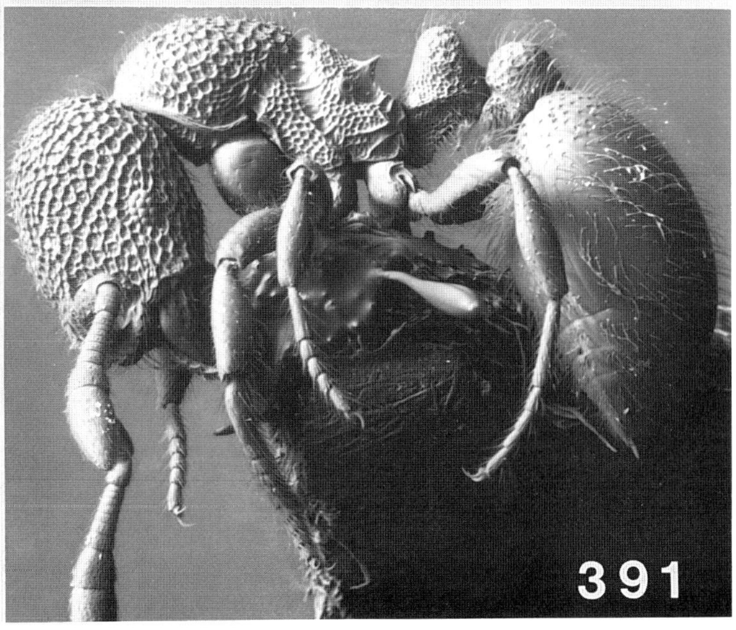
391

392

393

394

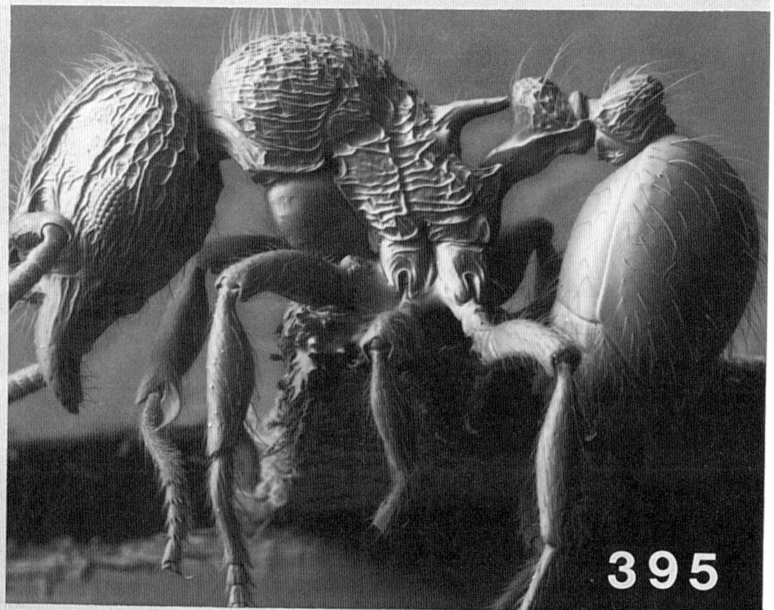
395

396

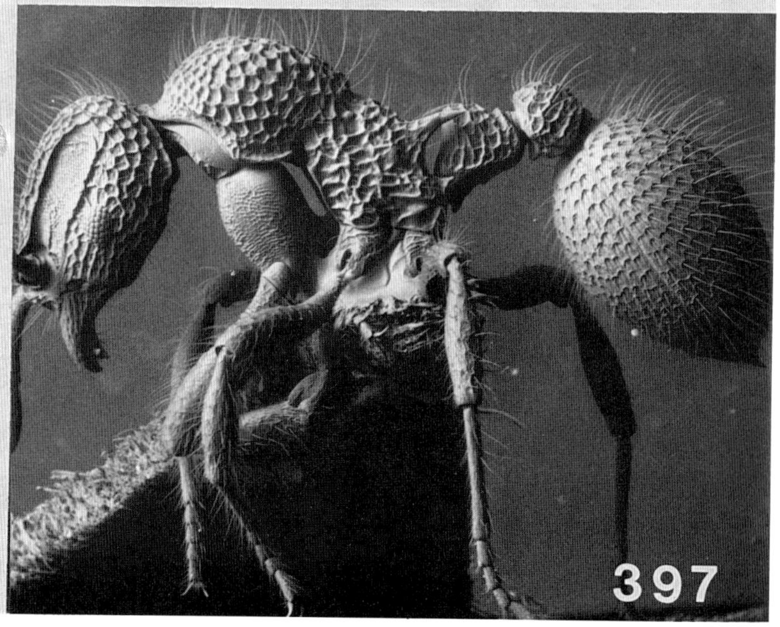
397

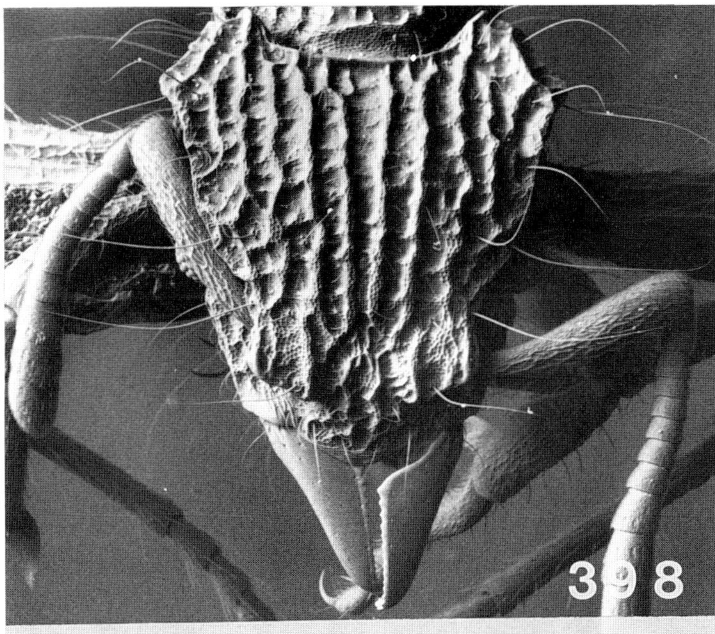

398

399

400

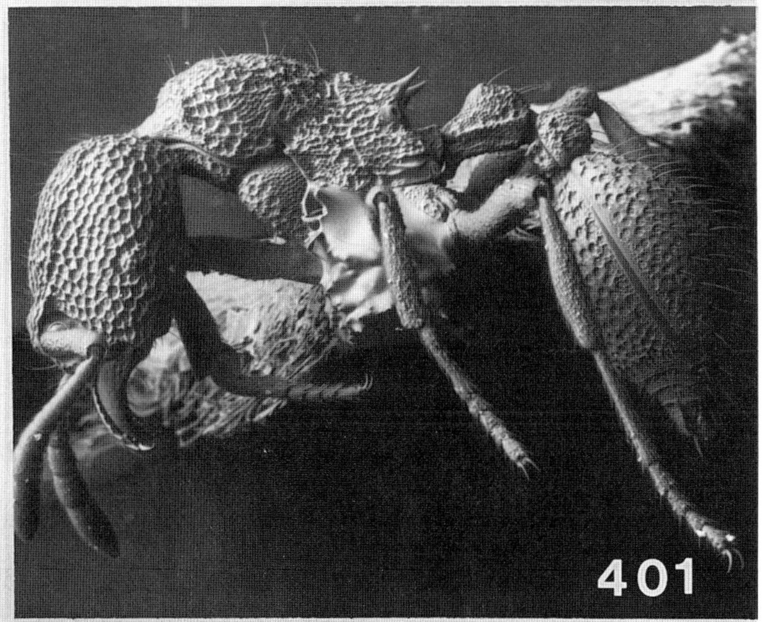

401

402

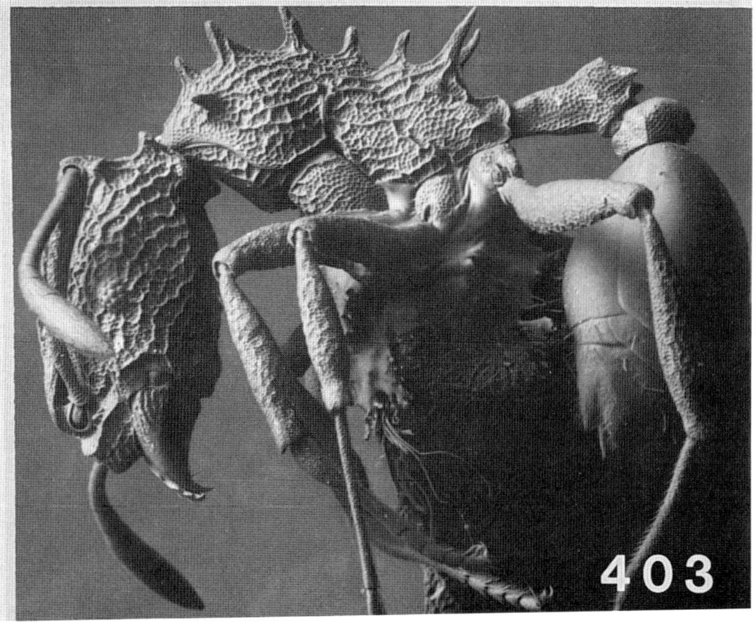

403

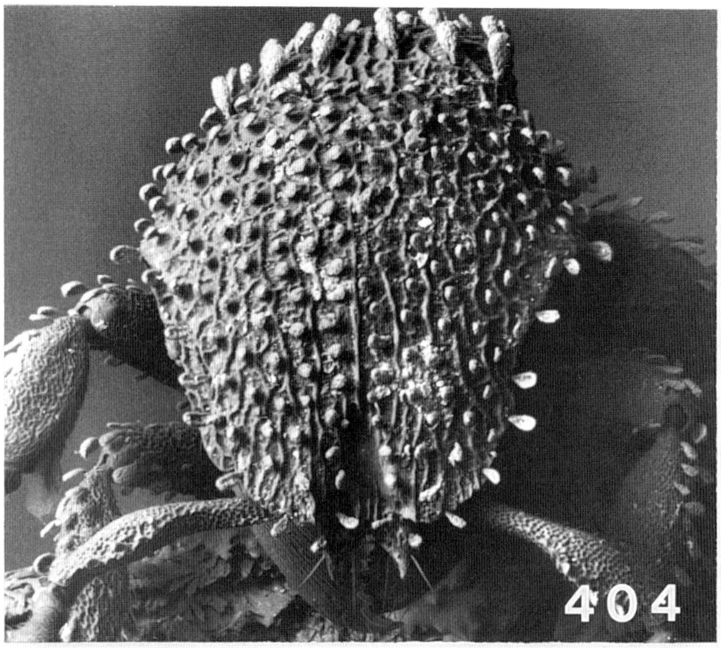

404

405

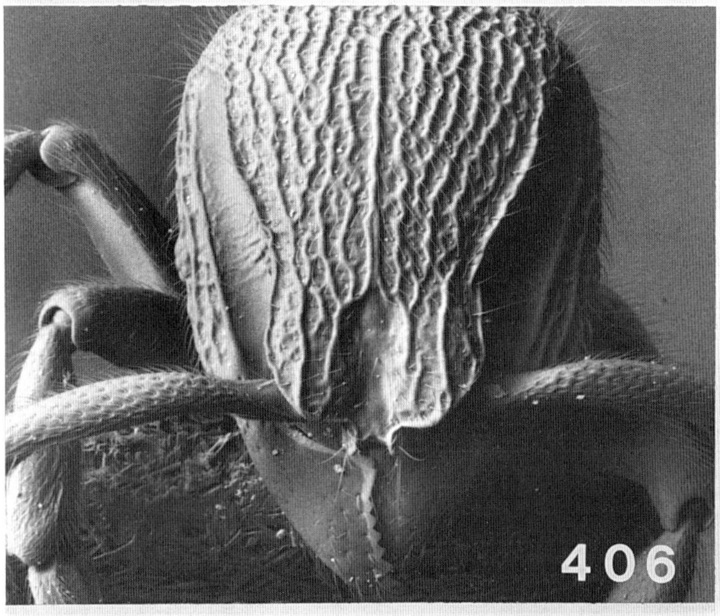

406

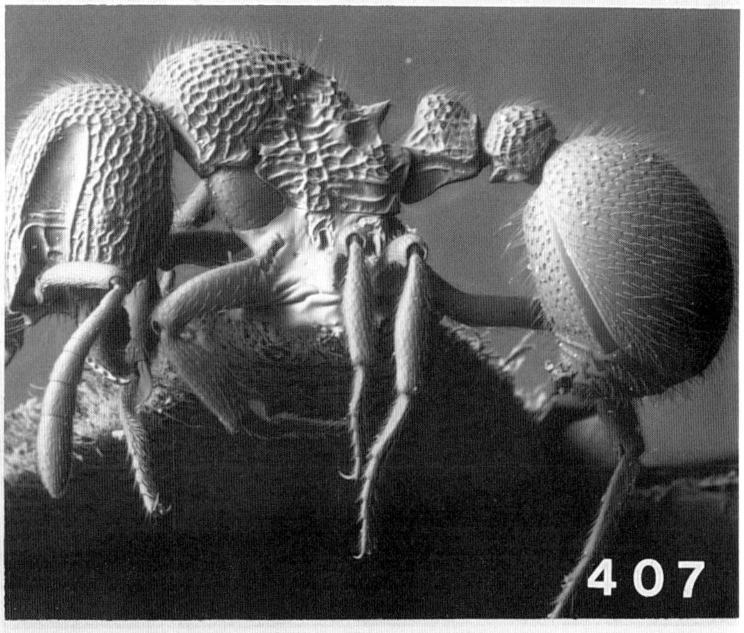

407

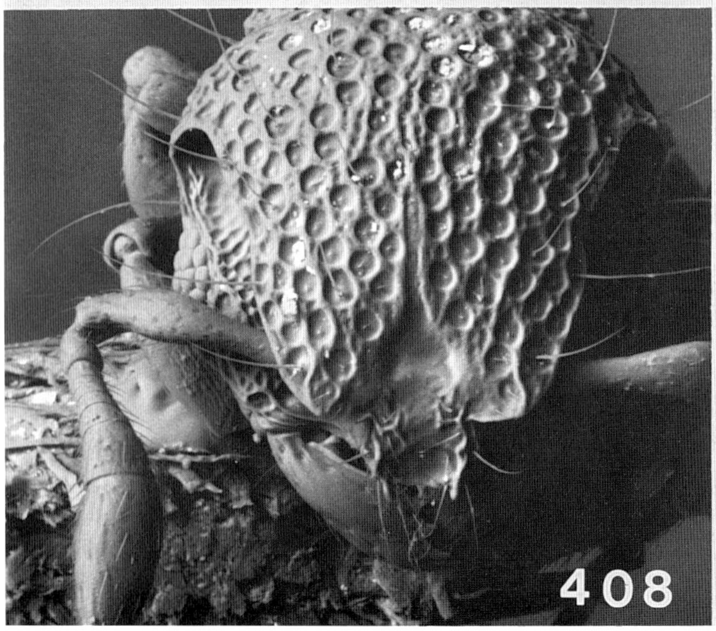

408

409

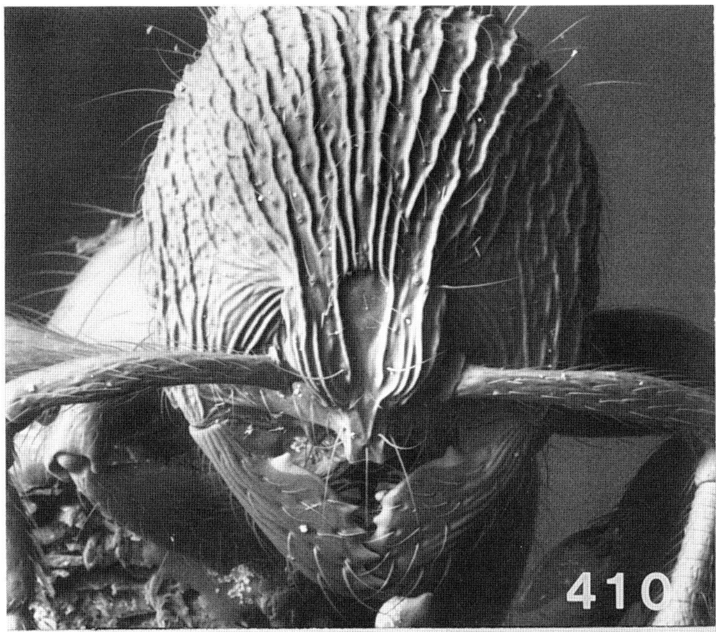
410

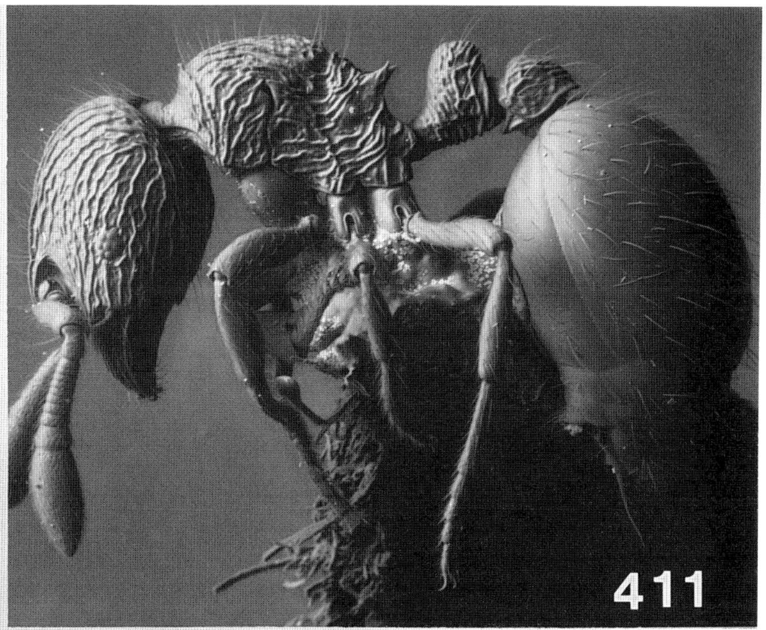
411

412

413

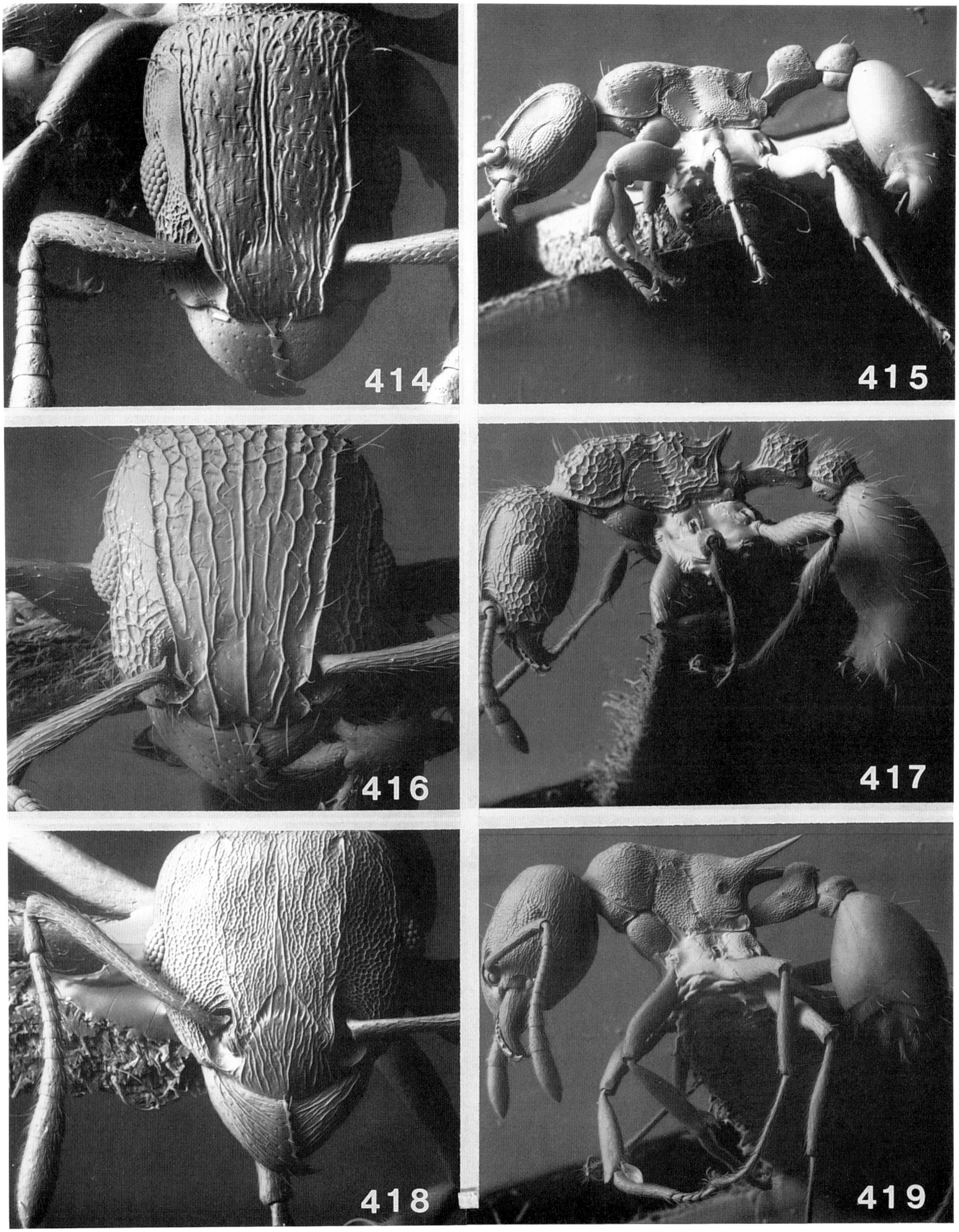
414
415
416
417
418
419

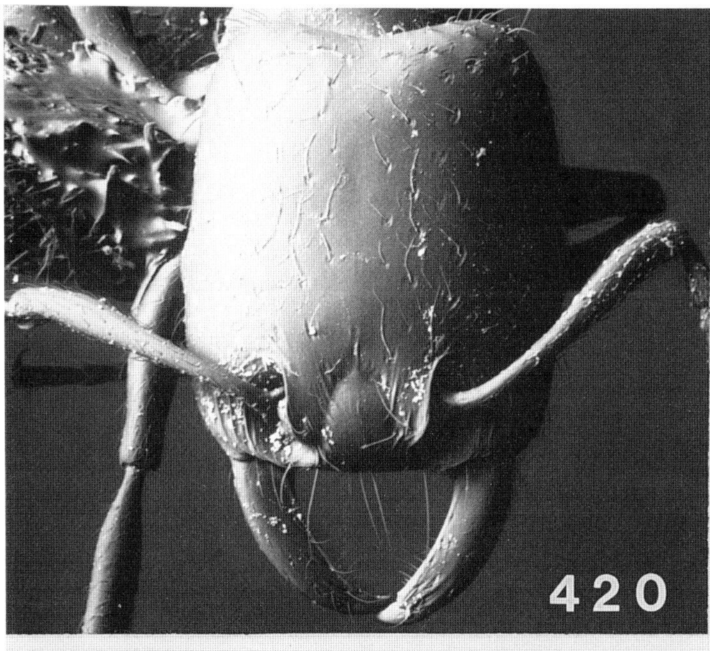
420
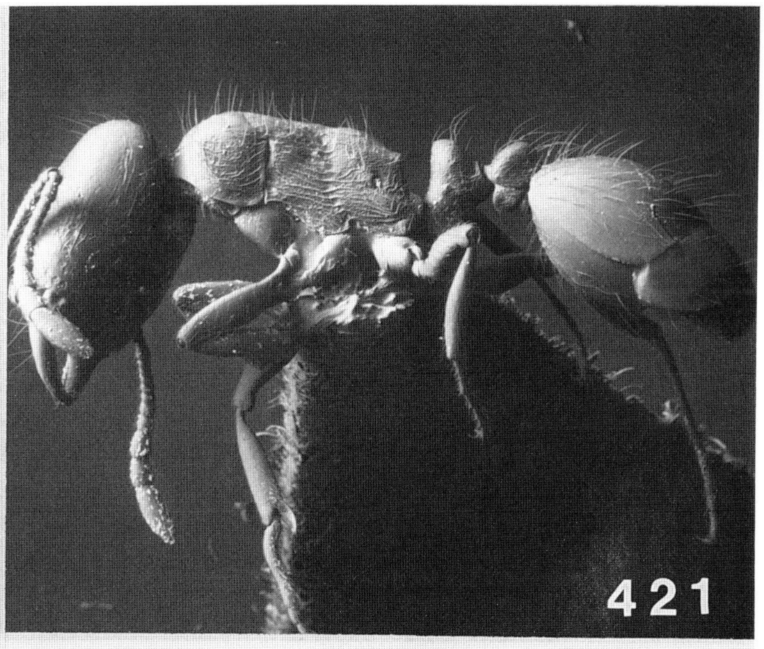
421
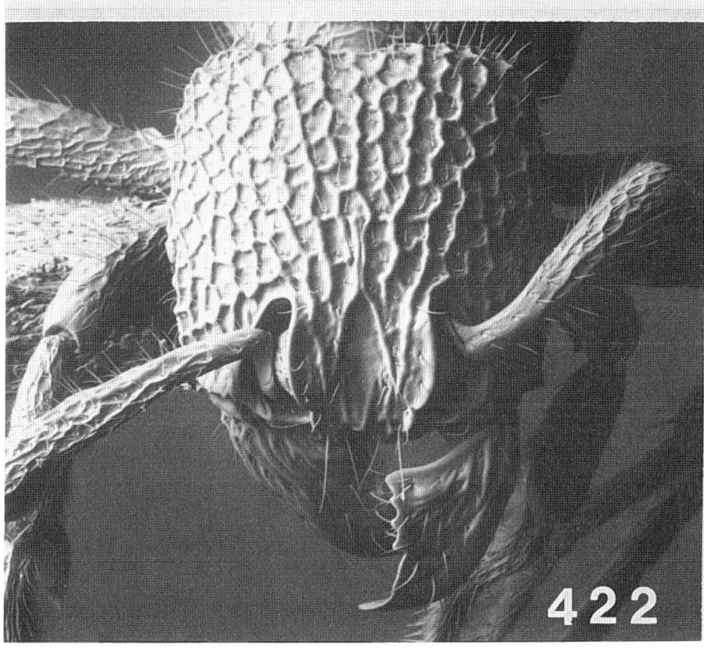
422

423

Subfamily NOTHOMYRMECIINAE

Diagnosis of Worker (Figs. 174, 175)

Ants with the following combination of characters together.

1 Clypeus broad from front to back so that antennal sockets are well behind anterior margin of head. Median portion of clypeus extended backwards between the frontal carinae.
2 Antennal sockets inclined; their margins and section of torulus closest to the midline of the head on a higher level than the margin most distant from the midline.
3 Frontal carinae present; frontal lobes absent but the antennal insertions partly concealed by the torular sclerites.
4 Narrow neck joining condylar bulb of antennal scape to shaft of scape proper straight.
5 Eyes present, large. Antenna with 12 segments.
6 Promesonotal suture present and flexible, the pronotum capable of movement relative to the mesonotum.
7 Metapleural gland orifice at lower posterior corner of metapleuron, opening laterally and not concealed by a cuticular flange or flap.
8 Mesonotum distinctly defined; metanotum present on dorsal alitrunk.
9 Metacoxal cavities open; cuticular annulus around each cavity with a wide break or interruption medially so that the coxal cavity is confluent with the cavity in which the petiole articulates.
10 Propodeal lobes present.
11 Waist of 1 segment, the petiole (= abdominal segment 2).
12 Abdominal stridulatory system present ventrally, the stridulitrum anterior on abdominal sternite 4 (= gastral sternite 2), the plectrum posterior on the preceding sternite.
13 Abdominal segment 4 (= gastral segment 2) without differentiated presclerites.
14 Abdominal spiracles 5–7 (= gastral 3–5) concealed by posterior margins of preceding tergites, not visible without distension of the gaster.
15 Helcium sternite relatively small, retracted, concealed by the tergite, and not visible in profile.
16 Abdominal segments 2–7 (petiole to apex of gaster) without tergosternal fusion.
17 Pygidium (tergite of abdominal segment 7 = gastral segment 5) simple, biconvex, unarmed.
18 Sting present, large, and strongly developed.

Synoptic Classification

Subfamily **NOTHOMYRMECIINAE.**
Tribe **Nothomyrmeciini.** Genus: ***Nothomyrmecia*** (Figs. 174, 175).

Distribution

The single extant species of this small subfamily is found only in the Australasian region, where it occurs in south and west Australia.

Taxonomic References

Nothomyrmeciinae: Taylor (1978a); Hölldobler and Wilson (1990); Shattuck (1992b); Baroni Urbani, Bolton, and Ward (1992).

Subfamily PONERINAE

Diagnosis of Worker (Figs. 424–516)

Ants with the following combination of characters together.

1 Clypeus usually broad from front to back so that antennal sockets are well back from anterior margin of head; much less commonly clypeus reduced so that the antennal sockets are close to the anterior margin of the head.

2 Antennal sockets horizontal, in the plane of the transverse axis of the head, usually concealed by the frontal lobes in full-face view.

3 Frontal lobes usually present and partially to entirely concealing the antennal insertions; rarely the frontal lobes absent and the antennal insertions exposed.

4 Narrow neck joining condylar bulb of antennal scape to shaft of scape proper generally sharply angled or bent downwards in frontal or full-face view, only rarely straight.

5 Eyes usually present, may be small to vestigial, only rarely absent; antenna with 6–12 segments.

6 Promesonotal suture usually present and flexible; fused and immobile in some, sometimes absent.

7 Metapleural gland orifice in lower posterior corner of metapleuron, opening laterally or posteriorly, the orifice not concealed by a cuticular flange or flap.

8 Metacoxal cavities variable; cuticular annulus around each cavity uncommonly broad and complete, the annulus usually interrupted by a flexible suture running from the coxal cavity to the cavity in which the petiole articulates or with a wide gap in the annulus medially.

9 Propodeal lobes usually present.

10 Waist of 1 segment, the petiole (= abdominal segment 2); usually also with a constriction between abdominal segments 3 and 4 (= gastral segments 1 and 2), which morphologically marks the boundary between the presclerites and postsclerites of abdominal segment 4.

11 Abdominal stridulatory system frequently present, less commonly absent; when present the stridulitrum located on the pretergite of abdominal segment 4 (= gastral segment 2), the plectrum posterior on the preceding tergite.

12 Abdominal spiracles 5–7 (= gastral 3–5) concealed by posterior margins of preceding tergites, not visible without distension of the gaster. Spiracle on abdominal tergite 5 (= gastral 3) may be very close to level of posterior margin of abdominal tergite 4, and thus may become visible with only slight distension.

13 Helcium sternite small, retracted, concealed by the tergite; not visible without dissection.

14 Abdominal segments 3 and 4 (= gastral segments 1 and 2) with tergosternal fusion; tergites and sternites of following abdominal segments (5–7, = gastral segments 3–5) not fused.

15 Abdominal segment 4 (= gastral segment 2) with presclerites sharply defined and differentiated from the postsclerites, the former fitting tightly within the posterior end of the third segment.

16 Pygidium (tergite of abdominal segment 7 = gastral segment 5) large, usually simple but very rarely with a few peg-like teeth or a pair of spines. Hypopygium sometimes with a marginal row of teeth but usually simple.

17 Sting present, usually large and strongly developed.

Key to Palaearctic PONERINAE (Workers)

1 Petiole broadly attached to first gastral segment, the two separated dorsally and laterally only by a constriction; petiole without a free posterior face and helcium located high on anterior face of first gastral segment in profile (Figs. 425, 427). Mandibles elongate, multidentate, and linear, articulated at corners of anterior margin of head, not closing against the clypeus (Figs. 424, 426) . . . . ***Amblyopone***

— Petiole narrowly attached to first gastral segment, the two joined via a slender articulatory junction; petiole usually with a free posterior face and helcium located at or below mid-height on anterior face of first gastral segment in profile (Figs. 445, 453, 457, 476, 490, 492, 500). Mandibles usually triangular or subtriangular, if linear then either not multidentate or articulated in middle of anterior margin of head (Figs. 444, 452, 456, 487, 489, 491, 499) **2**

2 Mandible long and linear, in full-face view inserted in the middle of the anterior margin of the head (Figs. 456, 459), with an apical armament of 3 teeth arranged in a vertical series (Fig. 457) **3**

— Mandible linear to triangular, in full-face view inserted at the anterolateral corner of head and not armed apically with a vertical series of 3 teeth (Figs. 444, 452, 467, 485, 487, 489, 491, 495, 499) **4**

3 Nuchal carina (separating dorsal from posterior surfaces of head) converging in a V at the midline, and also receiving a pair of prominent, dark, posterior apophyseal lines that converge to form the sharp median-dorsal groove of the vertex (Fig. 460) . . . . ***Odontomachus***

— Nuchal carina forming a broad, uninterrupted curve across the posterodorsal extremity of the head; posterior surface without paired, dark, apophyseal lines; on vertex median groove absent or ill-defined and shallow (Fig. 458) . . . . . . ***Anochetus***

4 Promesonotal suture absent (Figs. 443, 445, 455). With the head in full-face view horizontal frontal lobes vestigial to absent, the antennal sockets exposed (Figs. 442, 454) . . . **5**

— Promesonotal suture present (Figs. 453, 476, 486, 490, 496, 500). With the head in full-face view horizontal frontal lobes partially or entirely conceal the antennal sockets (Figs. 452, 474, 485, 487, 489, 495, 499) **7**

5 Tergite of second gastral segment strongly arched and vaulted so that the remaining segments point anteriorly (Figs. 443, 445). Sternite of second gastral segment small to minute in profile, roughly triangular in shape, and with the apex of the triangle directed ventrally or anteriorly. Eyes present, often small (Figs. 442–445) **6**

— Tergite of second gastral segment not arched and vaulted, the remaining segments directed posteriorly (Fig. 455). Sternite of second gastral segment large and elongate in profile, subrectangular to trapezoidal in shape. Eyes absent (Figs. 454, 455) . . . . ***Probolomyrmex***

6 Mandible edentate, overhung by the projecting clypeus (Fig. 442). Apical funicular segment strongly bulbous (Fig. 443) . . . . ***Discothyrea***

— Mandible with 3 or more teeth, not overhung by the clypeus (Fig. 444). Apical funicular segment enlarged but not strongly bulbous . . . . ***Proceratium***

7 Sides of head with broad, deep antennal scrobes present . . . . ***Aulacopone***

— Sides of head without antennal scrobes **8**

8 Pretarsal claws multidentate to pectinate on the inner curvature behind the apical point (Fig. 516) . . . . ***Leptogenys***

— Pretarsal claws unarmed or at most with a single tooth on the inner curvature behind the apical point (Figs. 514, 515) **9**

9 Petiole node armed dorsally with a pair of spines (Fig. 476). Alitrunk laterally with a conspicuous, pocket-like excavation above the mesopleuron . . . . ***Diacamma***

— Petiole node unarmed dorsally (Figs. 453, 469, 486, 488, 490, 496, 500). Alitrunk laterally without a pocket-like excavation above the mesopleuron **10**

10 Mandible elongate-triangular and armed with 5 long but slender spiniform teeth. Apical tooth particularly long, saber-like, broadly curved, and strongly crossing over its opposite number when the mandibles are closed (Fig. 487) . . . . ***Emeryopone***

— Mandible triangular and closing tightly against the clypeus, not armed with 5 spiniform teeth. Apical tooth not saber-like, overlapping but not strongly crossing over its opposite number when the mandibles are closed (Figs. 452, 467, 485, 489, 495, 499) **11**

11 Basal portion of mandible with a distinct circular or near-circular pit or fovea dorsolaterally **12**

— Basal portion of mandible without a dorsolateral pit or fovea **13**

12 Eyes absent (Figs. 485, 486). Dorsal (outer) surface of middle tibia and middle basitarsus with stout, peg-like setae or narrow, cuticular spines among the normal pilosity (Fig. 510) . . . . ***Cryptopone***

— Eyes present (Figs. 495, 496). Dorsal (outer) surface of middle tibia and middle basitarsus without stout, peg-like setae or narrow, cuticular spines among the normal pilosity . . . . ***Pachycondyla***

13 Ventral apex of hind tibia, when viewed from in front with the femur at right angle to the body, with a single, large, pectinate spur; without a second, smaller spur in front of the pectinate main spur in the direction of observation (Fig. 511) **14**

— Ventral apex of hind tibia, when viewed from in front with the femur at right angle to the body, with 2 spurs, consisting of a large, pectinate spur and a second, smaller spur in front of the main spur in the direction of observation (Figs. 512, 513) **15**

14 Subpetiolar process in profile with an acute angle posteroventrally and with a fenestra or tranducent thin spot anteriorly . . . . ***Ponera***

— Subpetiolar process in profile a simple lobe, without an acute posteroventral angle and lacking an anterior fenestra or thin spot . . . . ***Hypoponera***

15 Helcium located approximately at midheight of first gastral segment in profile (Fig. 453). Tibiae of middle and hind legs each with 2 pectinate spurs (Fig. 513). Sculpture universally of fine, dense shagreening with associated larger punctures. Pretarsal claws each with a tooth on its inner curvature, some distance from the apex (Fig. 515) . . . . ***Platythyrea***

— Helcium located at base of first gastral segment in profile (Figs. 469, 496). Tibiae of middle and hind legs each with 1 pectinate and 1 simple spur (Fig. 512). Sculpture not of fine, dense shagreening with associated larger punctures. Pretarsal claws lacking teeth on the inner curvature (Fig. 514) . . . . *Pachycondyla* (part)

Key to Afrotropical and Malagasy PONERINAE (Workers)

1 Petiole broadly attached to first gastral segment, the two separated dorsally and laterally only by a constriction (Figs. 425, 427, 431, 433); petiole without a free posterior face. In profile the helcium located high on the front of the first gastral segment, above the midheight of the anterior face **2**

— Petiole narrowly attached to first gastral segment, the two joined via a slender articulatory junction (Figs. 443, 453, 461, 472, 484, 490, 492, 494, 498); petiole usually with a free posterior face. In profile the helcium located low on the front of the first gastral segment, at or below the midheight of the anterior face **5**

2 Mandible elongate and usually linear, multidentate, and not closing tightly against the clypeus, always with more than 3 teeth (Figs. 424, 426, 430) **3**

— Mandible short, triangular or subtriangular, with relatively few teeth (1–3) (Fig. 432), and closing tightly against the clypeus **4**

3 Mandible pointed at apex, not as long as head (Figs. 424, 426); tooth row on inner margin of mandible single. Spatulate setae absent from head . . . . *Amblyopone*

— Mandible blunt at apex and very long, longer than head (Fig. 430); tooth row on inner margin of mandible double. Spatulate setae present on head . . . . *Mystrium*

4 Apical tooth of mandible followed by a short cleft, remainder of apical (masticatory) margin edentate. Alitrunk in dorsal view marginate anteriorly and strongly constricted from side to side in front of metanotal groove. Antenna 9-segmented with a strongly defined, 4-segmented club . . . . *Concoctio*

— Apical tooth of mandible followed by 2 more teeth, the second tooth small, the third larger than the second (Fig. 432). Alitrunk in dorsal view not marginate anteriorly nor constricted from side to side in front of the metanotal groove. Antenna 8–12 segmented with a 3- or 4-segmented club . . . . *Prionopelta*

5 Promesonotal suture absent (Figs. 443, 445, 455). Frontal lobes vestigial to absent (Figs. 442, 454); with the head in full-face view the antennal sockets entirely visible and directed vertically **6**

— Promesonotal suture present (Figs. 453, 461, 469, 472, 477, 482, 490, 494). Frontal lobes present (Figs. 452, 459, 467, 471, 475, 481, 489, 491, 495); with the head in full-face view the antennal sockets partially to entirely concealed by the frontal lobes **8**

6 Tergite of second gastral segment strongly arched and vaulted

so that the remaining segments point anteriorly (Figs. 443, 445). Sternite of second gastral segment small to minute in profile, triangular in shape, and with the apex of the triangle directed ventrally or anteriorly. Eyes present, often small **7**

— Tergite of second gastral segment not arched and vaulted, the remaining segments directed posteriorly (Fig. 455). Sternite of second gastral segment large and longitudinal in profile, subrectangular to trapezoidal in shape. Eyes absent . . . . *Probolomyrmex*

7 Mandible edentate, overhung by the projecting clypeus (Fig. 442). Apical funicular segment strongly bulbous (Fig. 443) . . . . *Discothyrea*

— Mandible with 3 or more teeth, not overhung by the clypeus (Fig. 444). Apical funicular segment moderately enlarged but not strongly bulbous . . . . *Proceratium*

8 Mandible long and linear, in full-face view inserted in the middle of the anterior margin of the head (Figs. 456, 459), with an apical armament of 3 teeth arranged in a vertical series (Fig. 457) **9**

— Mandible linear to triangular, in full-face view inserted at the anterolateral corner of the head and not armed apically with a vertical series of teeth (Figs. 452, 467, 470, 471, 475, 485, 489, 491) **10**

9 Nuchal carina (separating dorsal from posterior surfaces of head) converging in a V at the midline, and also receiving a pair of prominent, dark, posterior apophyseal lines that converge to form the sharp median-dorsal groove of the vertex (Fig. 460). Dorsalmost tooth of apical mandibular series truncated . . . . *Odontomachus*

— Nuchal carina forming a broad, uninterrupted curve across the posterodorsal extremity of the head; posterior surface without paired, dark apophyseal lines; on vertex the median groove absent or ill-defined and shallow (Fig. 458). Dorsalmost tooth of apical mandibular series acute . . . . *Anochetus*

10 Ventral apex of hind tibia, when viewed from in front with the femur at right angle to the body, with a single, large, pectinate spur; without a second, smaller spur in front of the main spur in the direction of observation (Fig. 511) **11**

— Ventral apex of hind tibia, when viewed from in front with the femur at right angle to the body, with two spurs, consisting of a large, pectinate spur and a second, smaller spur in front of the main spur in the direction of observation (Figs. 512, 513) **19**

11 Mandible elongate, linear, and weakly curved, blunt apically, the inner margin with 0–2 blunt teeth (Fig. 470). Mandibular articulation associated with a marked semicircular excavation of the dorsal anterior margin of the head in front of the eyes (Fig. 470) . . . . *Plectroctena*

— Mandible triangular to elongate-triangular and with a sharp apical tooth (Figs. 481, 483, 485, 489, 493, 499), the apical (masticatory) margin sometimes edentate (Fig. 471) but usually with several to many teeth. Mandibular articulation not associated with a semicircular excavation of the dorsal anterior margin of the head in front of the eyes **12**

12 Basal portion of mandible with a distinct circular or near-circular pit or fovea dorsolaterally .... ***Cryptopone***

— Basal portion of mandible without a dorsolateral pit or fovea **13**

13 Dorsal (outer) surfaces of middle tibiae and middle and hind basitarsi equipped with numerous strong cuticular spines or peg-like teeth which are very conspicuous (Fig. 509) .... ***Centromyrmex*** (part)

— Dorsal (outer) surfaces of middle tibiae and middle and hind basitarsi with setae but lacking cuticular spines or teeth **14**

14 Gaster in profile and in dorsal view with a distinct impression or girdling constriction between the first and second segments (Figs. 473, 490, 494, 500) **15**

— Gaster in profile and in dorsal view without an impression or girdling constriction between the first and second segments (Fig. 482) .... ***Asphinctopone***

15 Mandible elongate-falcate, with an extremely long apical tooth so that the tips cross over at rest (Fig. 471). Apical (masticatory) margin edentate or crenulate. Labrum prominent, in dorsal view projecting beyond the anterior clypeal margin as a striated lobe. Palp formula 3,4. Larger ants, total length 9–16 mm .... ***Psalidomyrmex***

— Mandible short and triangular, lacking an extremely long apical tooth (Figs. 493, 489, 499). Apical (masticatory) margin multidentate. Labrum not projecting beyond clypeus as a striated lobe in dorsal view. Palp formula less than 3,4 (unknown in *Dolioponera*). Smaller ants, total length less than 6 mm **16**

16 Frontal lobes massive, projecting anteriorly and overlapping the clypeus (Fig. 493). Median portion of clypeus projecting as a broad, truncated lobe **17**

— Frontal lobes small, not projecting anteriorly and not overlapping the clypeus (Figs. 489, 499). Median portion of clypeus not forming a broad, truncated lobe **18**

17 Second gastral segment with dorsum vaulted, strongly arched and down-curved posteriorly (Fig. 494). Sternite of second gastral segment much reduced and with a bluntly U-shaped outline in profile, very much smaller than the tergite (Fig. 494). Second gastral segment only slightly larger than first. Basal angle of mandible angulate, the apical (masticatory) margin with fewer than 8 teeth .... ***Loboponera*** [*Note:* This name is unavailable here; it will be formally described by W. L. Brown, Jr., in a forthcoming publication.]

— Second gastral segment barrel-shaped and longitudinal, the dorsum not vaulted, not arched and downcurved posteriorly. Sternite of second gastral segment longitudinal, without a bluntly U-shaped outline in profile, only slightly smaller than the tergite. Second gastral segment very much larger than the first. Basal angle of mandible evenly rounded, the apical (masticatory) margin with 8 teeth .... ***Dolioponera***

Afrotropical and Malagasy PONERINAE *(continued)*

18 Subpetiolar process in profile with an acute angle posteroventrally and with a fenestra or translucent thin spot anteriorly .... ***Ponera***

— Subpetiolar process in profile a simple lobe, without an acute posteroventral angle and lacking an anterior fenestra or thin spot .... ***Hypoponera***

19 Pretarsal claws of middle and hind legs armed on the inner curvature with a tooth, either close to the midlength or near the base (Fig. 515), or the entire inner curvature dentate to pectinate (Fig. 516) **20**

— Pretarsal claws of middle and hind legs simple (Fig. 514), the inner curvature without a tooth medially or near the base, never dentate or pectinate **22**

20 Pretarsal claws of middle and hind legs pectinate (Fig. 516) or with 1–3 small teeth behind the apex. If only 1 preapical tooth present on claw then mandible with only 1–3 teeth and clypeus with a sharp, median longitudinal carina (Fig. 491) .... ***Leptogenys*** (part)

— Pretarsal claws of middle and hind legs never pectinate, the claws always with only a single preapical tooth (Fig. 515). Mandible usually with more than 3 teeth but may be edentate, in which case the clypeus without a median longitudinal carina **21**

21 Helcium located approximately at midheight on the front of the first gastral segment so that the first gastral segment does not have a long, vertical anterior face in profile (Fig. 453). Tibiae of middle and hind legs each with 2 pectinate spurs (Fig. 513). Sculpture universally of fine, dense shagreening with associated larger punctures. Eyes never positioned well behind the midlength of the sides of the head .... ***Platythyrea***

— Helcium located very low on the front of the first gastral segment so that the first gastral segment has a long, vertical anterior face in profile (Fig. 496). Tibiae of middle and hind legs each with 1 large, pectinate spur and 1 small, simple spur (Fig. 512). Sculpture usually not of fine, dense shagreening with associated larger punctures, but if such is present then the eyes are positioned a considerable distance behind the midlength of the sides of the head .... ***Pachycondyla*** (part)

22 Eyes absent (Fig. 483). Dorsal (outer) surface of middle tibiae and middle and hind basitarsi with numerous cuticular spines or peg-like teeth (Fig. 509) .... ***Centromyrmex***

— Eyes present, varying from large to insignificant (Figs. 467, 475, 491, 495, 497). Middle tibiae and middle and hind basitarsi without cuticular spines or teeth, though stiff setae may be present **23**

23 Petiole dorsally with a comb of 5 long spines, which curve backwards over the base of the first gastral segment (Fig. 498) .... ***Phrynoponera***

— Petiole dorsally without a comb of 5 spines (Figs. 469, 477, 492, 496) **24**

24 Sides of petiole converging dorsally into a sharp, longitudinal crest, which runs the length of the segment. Posterolateral margins of petiole also sharply angulate in the dorsal half, these sharp angles meeting the dorsal crest at its posterior end (Fig. 477). Anterior clypeal margin broadly concave, the concavity terminating at each side in a prominent angle or tooth-like projection (Fig. 475) . . . . ***Streblognathus***

— Petiole scale-like to nodiform but without a sharp, longitudinal crest running the length of the dorsum (Figs. 469, 492, 496). Clypeus usually prominent but if shallowly concave medially then the concavity not terminating in prominent angles or teeth **25**

25 Mandible armed with only 1–3 teeth (usually 2) (Fig. 491) . . . . ***Leptogenys***

— Mandible armed with 5 or more teeth (Figs. 467, 495) . . . . ***Pachycondyla*** (part)

Key to Oriental and Indo-Australian PONERINAE (Workers)

1 Petiole broadly attached to first gastral segment, the two separated dorsally and laterally only by a constriction (Figs. 425, 427, 429, 431, 433); petiole without a free posterior face. In profile the helcium attached high on the anterior face of the first gastral segment **2**

— Petiole narrowly attached to first gastral segment, the two joined via a slender articulatory junction (Figs. 443, 447, 453, 461, 465, 468, 490, 492, 496, 500); petiole usually with a free posterior face. In profile the helcium attached at about the midheight or lower on the anterior face of the first gastral segment **5**

2 Mandible elongate and usually linear, multidentate, and not closing tightly against the clypeus, always with more than 3 teeth (Figs. 424, 426, 428, 430) **3**

— Mandible short and narrow, closing tightly against the clypeus, and armed with only 3 teeth, of which the median tooth is the smallest (Fig. 432) . . . . ***Prionopelta***

3 With head in full-face view the frontal lobes approximately even with, or slightly surpassing, the anterior clypeal margin beneath them (Fig. 428). Antennal funiculi markedly compressed . . . . ***Myopopone***

— With head in full-face view the frontal lobes distinctly posterior to the anterior clypeal margin (Figs. 424, 426, 430). Antennal funiculi not compressed, approximately round in section **4**

4 Mandible pointed at apex, in the form of an acute tooth (Figs. 424, 426). Spatulate setae absent from head . . ***Amblyopone***

— Mandible blunt at apex, rounded or subtruncate in full-face view (Fig. 430). Spatulate setae present on head . . . . ***Mystrium***

5 Mandible long and linear, in full-face view inserted in the middle of the anterior margin of the head (Figs. 456, 459), with an apical armament of 2 or 3 teeth arranged in a vertical series (Fig. 457) **6**

— Mandible linear to triangular, in full-face view inserted at the anterolateral corner of the head, and not armed apically with a vertical series of 2 or 3 teeth (Figs. 444, 449, 452, 462, 463, 466, 485, 495) **7**

6 Nuchal carina (separating dorsal from posterior surface of head) converging in a V at the midline, and also receiving a pair of prominent, dark apophyseal lines that converge to form the sharp mediodorsal groove of the vertex (Fig. 460) . . . . ***Odontomachus***

— Nuchal carina forming a broad, uninterrupted curve across the posterodorsal extremity of the head; posterior surface without paired, dark apophyseal lines; on vertex, median groove absent or ill-defined and shallow (Fig. 458) . . . . ***Anochetus***

7 With head in full-face view horizontal frontal lobes absent. Either the frontal lobes completely absent, in which case the antennal sockets are entirely visible and set on a shelf-like projection which overhangs the mandibles (Figs. 442, 454), or antennal sockets at extreme anterior margin of head; or (rarely) strongly elevated, strip-like frontal lobes are present (as Fig. 444), in which case the sockets are at the extreme anterior margin of the head and the head capsule has a median carina running its length **8**

— With head in full-face view horizontal frontal lobes present. These usually cover and conceal the antennal sockets (Figs. 446, 449, 452, 463, 467, 489, 495, 499) but if the sockets are partially visible then either the sockets are well behind the anterior margin of the head (Fig. 491), or a median longitudinal carina is absent from the head capsule, or both; the antennal sockets never on a shelf-like projection overhanging the mandibles **10**

8 Tergite of second gastral segment strongly arched and vaulted so that the remaining segments point anteriorly (Figs. 443, 445). Sternite of second gastral segment small to minute in profile, triangular in shape, and with the blunt apex of the triangle directed ventrally or anteriorly. Eyes present even if very small **9**

— Tergite of second gastral segment not arched and vaulted, the remaining segments directed posteriorly (Fig. 455). Sternite of second gastral segment large and elongate in profile, subrectangular to trapezoidal in shape. Eyes absent . . . . ***Probolomyrmex***

9 Mandible edentate, overhung by the projecting clypeus (Fig. 442). Apical funicular segment strongly bulbous (Fig. 443) . . . . ***Discothyrea***

— Mandible with 3 or more teeth, not overhung by the clypeus

(Fig. 444). Apical funicular segment moderately enlarged but not strongly bulbous . . . . ***Proceratium***

10 Metatibia (tibia of hind leg) in anterior view, with the leg at right angle to the long axis of the body, with 2 distinctly pectinate spurs, the posterior spur usually much larger than the anterior (Fig. 513) . . . . ***Platythyrea***

— Metatibia in anterior view, with the leg at right angles to the long axis of the body, either with only 1 spur, which is pectinate (Fig. 511), or with a large, pectinate, posterior spur and a much smaller, simple, anterior spur (Fig. 512) **11**

11 Frontal lobes widely separated throughout their length (Figs. 446, 449). Posterior clypeal margin usually broadly rounded or truncated between anterior ends of frontal lobes, the lobes themselves never separated only by a slender triangle anteriorly or a narrow median strip of cuticle throughout. Frontal lobes usually elongate and generally more or less straight-sided, not consisting of simple semicircles or blunt triangles and not having a distinct pinched-in appearance posteriorly **12**

— Frontal lobes closely approximated or even partially to entirely confluent (Figs. 462, 463, 466, 467, 474, 483, 485, 487, 489, 491, 495, 499). Frontal lobes separated only by a slender triangle of cuticle anteriorly or by a longitudinal line or narrow median strip of cuticle throughout. Frontal lobes usually consisting of simple, short semicircles or blunt triangles, and having a distinct pinched-in appearance posteriorly **13**

12 With alitrunk in profile the anteroventral pronotal angle, just in front of the anterior coxa, with a distinct and usually acute tooth (Fig. 451) (rarely this tooth may be reduced or missing in individual specimens). Posterior pretarsal claws always with a distinct median tooth (Fig. 515); posterior coxae unarmed above . . . . ***Rhytidoponera***

— With alitrunk in profile the anteroventral pronotal angle unarmed or forming an obtuse angle (Fig. 447). In rare cases where the angle is present and more nearly tooth-like then the posterior pretarsal claws lack a median tooth, or else the posterior coxae are toothed above . . . . ***Gnamptogenys***

13 Basal portion of mandible with a distinct circular, near-circular, or elongate pit or fovea dorsolaterally **14**

— Basal portion of mandible without a dorsolateral pit or fovea **15**

14 Dorsal (outer) surface of middle tibia and middle basitarsus with traction-enhancing thickened peg-like setae or narrow cuticular spines mixed with the normal finer pilosity (Fig. 510). Eyes usually absent (Figs. 485, 486), very rarely vestigially present . . . . ***Cryptopone***

— Dorsal (outer) surface of middle tibia and middle basitarsus without thickened peg-like setae or narrow cuticular spines mixed with the normal pilosity. Eyes always present and conspicuous (Figs. 495, 496) . . . . ***Pachycondyla*** (part)

15 Ventral apex of hind tibia, when viewed from in front with the femur at right angle to the body, with a single, large, pectinate spur; without a second, smaller spur in front of the main spur in the direction of observation (Fig. 511) **16**

— Ventral apex of hind tibia, when viewed from in front with the femur at right angle to the body, with 2 spurs, consisting of a large, pectinate spur and a second, smaller, simple spur in front of the main spur in the direction of observation (Fig. 512) **19**

16 Mandible elongate-triangular and armed with 5 long, slender, spiniform teeth. Apical tooth particularly long, saber-like, broadly curved, and strongly crossing over its opposite number when mandibles closed (Fig. 487) . . . . ***Emeryopone***

— Mandible triangular and closing tightly against the clypeus, not armed with 5 spiniform teeth. Apical tooth not saber-like, overlapping but not strongly crossing with its opposite number when mandibles closed (Figs. 483, 489, 499) **17**

17 Dorsal (outer) surfaces of middle tibiae and middle and hind basitarsi equipped with numerous strong cuticular spines or peg-like teeth (Fig. 509) . . . . ***Centromyrmex***

— Dorsal (outer) surfaces of middle tibiae, and middle and hind basitarsi, with setae but lacking cuticular spines or teeth. **18**

18 Subpetiolar process in profile with an acute angle posteroventrally and with a fenestra or translucent thin spot anteriorly . . . . ***Ponera***

— Subpetiolar process in profile a simple lobe, without an acute posteroventral angle and lacking an anterior fenestra or thin-spot . . . . ***Hypoponera***

19 Pretarsal claws of hind leg, on the inner curvature behind the apical point, either pectinate or equipped with 1 or more teeth (Figs. 515, 516) **20**

— Pretarsal claws of hind leg, on the inner curvature behind the apical point, unarmed, not pectinate and without teeth (Fig. 514) **21**

20 Ocelli present (Fig. 463). Pretarsal claws each with a large, stout preapical tooth (Fig. 515). Mandible forceps-like, each blade with a double longitudinal row of teeth and with more than 25 teeth in each row, so that each mandibular blade has >50 teeth (Fig. 463). Ventral surface of each mandible close to the base with a large, triangular flange whose inner margin forms an extension of the main gripping edge of the mandible . . . . ***Harpegnathos***

— Ocelli absent (Fig. 491). Pretarsal claws usually pectinate (Fig. 516), less commonly the pectination reduced to 1–3 small teeth on the basal half of the claw. Mandible very variable in shape but never with 2 rows of teeth on each blade and with <30 teeth on each mandible; often with only 1–3 teeth on each blade. Ventral surface of mandible close to base without a triangular flange . . . . ***Leptogenys***

21 Alitrunk laterally with a conspicuous, pocket-like excavation above the mesopleuron (Fig. 476). Petiole a node, armed dorsally with a pair of spines . . . . ***Diacamma***

— Alitrunk laterally without a pocket-like excavation above the mesopleuron (Figs. 464, 468, 469, 496). Petiole usually unarmed but sometimes a scale, which is emarginate dorsally, sometimes a node with a tridentate to multidentate posterodorsal margin **22**

22 Antennal sockets very close to or at the anterior clypeal margin. With head in full-face view frontal lobes reaching or overhanging the anterior clypeal margin on each side (Fig. 462). Medially the frontal lobes usually with a narrow, truncated clypeal lobe projecting freely in front of them; only very rarely is this lobe absent. Mandible usually linear . . . . ***Myopias***

— Antennal sockets well behind anterior clypeal margin. With head in full-face view frontal lobes far behind anterior clypeal margin on each side (Figs. 466, 467, 495). Medially the frontal lobes usually without a narrow, truncated, freely projecting clypeal lobe in front of them; only very rarely is such a lobe present. Mandible never linear **23**

23 Pronotum with a pair of laterally directed triangular teeth (Fig. 468). Mandible with 5 large, stout teeth and usually also with a minute basal denticle. Anterior clypeal margin with 7–9 acute to blunt projecting teeth (Fig. 466) . . . . ***Odontoponera***

— Pronotum unarmed (Figs. 469, 496). Mandible usually with 7 or more teeth, rarely with 6. Anterior clypeal margin unarmed, without projecting teeth (Figs. 467, 495) . . . . ***Pachycondyla*** (part)

Key to Australasian PONERINAE (Workers)

1 Petiole broadly attached to first gastral segment, the two separated dorsally and laterally only by a constriction (Figs. 425, 427, 429, 431, 433, 435); petiole without a free posterior face. In profile the helcium attached high on the anterior face of the first gastral segment **2**

— Petiole narrowly attached to first gastral segment, the two joined via a slender articulatory junction (Figs. 439, 443, 447, 453, 461, 476, 490, 492, 496, 500); petiole with a free posterior face. In profile the helcium attached at about the midheight or lower on the anterior face of the first gastral segment **6**

2 Mandible elongate and generally linear, always multidentate, with more than 3 teeth (Figs.. 424, 426, 428, 430, 434) . **3**

— Mandible short and narrow, armed with just 3 teeth of which the median tooth is the smallest (Fig. 432) . . . . ***Prionopelta***

3 With head in full-face view the frontal lobes approximately even with, or slightly surpassing, the anterior clypeal margin beneath them (Fig. 428). Antennal funiculi markedly compressed . . . . ***Myopopone***

— With head in full-face view the frontal lobes distinctly posterior to the anterior clypeal margin (Figs. 424, 426, 430, 434). Antennal funiculi not compressed, approximately round in section **4**

4 Mandible pointed at apex, in the form of an acute tooth (Figs. 424, 426, 434). Spatulate setae absent from head **5**

— Mandible blunt at apex, rounded or subtruncate in full-face view (Fig. 430). Spatulate setae present on head . ***Mystrium***

5 Hind tibiae each with a large, pectinate apical spur. Pretarsal claws small . . . . ***Amblyopone***

— Hind tibiae lacking pectinate apical spurs; either spur completely absent or reduced to a small setiform vestige. Pretarsal claws very large and stout . . . . ***Onychomyrmex***

6 Mandible long and linear, in full-face view inserted in the middle of the anterior margin of the head (Figs. 456, 459), with an apical armament of 2 or 3 teeth arranged in a vertical series (Fig. 457) **7**

— Mandible linear to triangular, in full-face view inserted at the anterolateral corner of the head and not armed apically with a vertical series of 2 or 3 teeth (Figs. 438, 444, 449, 452, 462, 467, 485, 489, 495, 499) **8**

7 Nuchal carina (separating dorsal from posterior surface of head) converging in a V at the midline, and also receiving a pair of prominent, dark apophyseal lines that converge to form the sharp, median-dorsal groove of the vertex (Fig. 460) . . . . ***Odontomachus***

— Nuchal carina forming a broad, uninterrupted curve across the posterodorsal extremity of the head; posterior surface without paired, dark apophyseal lines; on vertex, median groove absent or ill-defined and shallow (Fig. 458) . . . . ***Anochetus***

8 With head in full-face view horizontal frontal lobes absent. Either the frontal lobes completely absent, in which case the antennal sockets are entirely visible and set on a shelf-like projection which overhangs the mandibles (Figs. 442, 454), or the sockets at the extreme anterior margin of the head; or (rarely) strongly elevated, strip-like frontal lobes are present (as Fig. 444), in which case the sockets are at the extreme anterior margin of the head and the head capsule has a median carina running its length **9**

— With head in full-face view horizontal frontal lobes present. These usually cover and conceal the antennal sockets (Figs. 438, 449, 452, 462, 467, 474, 485, 499), but if the sockets are partially visible then either they are well behind the anterior margin of the head (Fig. 491), or a median longitudinal carina is absent from the head capsule, or both; antennal sockets never on a shelf-like projection overhanging the mandibles **11**

9 Tergite of second gastral segment strongly arched and vaulted so that the remaining segments point anteriorly (Figs. 443, 445). Sternite of second gastral segment small to minute in profile, triangular in shape, and with the blunt apex of the triangle directed ventrally or anteriorly. Eyes present, even if very small **10**

— Tergite of second gastral segment not arched and vaulted, remaining segments directed posteriorly (Fig. 455). Sternite of second gastral segment large and elongate in profile, subrectangular to trapezoidal in shape. Eyes absent . . . . ***Probolomyrmex***

10 Mandible edentate, overhung by the projecting clypeus (Fig. 442). Apical funicular segment strongly bulbous (Fig. 443) . . . . ***Discothyrea***

— Mandible with 3 or more teeth, not overhung by the clypeus (Fig. 444). Apical funicular segment moderately enlarged but not strongly bulbous . . . . ***Proceratium***

11 Helcium in profile broad and attached at or close to the mid-height of the first gastral segment (Figs. 447, 451, 453). Frontal lobes widely separated throughout their length. Posterior clypeal margin broadly rounded, truncated, or broadly triangular between anterior ends of frontal lobes (Figs. 438, 446, 449, 452) **12**

— Helcium in profile narrow and attached at or close to the ventralmost point of the first gastral segment (Figs. 476, 490, 492, 496, 500). Frontal lobes closely approximated or even partially to entirely confluent. Frontal lobes separated only by a slender triangle of cuticle anteriorly or by a longitudinal line or narrow median strip of cuticle throughout (Figs. 462, 467, 474, 485, 489, 491, 495, 499) **15**

12 Promesonotal suture absent from dorsal alitrunk (Fig. 447). Metacoxa dorsally with a spine . . . . ***Gnamptogenys***

— Promesonotal suture present and conspicuous on dorsal alitrunk (Figs. 439, 451, 453). Metacoxa without a dorsal spine **13**

13 Metatibia (tibia of hind leg) in anterior view, with the leg at right angle to the long axis of the body, with 2 distinctly pectinate spurs, the posterior spur usually much larger than the anterior (Fig. 513) . . . . ***Platythyrea***

— Metatibia in anterior view, with the leg at right angle to the long axis of the body, usually with a single, simple to pectinate spur (Fig. 511); sometimes a second, smaller, simple spur is also present, anterior to the main spur **14**

14 Pretarsal claws of hind legs each with a single preapical tooth at about the midlength of the inner curvature (Fig. 515) . . . . ***Rhytidoponera***

— Pretarsal claws of hind legs simple, without preapical teeth on the inner curvature (Fig. 514) . . . . ***Heteroponera***

15 Basal portion of mandible with a circular to elongate pit or fovea dorsolaterally **16**

— Basal portion of mandible without a dorsolateral pit or fovea **17**

16 Dorsal (outer) surface of middle tibia, and usually also middle basitarsus, with traction-enhancing thickened peg-like setae or narrow cuticular spines mixed in with the normal finer pilosity (Fig. 512). Eyes present but very small . . . . ***Cryptopone***

— Dorsal (outer) surface of middle tibia and middle basitarsus without traction-enhancing thickened peg-like setae or narrow cuticular spines mixed in with the normal pilosity; often all pilosity lacking. Eyes present, moderate to large, conspicuous (Figs. 495, 496) . . . . ***Pachycondyla*** (part)

17 Ventral apex of hind tibia, when viewed from in front with the femur at right angle to the body, with a single, large, pectinate spur; without a second, smaller spur in front of the main spur in the direction of observation (Fig. 511) **18**

— Ventral apex of hind tibia, when viewed from in front with the femur at right angle to the body, with 2 spurs, consisting of a large, pectinate spur and a second, smaller spur in front of the main spur in the direction of observation (Fig. 512) **19**

18 Subpetiolar process in profile with an acute angle posteroventrally and with a fenestra or translucent thin spot anteriorly . . . . ***Ponera***

— Subpetiolar process in profile a simple lobe, without an acute posteroventral angle and lacking an anterior fenestra or thin spot . . . . ***Hypoponera***

19 Pretarsal claws of hind leg, on the inner curvature behind the apical point, usually pectinate (Fig. 516), at least equipped with 1 or more teeth (Fig. 515) . . . . ***Leptogenys***

— Pretarsal claws of hind leg, on the inner curvature behind the apical point simple, unarmed, not pectinate, and without teeth (Fig. 514) **20**

20 Alitrunk laterally with a conspicuous pocket-like excavation above the mesopleuron (Fig. 476). Petiole a node, armed dorsally with a pair of spines . . . . ***Diacamma***

— Alitrunk laterally without a pocket-like excavation above the mesopleuron (Figs. 464, 469, 496). Petiole usually unarmed but sometimes multidenticulate along its posterodorsal margin **21**

21 Mandible triangular and closing against the anterior clypeal margin (Figs. 467, 495). Antennal sockets well behind anterior clypeal margin. With head in full-face view frontal lobes behind anterior clypeal margin on each side. Medially the frontal lobes usually without a narrow, truncated, freely projecting clypeal lobe in front of them . . . ***Pachycondyla*** (part)

— Mandible elongate to linear, slender, not closing against the anterior clypeal margin (Fig. 462). Antennal sockets very close to or at the anterior clypeal margin. With head in full-face view frontal lobes reaching or overhanging the anterior clypeal margin on each side. Medially the frontal lobes usually with a narrow, truncated, clypeal lobe freely projecting in front of them . . . . ***Myopias***

Key to Nearctic PONERINAE (Workers)

1 Petiole broadly attached to first gastral segment, the two separated dorsally and laterally only by a constriction (Figs. 425, 427, 433); petiole without a free posterior face. Helcium very broad and in profile attached high on anterior face of first gastral segment **2**

— Petiole narrowly attached to first gastral segment, the two joined via a slender articulatory junction (Figs. 443, 447, 450, 453, 461, 490, 492, 496, 500); petiole usually with a free posterior face. Helcium narrow, in profile attached near to or below the midheight of the first gastral segment **3**

2 Mandible elongate and usually linear, multidentate, and not closing tightly against the clypeus; always with more than 3 teeth (Figs. 424, 426) . . . . ***Amblyopone***

— Mandible short and narrow, closing tightly against the clypeus, and armed with only 3 teeth, of which the median tooth is the smallest (Fig. 432) . . . . ***Prionopelta***

3 Mandible long and linear, in full-face view inserted in the middle of the anterior margin of the head (Figs. 456, 459), with an apical armament of 2 or 3 teeth arranged in a vertical series (Fig. 457) **4**

— Mandible linear to triangular, in full-face view inserted at the anterolateral corner of the head, and not armed apically with a vertical series of 2 or 3 teeth (Figs. 444, 446, 448, 452, 467, 489, 491, 495) **5**

4 Nuchal carina (separating dorsal from posterior surface of head) converging in a V at the midline, and also receiving a pair of prominent, dark apophyseal lines that converge to form the sharp median-dorsal groove of the vertex (Fig. 460) . . . . ***Odontomachus***

— Nuchal carina forming a broad, uninterrupted curve across the posterodorsal extremity of the head; posterior surface without paired dark apophyseal lines; on vertex, median groove absent or ill-defined and shallow (Fig. 458) . . . . . . ***Anochetus***

5 With head in full-face view horizontal frontal lobes absent. Either the frontal lobes completely absent, in which case the antennal sockets are entirely visible and set on a shelf-like projection which overhangs the mandibles (Figs.. 442), or sockets at extreme anterior margin of head; or (rarely) strongly elevated, strip-like frontal lobes are present (as Fig. 444), in which case the sockets are at the anterior margin of the head. Promesonotal suture always absent **6**

— With head in full-face view horizontal frontal lobes present. These usually cover and conceal the antennal sockets (Figs. 446, 448, 452, 467, 485, 489, 495, 499), but if the sockets are partially visible (Fig. 491) then the sockets are well behind the anterior margin of the head and never on a shelf-like projection overhanging the mandibles. Promesonotal suture usually present and distinct, only very rarely obliterated . **7**

6 Mandible edentate, overhung by the projecting clypeus (Fig. 442). Apical funicular segment strongly bulbous (Fig. 443) . . . . ***Discothyrea***

— Mandible with 3 or more teeth, not overhung by the clypeus (Fig. 444). Apical funicular segment moderately enlarged but not strongly bulbous . . . . ***Proceratium***

7 Helcium in profile very low on anterior face of first gastral segment so that the petiole-gaster articulation is located at the extreme base of the segment (Figs. 490, 492, 496, 500). Frontal lobes closely approximated or confluent, the median portion of the clypeus posteriorly a very narrow triangle or merely a narrow, longitudinal strip between them (Figs. 467, 489, 491, 495, 499) **8**

— Helcium in profile at or close to midheight of anterior face of first gastral segment so that the petiole-gaster articulation is not located at the extreme base of the segment (Figs. 447, 450, 453). Frontal lobes widely separated, the median portion of the clypeus posteriorly rounded, broadly triangular, or truncated between them (Figs. 446, 448, 452) **12**

8 Ventral apex of hind tibia, when viewed from in front with the femur at right angle to the body, with a single spur, the spur large and pectinate (Fig. 511); without a second, smaller spur in front of the main spur in the direction of observation **9**

— Ventral apex of hind tibia, when viewed from in front with the femur at right angle to the body, with 2 spurs, consisting of a larger, pectinate spur and a smaller, simple spur in front of the main spur in the direction of observation (Fig. 512) . **10**

9 Subpetiolar process in profile with an acute angle posteroventrally and with a fenestra or translucent thin spot anteriorly . . . . ***Ponera***

— Subpetiolar process in profile a simple lobe, without an acute posteroventral angle and lacking an anterior fenestra or thin spot . . . . ***Hypoponera***

10 Basal portion of mandible with a distinct circular or near-circular pit or fovea dorsolaterally . . . . ***Cryptopone***

— Basal portion of mandible without a dorsolateral pit or fovea **11**

11 Pretarsal claws with inner curvatures usually pectinate (Fig. 516), more rarely the claws with only 1 or 2 small teeth behind the apical point (Fig. 515). If only 1 preapical tooth present on the claw then mandible with only 1–3 teeth and the clypeus with a sharp median longitudinal carina (Fig. 491) . . . . ***Leptogenys***

— Pretarsal claws with inner curvatures smooth and unarmed (Fig. 514), or at most with a single tooth behind the apical point (Fig. 515). If a single preapical tooth present on the claw then mandible distinctly with more than 3 teeth, or the clypeus lacking a median longitudinal carina, or usually both of these (Figs. 467, 495) . . . . ***Pachycondyla***

12 Hind tibiae each with 2 apical spurs, one larger than the other but both finely pectinate (Fig. 513). Dorsal surfaces of head and body without standing setae. Sculpture of entire body of fine, dense shagreening with associated larger punctures (Figs. 452, 453) . . . . ***Platythyrea***

— Hind tibiae each with either 1 apical spur only, or with 2 spurs; in the latter case 1 spur is larger and pectinate to barbulate, the other is smaller and simple (as Fig. 512). Dorsal surfaces of head and body with standing setae. Sculpture of entire body not of fine, dense shagreening with associated larger punctures (Figs. 446–448, 450) **13**

13 Mesonotum forming a prominent convexity surrounded by deeply impressed sutural lines (Fig. 450). Mesonotum and propodeum forming distinct convexities in front of and behind the metanotal groove .... ***Ectatomma***

— Mesonotum not forming a prominent convexity surrounded by deeply impressed sutural lines (Fig. 447). Metanotal groove present or absent; if present then mesonotum and propodeum not forming distinct convexities in front of and behind it .... ***Gnamptogenys***

Key to Neotropical PONERINAE (Workers)

1 Petiole broadly attached to first gastral segment, the two separated dorsally and laterally only by a constriction (Figs. 425, 427, 433); petiole without a free posterior face. Helcium very broad and in profile attached high on the anterior face of the first gastral segment **2**

— Petiole narrowly attached to first gastral segment, the two joined via a slender articulatory junction (Figs. 439, 443, 450, 453, 461, 490, 496, 500, 504, 506); petiole usually with a free posterior face. Helcium narrow and in profile attached near to or below the midheight on the anterior face of the first gastral segment **3**

2 Mandible elongate and usually linear, multidentate, and not closing tightly against the clypeus, always with more than 3 teeth (Figs. 424, 426) .... ***Amblyopone***

— Mandible short and subtriangular, closing tightly against the clypeus, and with only 3 teeth, of which the middle tooth is the smallest (Fig. 432) .... ***Prionopelta***

3 Mandible long and linear, in full-face view inserted in the middle of the anterior margin of the head (Figs. 456, 459), with an apical armament of 2 or 3 teeth arranged in a vertical series (Fig. 457) **4**

— Mandible linear to triangular, in full-face view inserted at the anterolateral corner of the head, and not armed apically with a vertical series of 2 or 3 teeth (Figs. 436, 444, 448, 467, 479, 487, 491, 499, 501, 503, 505) **5**

4 Nuchal carina (separating dorsal from posterior surfaces of head) converging in a V at the midline, and also receiving a pair of prominent, dark apophyseal lines that converge to form the sharp median-dorsal groove of the vertex (Fig. 460) .... ***Odontomachus***

— Nuchal carina forming a broad, uninterrupted curve across the posterodorsal extremity of the head; posterior surface without paired, dark apophyseal lines; on vertex the median groove absent or ill-defined and shallow (Fig. 458) .... ***Anochetus***

5 With head in full-face view horizontal frontal lobes absent. Either the frontal lobes completely absent, in which case the antennal sockets are entirely visible and set on a shelf-like projection which overhangs the mandibles (Figs. 442, 454), or sockets at extreme anterior margin of the head; or (rarely) strongly elevated, strip-like frontal lobes are present (as Fig. 444), in which case the sockets are at the extreme anterior margin of the head and the head capsule has a median carina running its length **6**

— With head in full-face view horizontal frontal lobes present. These usually cover and conceal the antennal sockets (Figs. 438, 440, 448, 452, 467, 479, 483, 489, 501, 503, 505), but if the sockets are partially visible then either the sockets are well behind the anterior margin of the head (Fig. 491), or a median longitudinal carina is absent from the head capsule, or both; antennal sockets never on a shelf-like projection overhanging the mandibles **8**

6 Tergite of second gastral segment strongly arched and vaulted so that the remaining segments point anteriorly (Figs. 443, 445). Sternite of second gastral segment small to minute in profile, triangular in shape, and with the apex of the triangle directed ventrally or anteriorly. Eyes usually present though they may be very small **7**

— Tergite of second gastral segment not arched and vaulted, the remaining segments directed posteriorly (Fig. 455). Sternite of second gastral segment large and elongate in profile, subrectangular to trapezoidal in shape. Eyes absent .... ***Probolomyrmex***

7 Mandible edentate, overhung by the projecting clypeus (Fig. 442). Apical funicular segment strongly bulbous (Fig. 443) .... ***Discothyrea***

— Mandible with 3 or more teeth, not overhung by the clypeus (Fig. 444). Apical funicular segment moderately enlarged but not strongly bulbous .... ***Proceratium***

8 With head in full-face view the median portion of the clypeus posteriorly either a very elongate, narrowly acute triangle which projects back between the closely approximated frontal lobes; or the frontal lobes confluent and separated only by a narrow, median cuticular strip; or a combination of these, the clypeus forming a narrowly acute, long triangle between the closely approximated lobes anteriorly, then petering out to a narrow strip behind (Figs. 467, 479, 483, 485, 487, 489, 491, 495, 499, 501). Frontal lobes always closely approximated or confluent, their outer margins always with a pinched-in appearance posteriorly **9**

— With head in full-face view the median portion of the clypeus posteriorly either rounded, broadly triangular, or truncated between the frontal lobes (Figs. 436, 438, 440, 446, 448, 452, 503). Frontal lobes distinctly and usually widely separated, never closely approximated, and never confluent, their outer margins variously shaped but only rarely with a pinched-in appearance posteriorly **18**

9 Ventral apex of hind tibia, when viewed from in front with the femur at right angle to the body, with a single spur; this spur usually large and pectinate (Fig. 511) but sometimes simple; without a second, smaller spur in front of the main spur in the direction of observation **10**

— Ventral apex of hind tibia, when viewed from in front with the femur at right angle to the body, with 2 spurs, consisting of a larger, pectinate spur and a smaller, simple spur in front of the main spur in the direction of observation (Fig. 512) **15**

10 Dorsal (outer) surface of middle tibiae and middle and hind basitarsi equipped with numerous strong, cuticular spines or peg-like teeth (Fig. 509) . . . . ***Centromyrmex***

— Dorsal (outer) surface of middle tibiae and middle and hind basitarsi with setae but without cuticular spines or teeth **11**

11 Mandible elongate, not triangular, each blade having 3–6 large, prominent teeth on the oblique apical (masticatory) margin (Figs. 487, 501) **12**

— Mandible triangular, each blade with 1 or 2 larger teeth apically, which are followed by numerous small teeth, denticles, or crenulations down the length of the straight apical (masticatory) margin (Figs. 489, 499, 505) **13**

12 Mandible with 5 or rarely 6 large, sharp teeth, the apical tooth narrow, very long, and curved (as Fig. 487). Apical and preapical teeth not forming a pair, not separated from the third large tooth by a long diastema. Frontal lobes not raised into a platform above the plane of the anterior portion of the head . . . . ***Belonopelta***

— Mandible with 3 or sometimes 4 large teeth (Fig. 501), sometimes also with 1–3 minute denticles on the diastema between teeth 2 and 3, the apical tooth short and stout. Apical and preapical teeth forming a distinct pair, separated from the third large tooth by a long diastema. Frontal lobes raised into a platform above the plane of the anterior portion of the head . . . . ***Simopelta***

13 In profile the helcium projecting at approximately the mid-height of the anterior face of the first gastral segment, so that the first gastral tergite is without a deep vertical face above the helcium (Fig. 506) . . . . ***Typhlomyrmex*** (part)

— In profile the helcium projecting very low on the anterior face of the first gastral segment, so that the first gastral tergite has a deep vertical face above the helcium (Figs. 490, 500) . **14**

14 Subpetiolar process in profile with an acute angle posteroventrally and with a fenestra or translucent thin spot anteriorly . . . . ***Ponera***

— Subpetiolar process in profile a simple lobe, without an acute posteroventral angle and lacking an anterior fenestra or thin spot . . . . ***Hypoponera***

15 Basal portion of mandible with a distict circular or near-circular pit or fovea dorsolaterally . . . . ***Cryptopone***

— Basal portion of mandible without a dorsolateral pit or fovea **16**

16 Pretarsal claws with their inner curvatures usually pectinate (Fig. 516), more rarely the claws with only 1 or 2 small teeth behind the apical point. If only 1 preapical tooth present on each claw then mandible with only 1–3 teeth and the clypeus with a sharp median longitudinal carina (Fig. 491) . . . . ***Leptogenys***

— Pretarsal claws with their inner curvatures smooth and unarmed (Fig. 514) or at most with a single tooth behind the apical point (Fig. 515). If a single preapical tooth present on each claw then the mandible distinctly with more than 3 teeth, or the clypeus lacking a median longitudinal carina, or usually both of these (Figs. 467, 479, 495) **17**

17 Anterior clypeal margin with a widely separated pair of prominent, usually blunted, large teeth which project forward (Fig. 479). Giant ants, with head width greater than 4.00 mm . . . . ***Dinoponera***

— Anterior clypeal margin without a pair of prominent, large teeth (Figs. 467, 495). Smaller ants, head width considerably less than 4.00 mm . . . . ***Pachycondyla***

18 Hind tibiae each with 2 distinctly pectinate spurs, the median spur usually much larger than the lateral (Fig. 513). Dorsal surfaces of head and body without standing setae. Sculpture everywhere of fine, dense shagreening with associated larger punctures (Figs. 452, 453) . . . . ***Platythyrea***

— Hind tibiae either with only 1 spur, which may be simple or pectinate (Fig. 511), or with a large, pectinate median spur and a much smaller, simple lateral spur (Fig. 512). Dorsal surfaces of head and body usually with standing setae, at least in part. Sculpture not of fine, dense shagreening with associated larger punctures **19**

19 Mandible armed with 3 extremely long, curved spiniform teeth, the apical tooth so long and curved that when the mandible is closed it surpasses the anterolateral corner of the head opposite from its insertion and points posteriorly almost toward the eye (Fig. 503) . . . . ***Thaumatomyrmex***

— Mandible triangular to linear, always with more than 3 teeth, denticles, or crenulations on the apical (masticatory) margin, the teeth never long and spiniform. Apical tooth not arching across front of head to opposite corner when the mandibles are closed (Figs. 436, 438, 440, 446, 448, 505) **20**

20 Dorsal margin of hypopygium with a comb-like row of vertical teeth on each side, the teeth projecting outside the pygidium (Fig. 508). Antennal scrobes present which run posteriorly above the eye, then reverse direction and run anteriorly below it (Fig. 440). Giant ants, with head width greater than 4.00 mm . . . . ***Paraponera***

— Dorsal margin of hypopygium smooth, without a row of teeth on each side. Antennal scrobes usually absent (Figs. 446, 448, 505) but if present then running only above the eye (Figs. 436, 438). Smaller ants, with head width less than 4.00 mm **21**

21 Mesonotum forming a prominent convexity surrounded by deeply impressed sutural lines (Fig. 450). Mesonotum and propodeum forming distinct convexities in front of and behind the metanotal groove . . . . ***Ectatomma***

— Mesonotum not forming a prominent convexity surrounded by deeply impressed sutural lines (Figs. 437, 439, 447, 506). Metanotal groove present or absent; if the former then mesonotum and propodeum not forming distinct convexities in front of and behind it **22**

22 Dorsum of head with a conspicuous median longitudinal costa, which is distinctly differentiated from any surrounding sculpture and which runs from clypeus to vertex (Figs. 436, 438); costa continuous across the frontal triangle **23**

— Dorsum of head lacking a median longitudinal costa running from clypeus to vertex and continuous across the frontal triangle (Figs. 446, 505) **24**

23 Palp formula 6,4. Propodeum bispinose (Fig. 437). Pretarsal claws each with a prominent basal lobe and a large submedian tooth . . . . ***Acanthoponera***

— Palp formula 4,3 or less. Propodeum bidentate to unarmed (Fig. 439). Pretarsal claws without prominent basal lobes and submedian tooth often absent from claws of hind legs . . . . ***Heteroponera***

24 With gaster in profile the visible sternite of the second segment reduced, usually roughly, bluntly triangular in shape, with the apex directed ventrally (Fig. 447). Tergite of second gastral segment arched and down-curved posteriorly. Propodeal lobes present. Petiole sessile or at most with a short anterior peduncle. Eyes usually large and conspicuous, only rarely minute . . . . ***Gnamptogenys***

— With gaster in profile the visible sternite of the second segment not reduced, roughly rectangular to trapezoidal in shape (Fig. 506). Tergite of second gastral segment not arched nor down-curved posteriorly. Propodeal lobes vestigial to absent. Petiole with a long anterior peduncle. Eyes always vestigial . . . . ***Typhlomyrmex*** (part)

Synoptic Classification

A name prefixed by * indicates an extinct taxon.

Subfamily **PONERINAE.**

Tribe **Amblyoponini** (= Ericapeltini, = Examblyoponini, = Onychomyrmicini, = Reneini). Genera: ***Amblyopone*** (Figs. 424–427, 507) (= *Amblyopopona,* = *Amblyopopone,* = *Arotropus,* = *Ericapelta,* = *Fulakora,* = *Lithomyrmex,* = *Neoamblyopone,* = *Protamblyopone,* = *Stigmatomma,* = *Xymmer*), ****Casaleia*** (= **Protamblyopone* (homonym)), ***Concoctio, Myopopone*** (Figs. 428, 429), ***Mystrium*** (Figs. 430, 431), ***Onychomyrmex*** (Figs. 434, 435), ***Prionopelta*** (Figs. 432, 433) (= *Examblyopone,* = *Renea*), ***Paraprionopelta*** (male only).

Tribe **Ectatommini** (= Paraponerini, = Proceratiini, = Stictoponerini, = Discothyrinae). Genera: ***Acanthoponera*** (Figs. 436, 437), ***Aulacopone,*** ****Bradoponera, Discothyrea*** (Figs. 442, 443) (= *Prodiscothyrea,* = *Pseudosphincta* (misspelling), = *Pseudosysphincta*), ***Ectatomma*** (Figs. 448, 450), ****Electroponera, Gnamptogenys*** (Figs. 446, 447) (= *Alfaria,* = *Barbourella,* = *Commateta,* = *Emeryella,* = *Holcoponera,* = *Mictoponera,* = *Opisthoscyphus,* = *Parectatomma,* = *Poneracantha,* = *Rhopalopone,* = *Spaniopone,* = *Stictoponera,* = *Tammoteca,* = *Wheeleripone*), ***Heteroponera*** (Figs. 438, 439) (= *Anacanthoponera,* = *Paranomopone*), ***Paraponera*** (Figs. 440, 441, 508), ***Proceratium*** (Figs. 444, 445) (= *Sysphincta* (misspelling), = *Sysphingta*), ***Rhytidoponera*** (Figs. 449, 451) (= *Chalcoponera*), ****Syntaphus.***

Tribe **Platythyreini**. Genera: ***Platythyrea*** (Figs. 452, 453, 513, 515) (= *Eubothroponera*), ***Probolomyrmex*** (Figs. 454, 455) (= *Escherichia*).

Tribe **Ponerini** (= *Archiponerini, = Centromyrmicini, = Dorylozelini, = Drepanognathini, = Euponerinae, = Hargepnathini, = Leptogenyini, = Odontomachini, = Pachycondylinae, = Plectroctenini). Genera: ***Anochetus*** (Figs. 456–458) (= *Myrmapatetes,* = *Stenomyrmex*), ****Archiponera, Asphinctopone*** (Figs. 481, 482) (= *Lepidopone*), ***Belonopelta*** (= *Leiopelta*), ***Centromyrmex*** (Figs. 483, 484, 509) (= *Glyphopone,* = *Leptopone,* = *Promyopias,* = *Spalacomyrmex,* = *Typhloteras*), ***Cryptopone*** (Figs. 485, 486, 510), ***Diacamma*** (Figs. 474, 476), ***Dinoponera*** (Figs. 478–480), ***Dolioponera, Emeryopone*** (Figs. 487, 488), ****Emplastus, Harpegnathos*** (Figs. 463, 465) (= *Drepanognathus*), ***Hypoponera*** (Figs. 489, 490), ***Leptogenys*** (Figs. 491, 492, 516) (= *Dorylozelus,* = *Lobopelta,* = *Machaerogenys,* = *Microbolbos,* = *Odontopelta,* = *Prionogenys*), ***Loboponera*** (Figs. 493, 494), ***Myopias*** (Figs. 462, 464) (= *Bradyponera,* = *Trapeziopelta*), ***Odontomachus*** (Figs. 459–461) (= *Champsomyrmex,* = *Myrtoteras,* = *Pedetes*), ***Odontoponera*** (Figs. 466, 468), ***Pachycondyla*** (Figs. 467, 469, 495, 496, 512) (= *Bothroponera,* = *Brachyponera,* = *Ectomomyrmex,* = *Eumecopone,* = *Euponera,* = *Hagensia,* = *Hiphopelta* (misspelling), = *Megaloponera* (misspelling), = *Megaponera,* = *Mesoponera,* = *Neoponera,* = *Ophthalmopone,* = *Paltothyreus,* = *Pseudoneoponera,* = *Pseudoponera,* = *Syntermitopone,* = *Termitopone,* = *Trachymesopus,* = *Trachyponera* (misspelling), = *Wadeura,* = *Xiphopelta*), *Pergandea* (nomen nudum), ***Phrynoponera*** (Figs. 497, 498), ***Plectroctena*** (Figs. 470, 472, 511) (= *Cacopone*), ***Ponera*** (Figs. 499, 500) (= *Pseudocryptopone,* = *Pteroponera,* = *Selenopone*), ****Poneropsis,*** ****Protopone, Psalidomyrmex*** (Figs. 471, 473), ***Simopelta*** (Figs. 501, 502), ***Streblognathus*** (Figs. 475, 477, 514), *Titusia* (nomen nudum).

Tribe **Thaumatomyrmecini.** Genus: ***Thaumatomyrmex*** (Figs. 503, 504).

Tribe **Typhlomyrmecini.** Genus: ***Typhlomyrmex*** (Figs. 505, 506).

Genus *incertae sedis* in Ponerinae: *Condylodon.*

[Material of the unavailable names Proponerinae and Taraxoponerinae is referable to Ponerinae; that of the unavailable name Exeuponerinae is referable to Ponerini.]

[*Note:* The unfamiliar and extensive synonymy under *Pachycondyla* is the result of a nearly completed revision of the group by William L. Brown, Jr., to whom I am indebted for permission to include them here.]

Distribution

The subfamily Ponerinae is found in all zoogeographical regions, as shown in the table given in the Introduction. The total number of ponerine genera shared by 2 or more regions is as follows, where PAL = Palaearctic, AFR = Afrotropical, MAL = Malagasy, ORI = Oriental, INA = Indo-Australian, AUS = Australasian, NEA = Nearctic, NEO = Neotropical. The following table excludes endemic genera and those accidentaly introduced by human activities.

	PAL	AFR	MAL	ORI	INA	AUS	NEA
AFR	9						
MAL	6	10					
ORI	10	13	9				
INA	10	15	10	21			
AUS	9	14	10	17	19		
NEA	7	10	7	11	10	11	
NEO	8	13	9	13	14	14	10

Taxonomic References

Identification of extant species

Some older references have a suffixed comment "[out of date]." These references are included as they contain the only identification keys ever attempted for the taxon in question. They should be used with great caution as, for the most part, they contain numerous infraspecific and infrasubspecific taxa that are no longer recognized. Older references that have been superseded, or those rendered useless by the volume of later descriptions and synonymies, are omitted.

Acanthoponera: Brown (1958b). ***Amblyopone:*** Brown (1960), Lattke (1991) [New World]; Baroni Urbani (1978a) [Mediterranean lands]; Taylor (1979) [Melanesia]; Terayama (1989) [Taiwan]; Tinaut (1990) [Iberian Peninsula]. ***Anochetus:*** Kempf (1964d) [partial, Neotropical]; Brown (1978b) [world]. ***Aulacopone:*** Taylor (1980c). ***Belonopelta:*** Baroni Urbani (1975b). ***Centromyrmex:*** Kempf (1967b) [Neotropical]. ***Concoctio:*** Brown (1974b). ***Cryptopone:*** W. M. Wheeler (1933) [out of date]; Brown (1963). ***Dinoponera:*** Kempf (1971). ***Dolioponera:*** Brown (1974c). ***Ectatomma:*** Brown (1958b); C. Kugler and Brown (1982). ***Emeryopone:*** Baroni Urbani (1975b). ***Gnamptogenys:*** Brown (1958b) [world]; Lattke (1990b) [Venezuela]; Lattke (1992) [*minuta*-group, Neotropical]. ***Heteroponera:*** Brown (1958b) [world]; Kempf (1962a) [Neotropical]. ***Hypoponera:*** Onoyama (1989) [Japan]. ***Leptogenys:*** Bolton (1975a) [Afrotropical, Malagasy]. ***Myopias:*** Willey and Brown (1983) [Australasian]. ***Myopopone:*** Brown (1960). ***Mystrium:*** Menozzi (1929) [out of date]. ***Odontomachus:*** Kempf (1962b) [Neotropical]; Brown (1976a) [world]; Deyrup, Trager, and Carlin (1985) [southern Nearctic]. ***Onychomyrmex:*** Brown (1960). ***Pachycondyla:*** W. M. Wheeler (1922) [partial, Afrotropical, out of date]; Arnold (1951, 1952) [partial, Afrotropical, out of date]; Kempf (1961b, 1964c) [partial, Neotropical]; Brown (1963) [notes]. ***Phrynoponera:*** W. M. Wheeler (1922) [out of date]. ***Platythyrea:*** Brown (1975) [world]. ***Plectroctena:*** Bolton (1974b). ***Ponera:*** Wilson (1957) [*tenuis*- and *selenophora*-groups]; Taylor (1967) [world]. ***Prionopelta:*** Brown (1960) [Indo-Australian, Neotropical]; Terron (1974) [Afrotropical]. ***Probolomyrmex:*** Taylor (1965b); Brown (1975). ***Proceratium:*** Snelling (1967), Brown (1980b), Ward (1988) [New World]; Baroni Urbani (1977b) [Europe]; Terron (1981) [Afrotropical]. ***Psalidomyrmex:*** Bolton (1975b). ***Rhytidoponera:*** Clark (1936) [Australia]; Brown (1958b), Ward (1984) [New Caledonia]; Ward (1980) [*impressa*-group, Australia, New Guinea]. ***Simopelta:*** Gotwald and Brown (1967). ***Thaumatomyrmex:*** Kempf (1975); Longino (1988). ***Typhlomyrmex:*** Brown (1965).

Other taxonomic references

Amblyoponini: Brown (1960). **Ectatommini:** Brown (1958b). **Platythyreini:** Brown (1975). **Ponerinae:** Brown (1954a, 1973); Snelling (1981); Dlussky and Fedoseeva (1988); Hölldobler and Wilson (1990); Baroni Urbani, Bolton and Ward (1992); Ogata (1987) [Japan]. **Ponerini:** Brown (in preparation). **Typhlomyrmecini:** Brown (1965).

See also References to Faunistic Studies.

Figures 424–516 PONERINAE workers. Figs. 424–506, heads in full-face view and bodies in profile (exceptions noted):

424–435, **Amblyoponini:** 424–427, *Amblyopone;* 428–429, *Myopopone;* 430–431, *Mystrium;* 432–433, *Prionopelta;* 434–435, *Onychomyrmex*

436–451, **Ectatommini:** 436–437, *Acanthoponera;* 438–439, *Heteroponera;* 440–441, *Paraponera;* 442–443, *Discothyrea;* 444–445, *Proceratium;* 446–447, *Gnamptogenys;* 448, 450, *Ectatomma;* 449, 451, *Rhytidoponera*

452–455, **Platythyreini:** 452–453, *Platythyrea;* 454–455, *Probolomyrmex*

456–461, **Ponerini:** 456–457, *Anochetus;* 459, 461, *Odontomachus;* 458, 460, posterodorsal view of occipital region (458, *Anochetus;* 460, *Odontomachus*)

462–502, **Ponerini:** 462, 464, *Myopias;* 463, 465, *Harpegnathos;* 466, 468, *Odontoponera;* 467, 469, *Pachycondyla;* 470, 472, *Plectroctena;* 471, 473, *Psalidomyrmex;* 474, 476, *Diacamma;* 475, 477, *Streblognathus;* 478–480, *Dinoponera;* 481–482, *Asphinctopone;* 483–484, *Centromyrmex;* 485–486, *Cryptopone;* 487–488, *Emeryopone;* 489–490, *Hypoponera;* 491–492, *Leptogenys;* 493–494, *Loboponera;* 495–496, *Pachycondyla;* 497–498, *Phrynoponera;* 499–500, *Ponera;* 501–502, *Simopelta*

503–504, **Thaumatomyrmecini,** *Thaumatomyrmex*

505–506, **Typhlomyrmecini,** *Typhlomyrmex*

507–508, development of hypopygial spines: 507, *Amblyopone;* 508, *Paraponera*

509–510, mesotibial and basitarsal spines and tractor setae: 509, *Centromyrmex;* 510, *Cryptopone*

511–513, development of metatibial (= hind tibial) spurs: 511, *Plectroctena;* 512, *Pachycondyla;* 513, *Platythyrea*

514–516, pretarsal claws of metathoracic (= hind) leg: 514, simple claws in *Streblognathus;* 515, toothed claws in *Platythyrea;* 516, pectinate claws in *Leptogenys.*

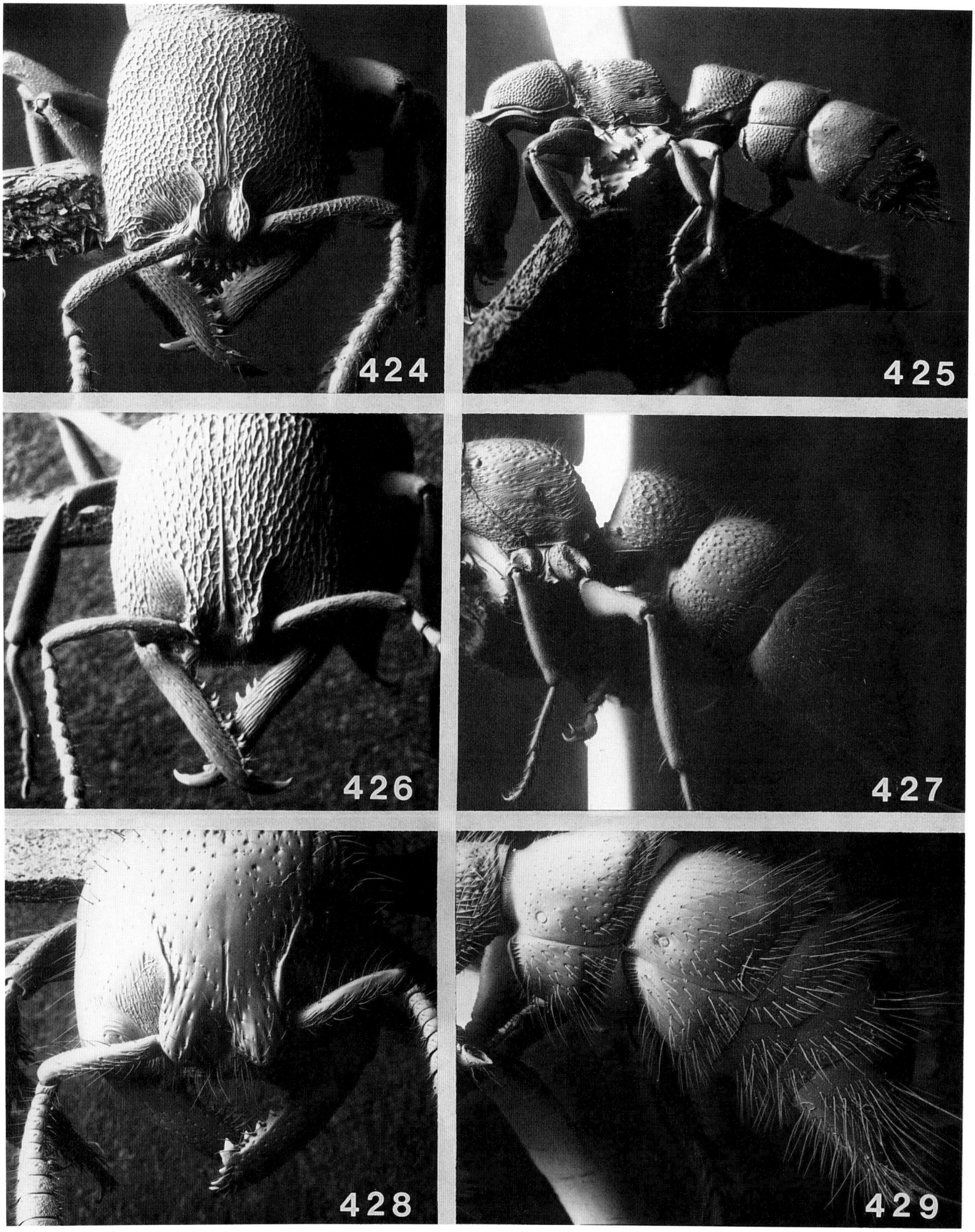
424
425
426
427
428
429

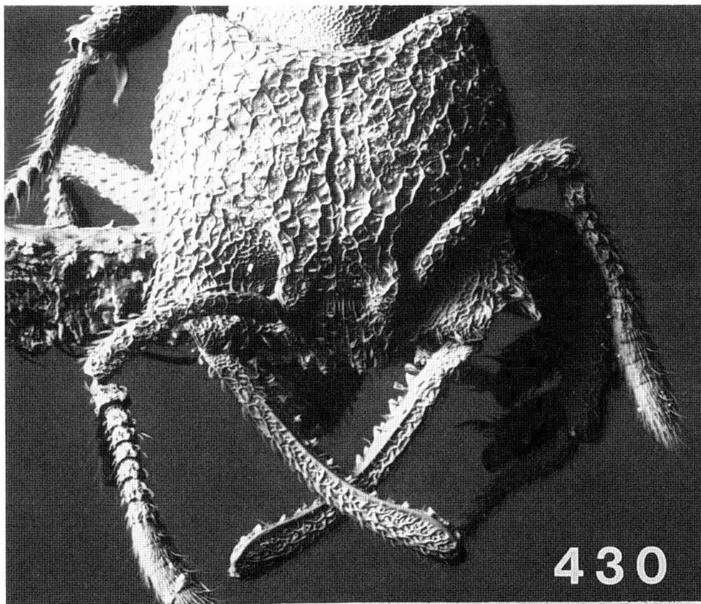
430

431

432

433

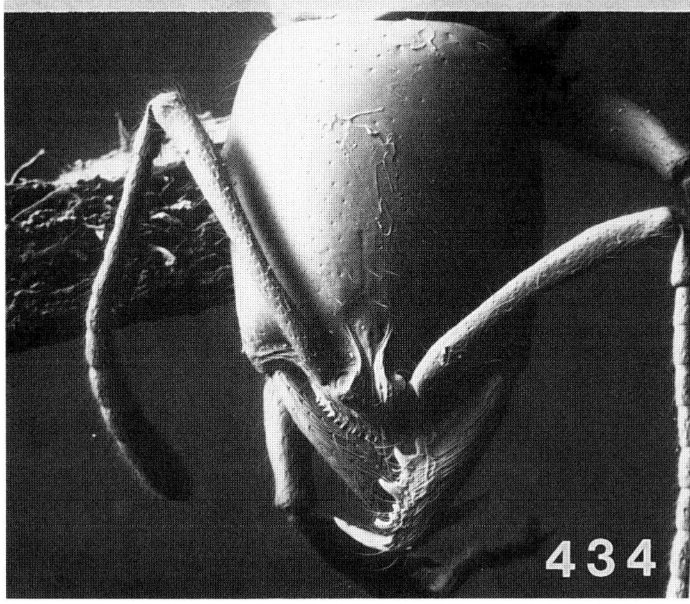
434

435

436

437

438

439

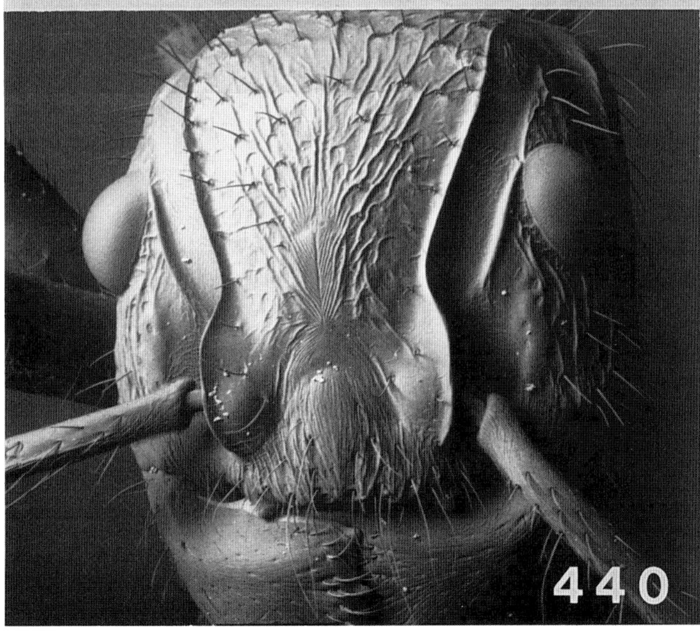

440

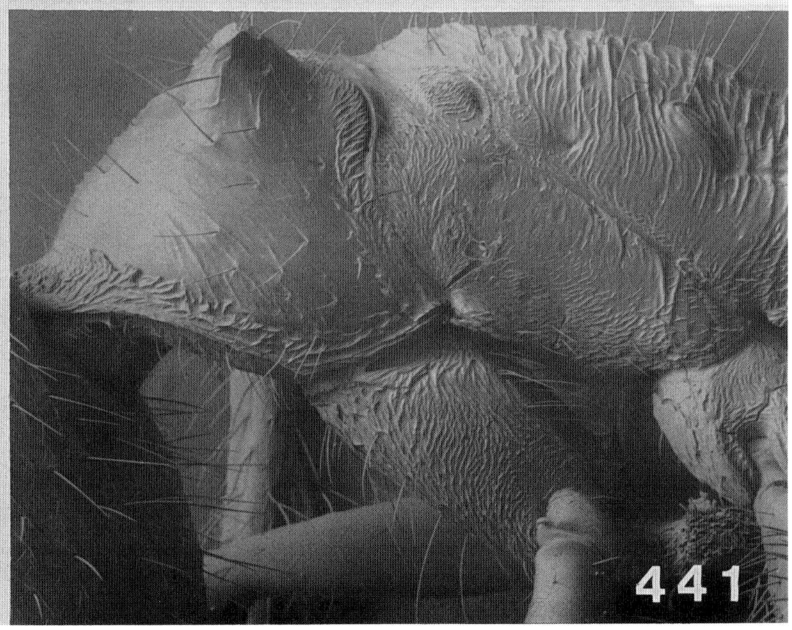

441

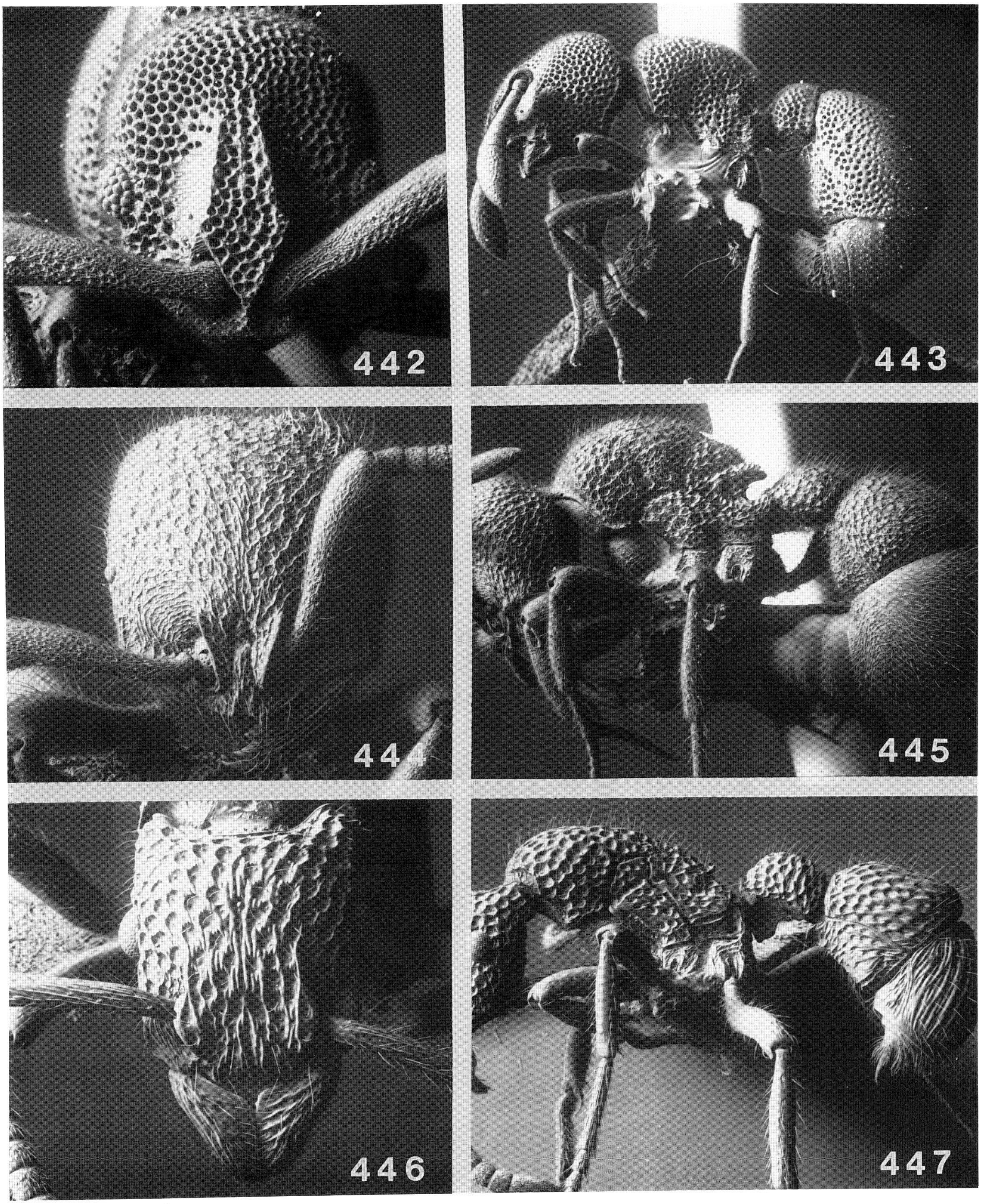
442
443
444
445
446
447

448
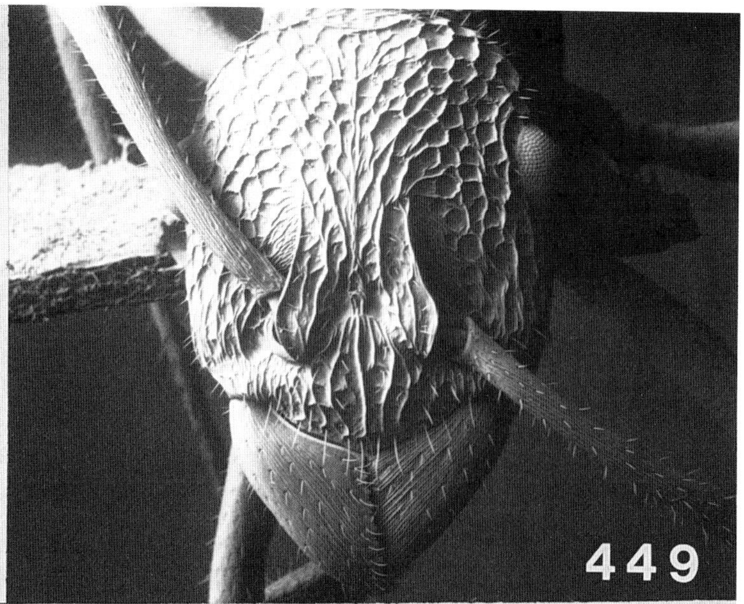
449
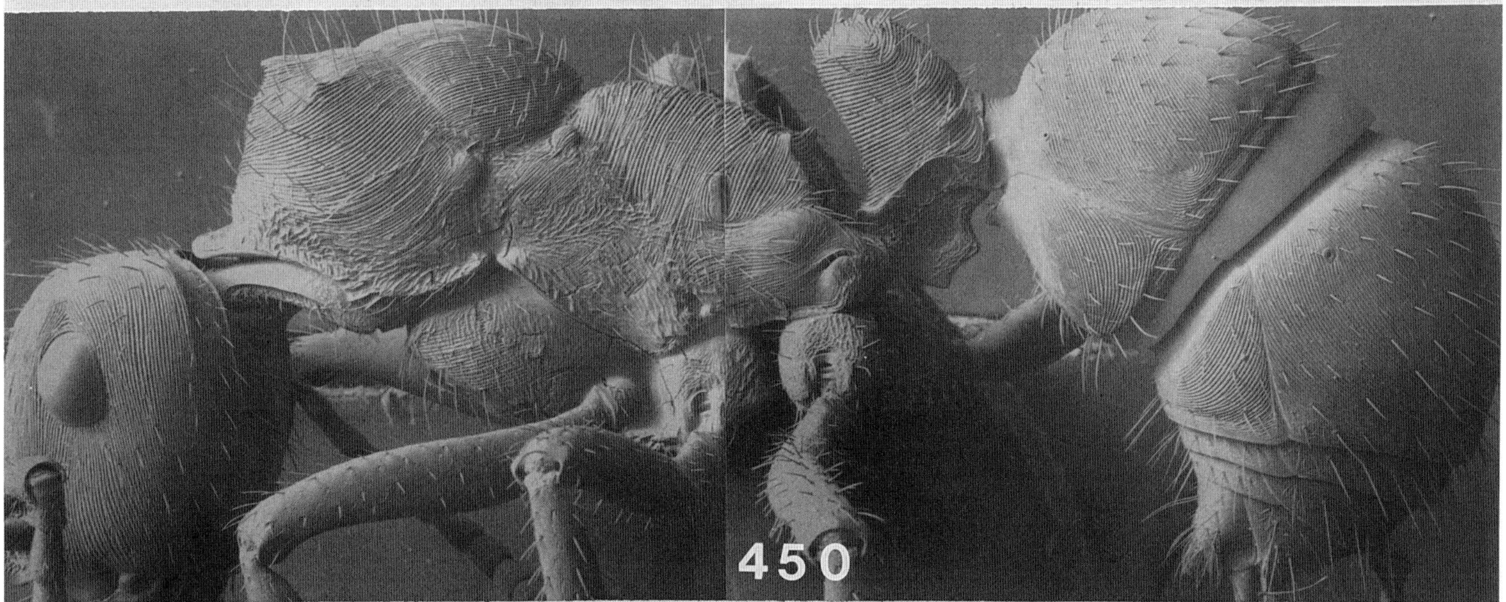
450

451

452

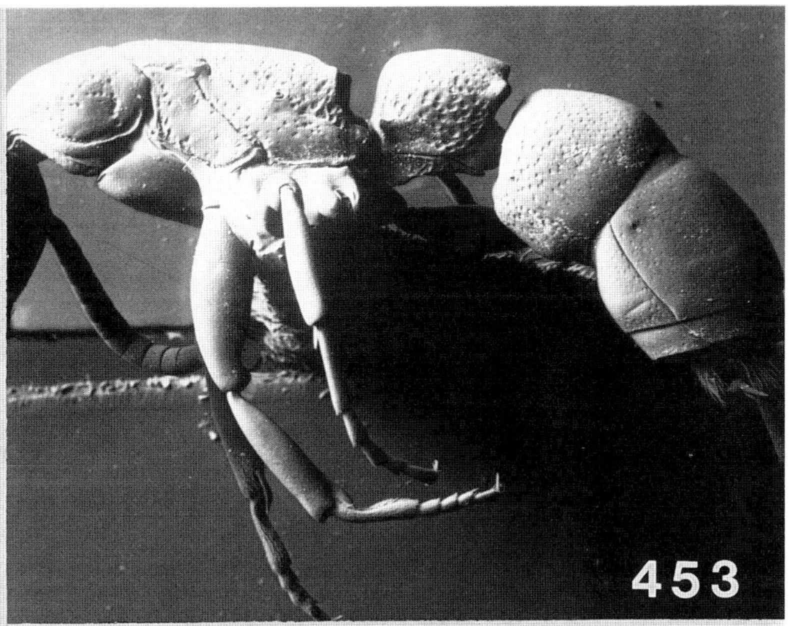
453

454

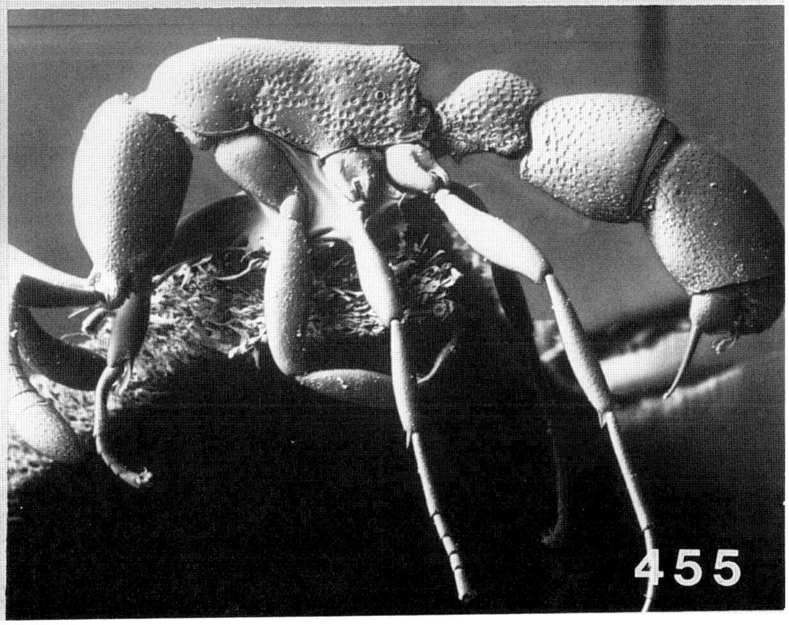
455

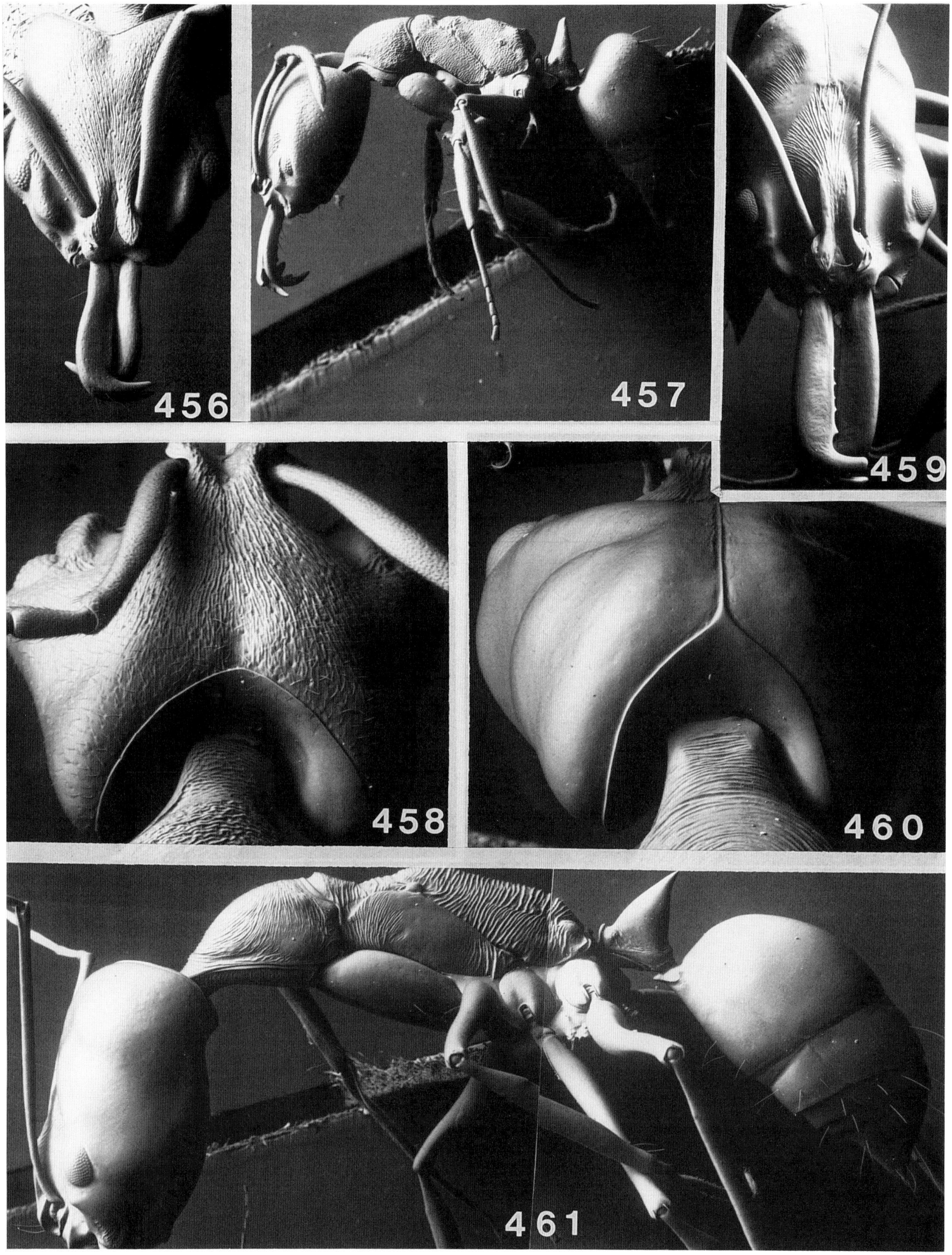
456
457
459
458
460
461

462

463

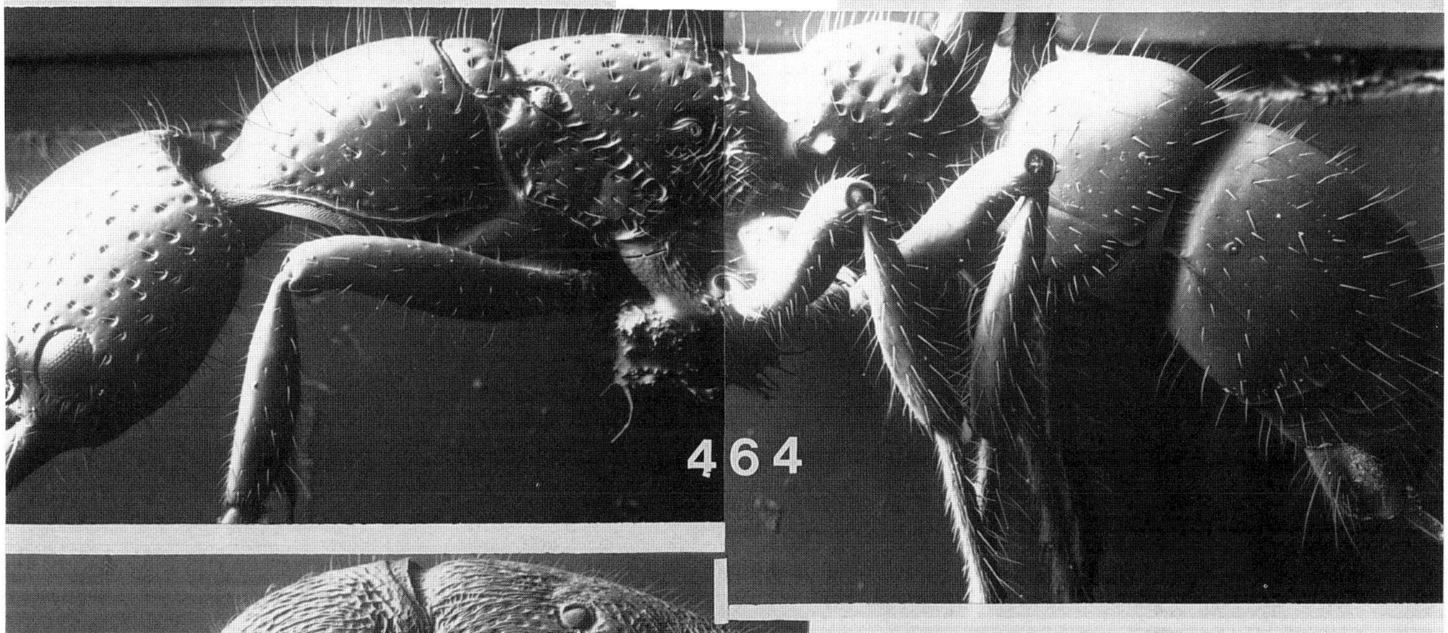
464

465

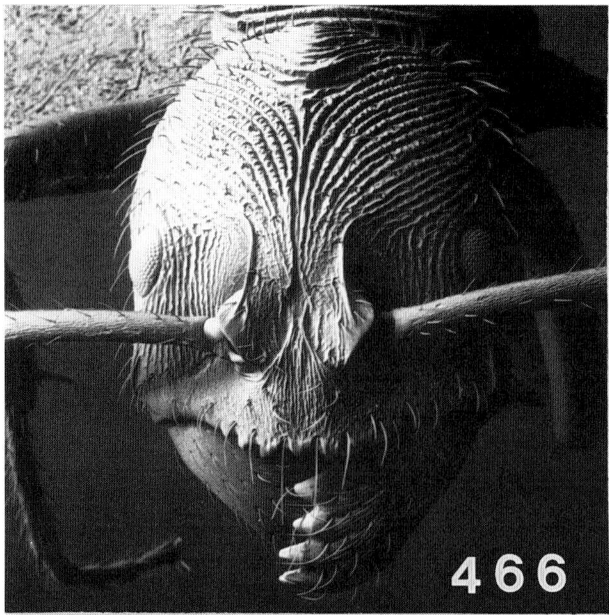
466

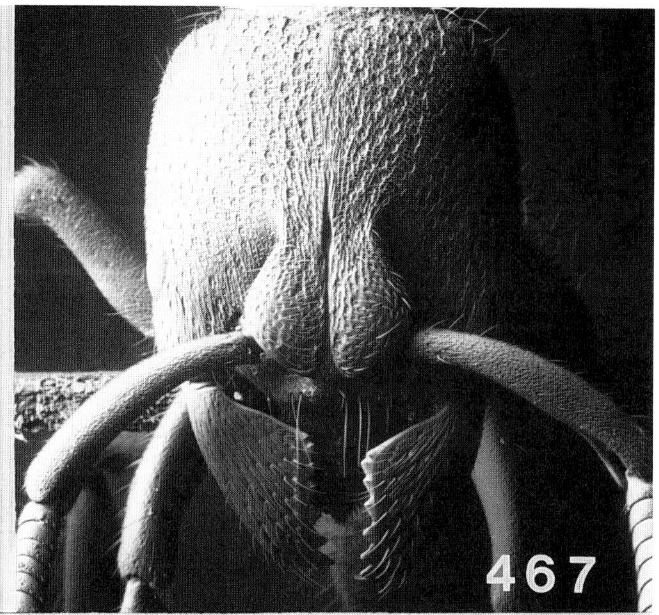
467

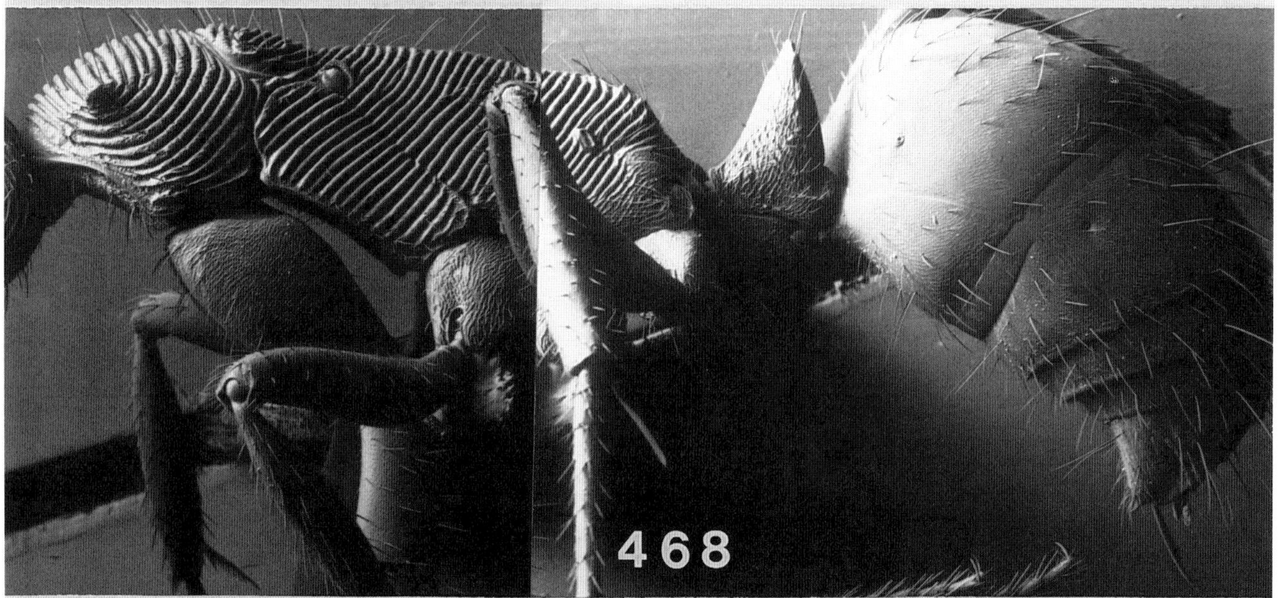
468

469

470

471

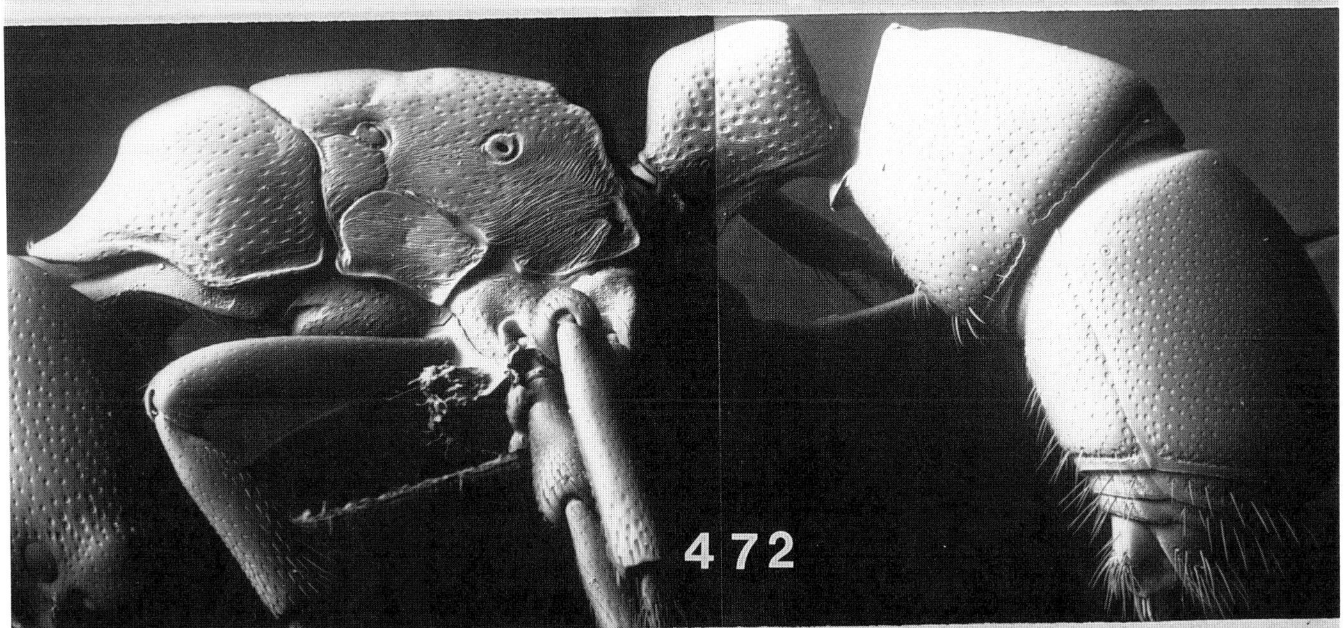
472

473

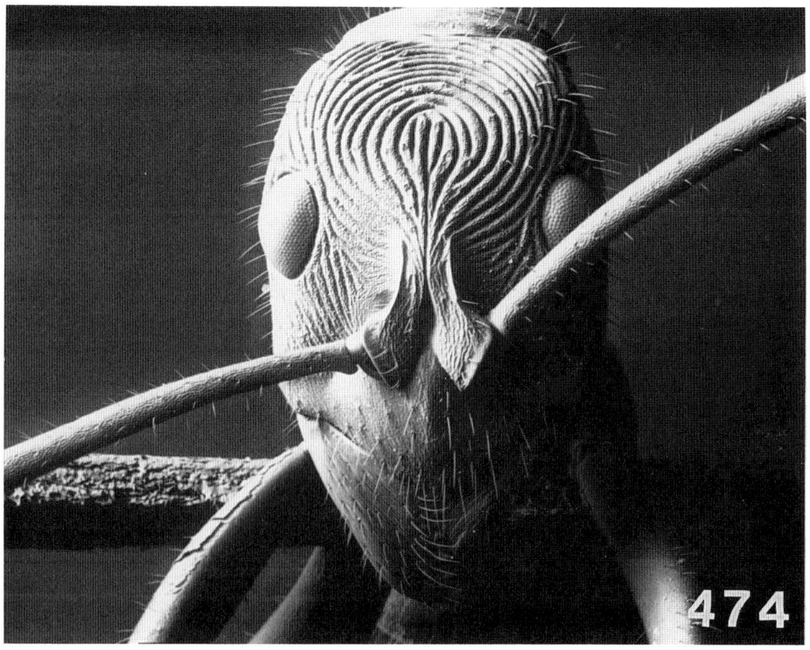
474

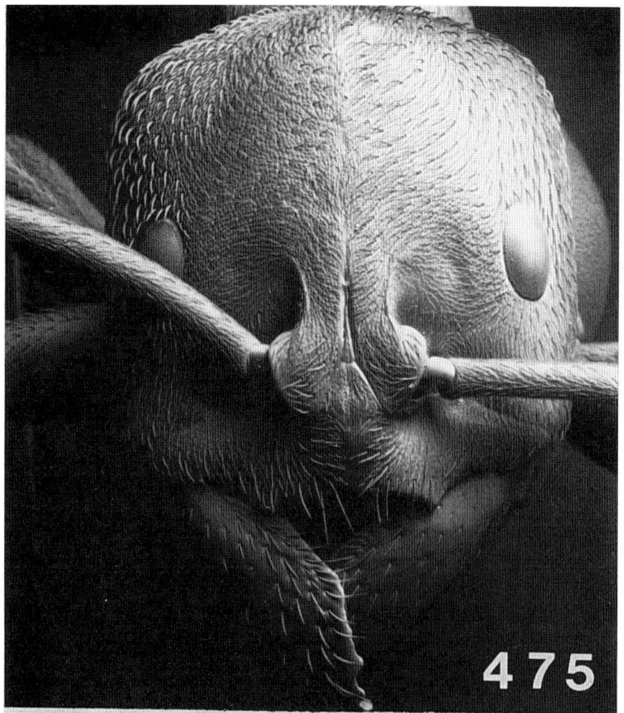
475

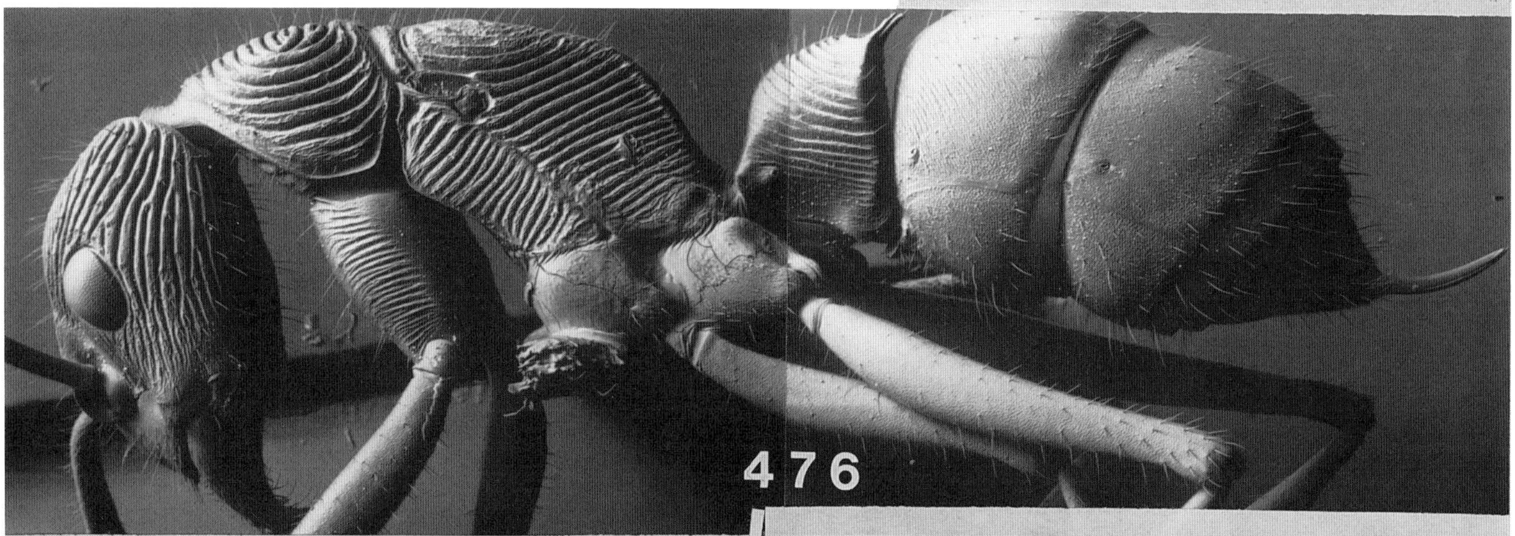
476

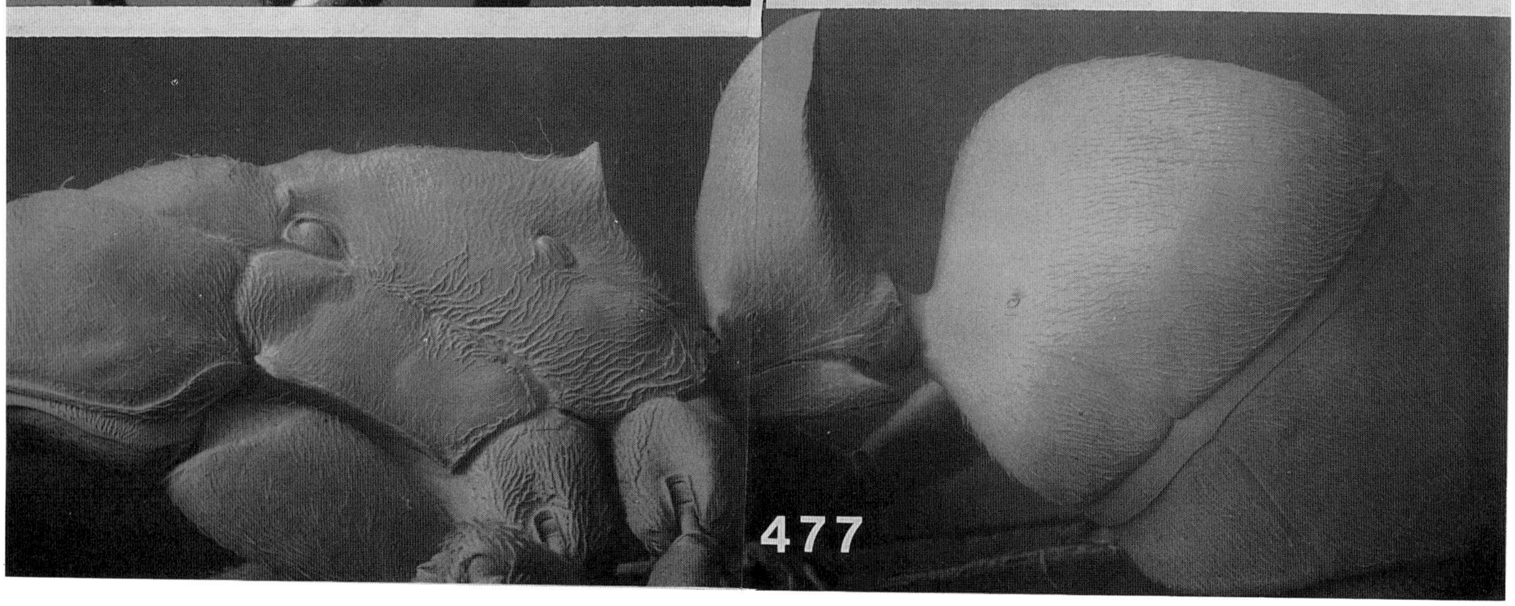
477

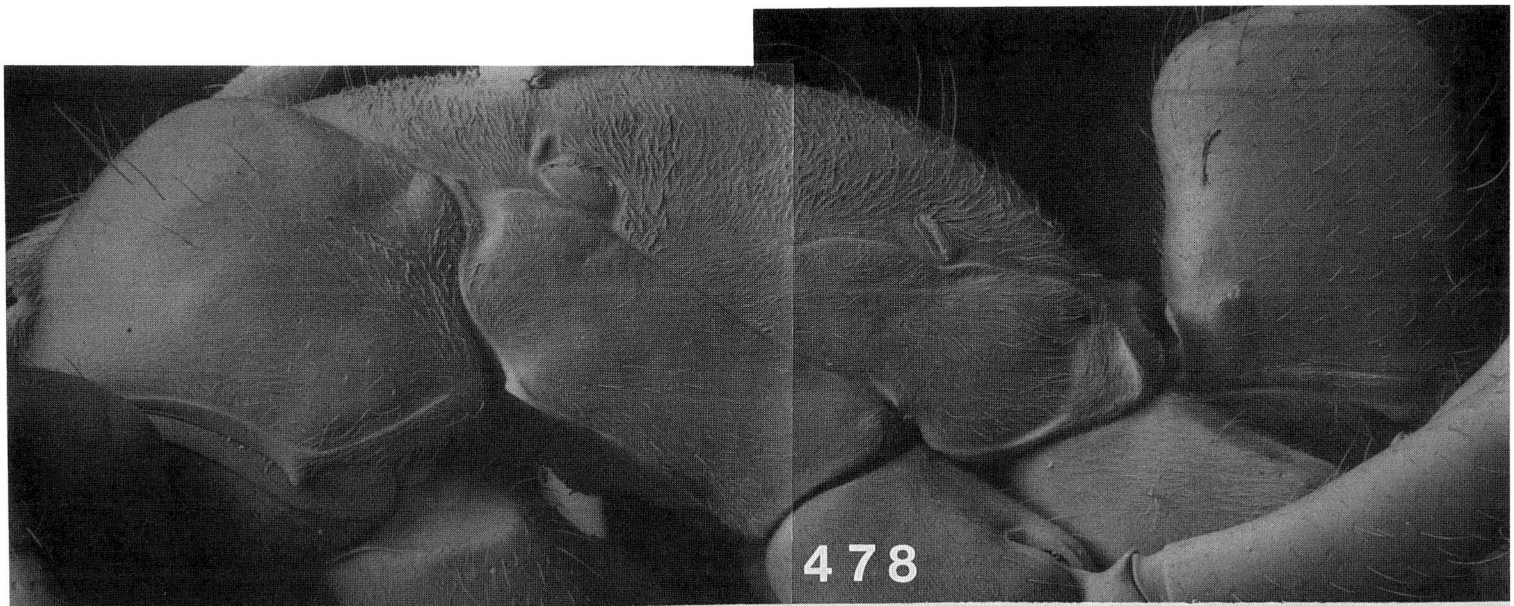

478

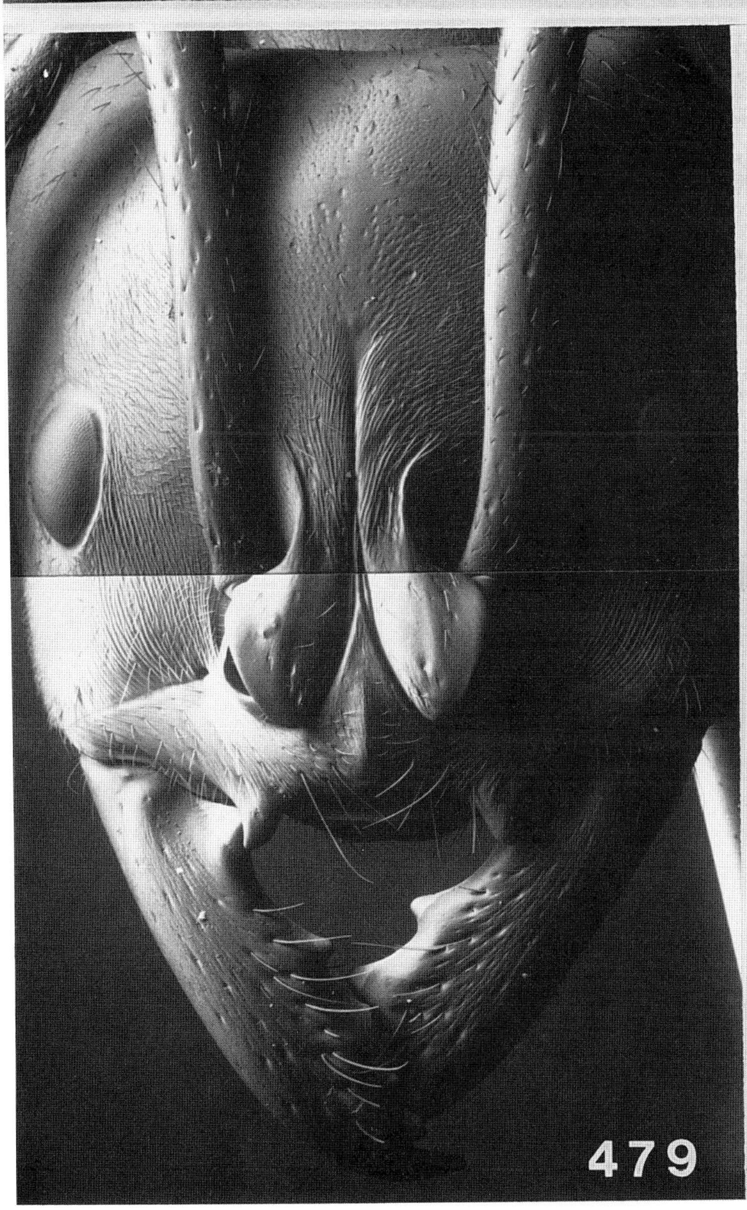

479

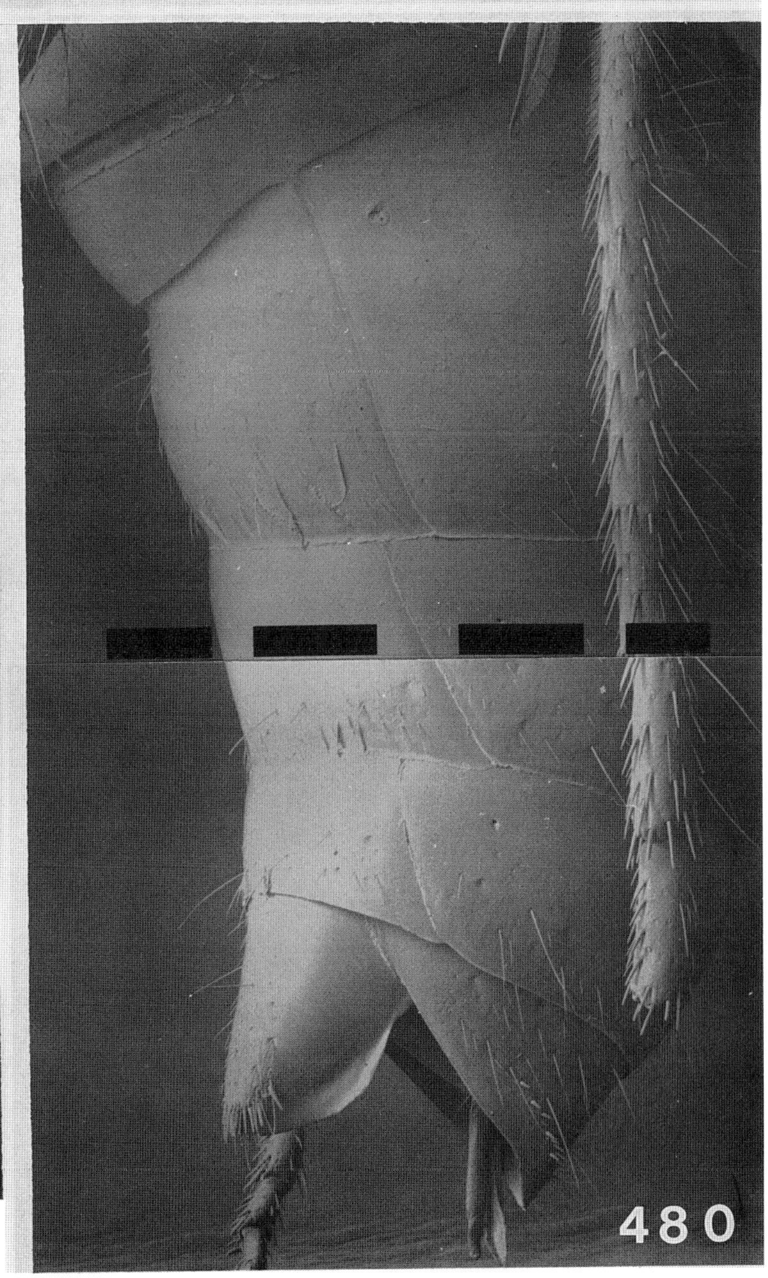

480

481

482

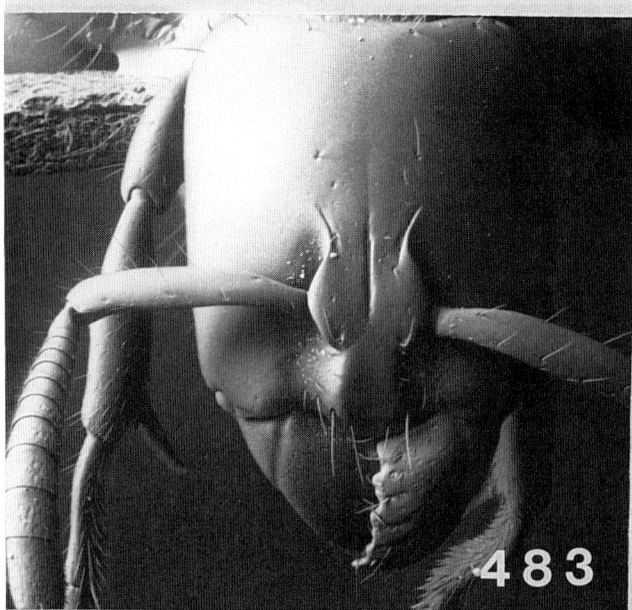

483

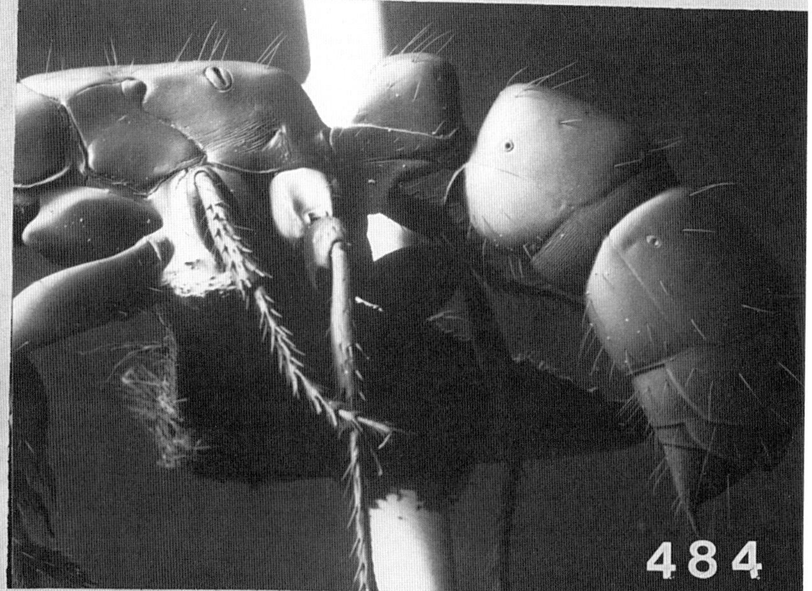

484

485

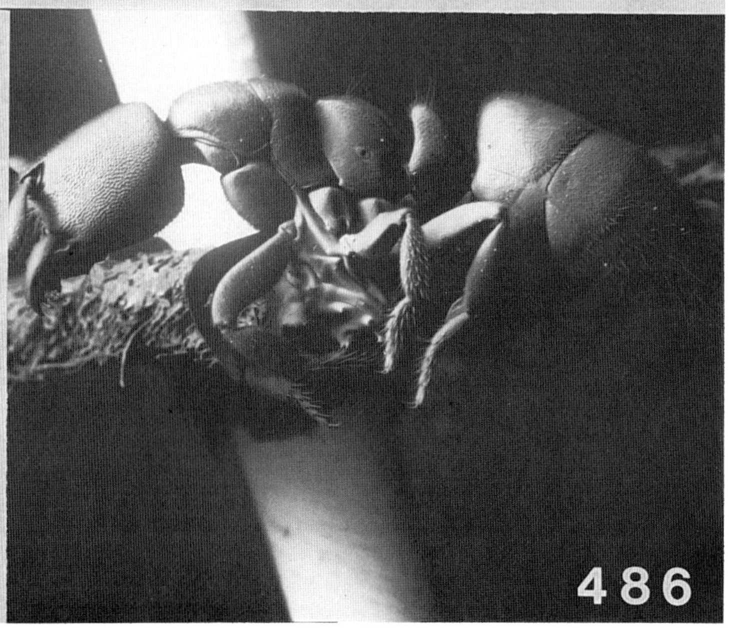

486

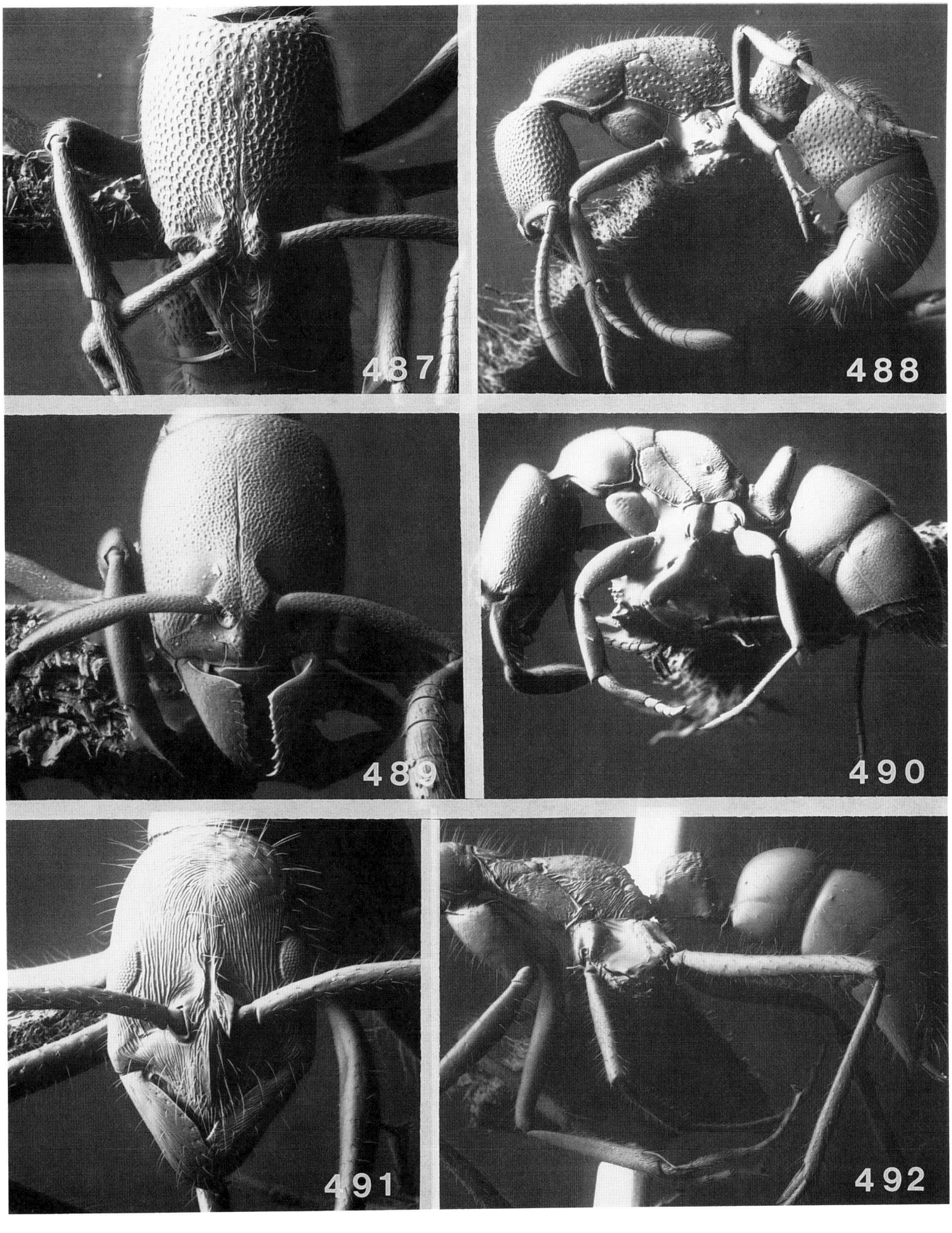
487
488
489
490
491
492

493

494

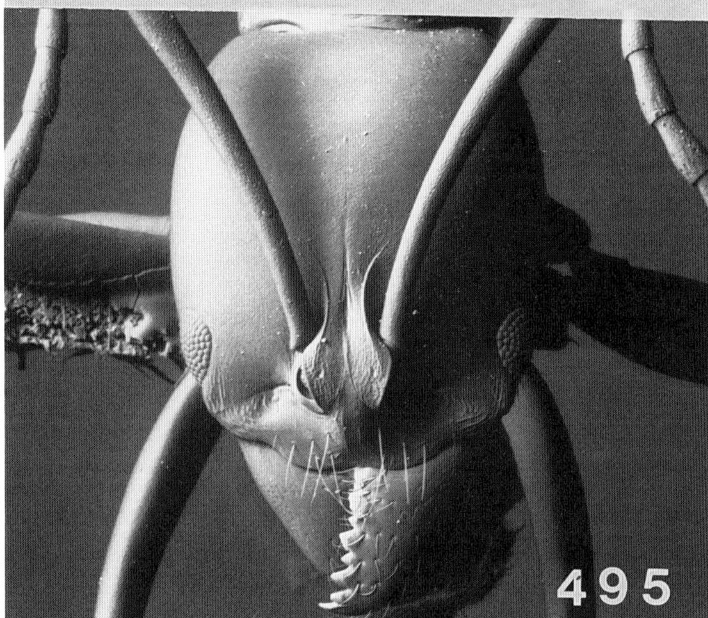
495

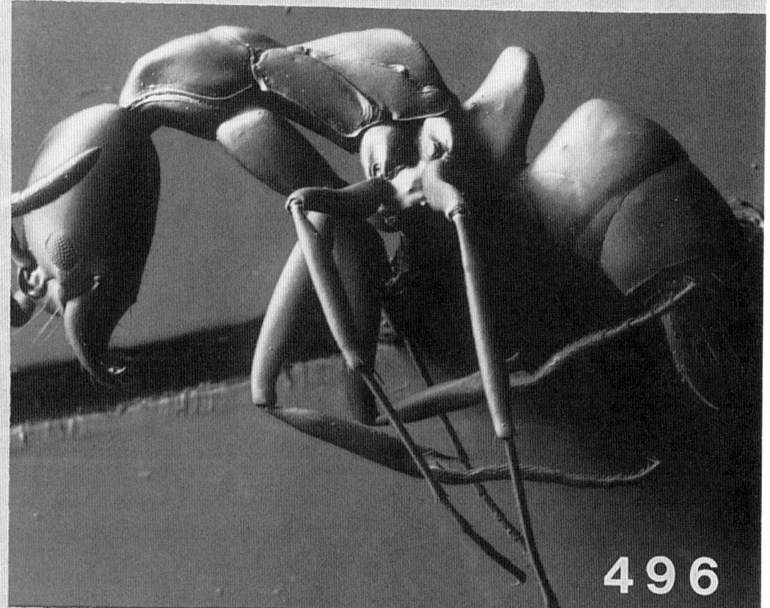
496

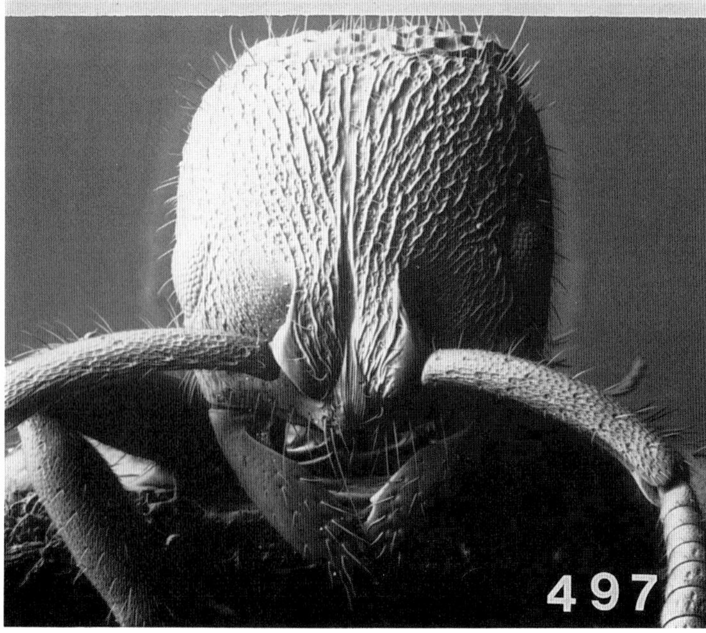
497

498

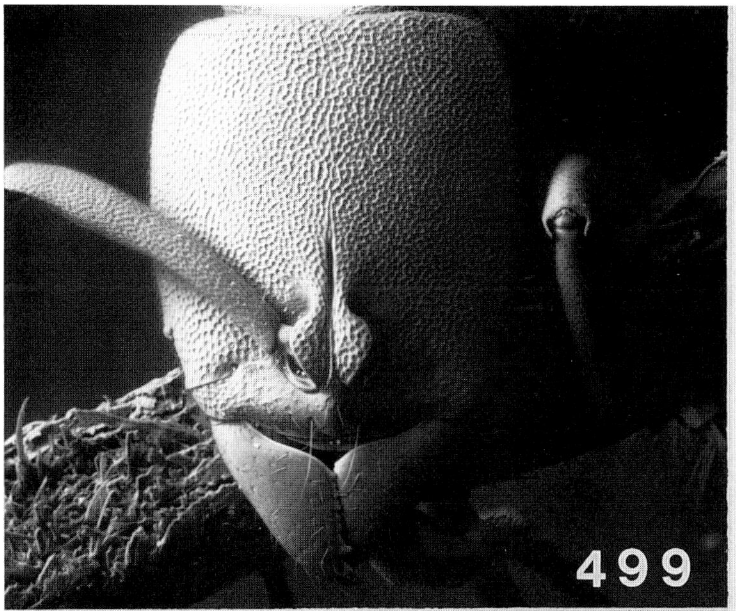

499

500

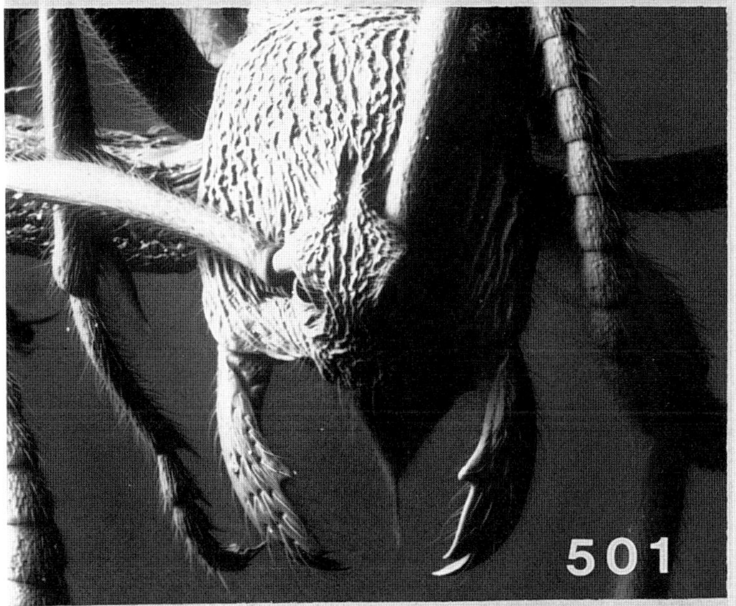

501

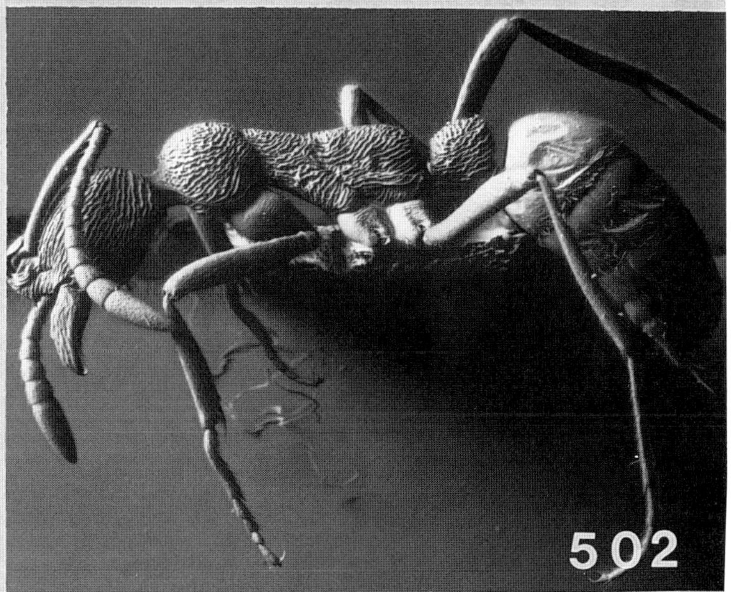

502

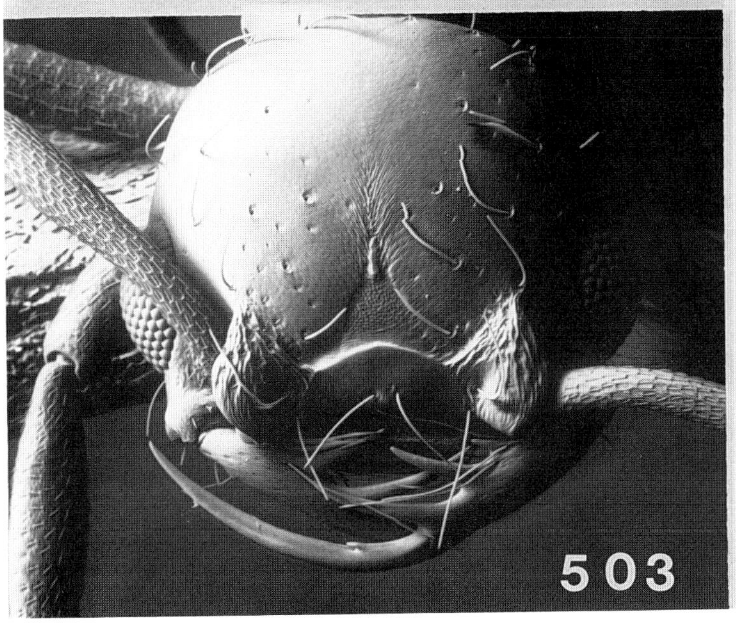

503

504

505

506

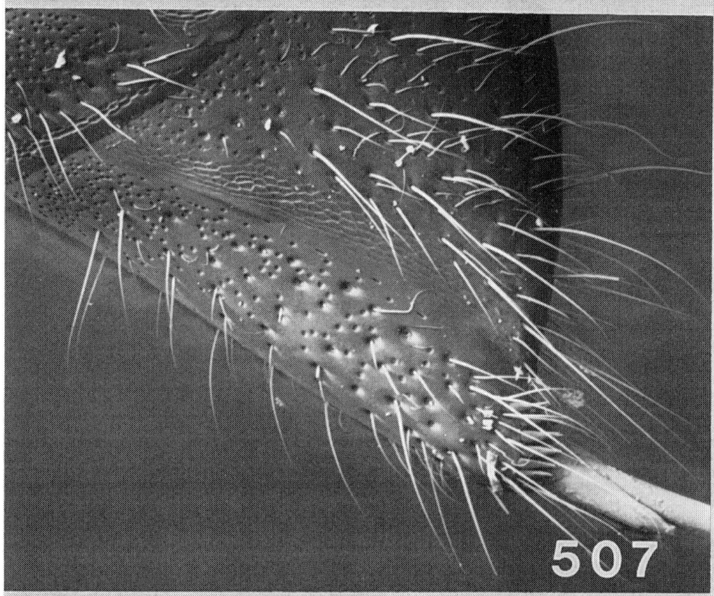
507

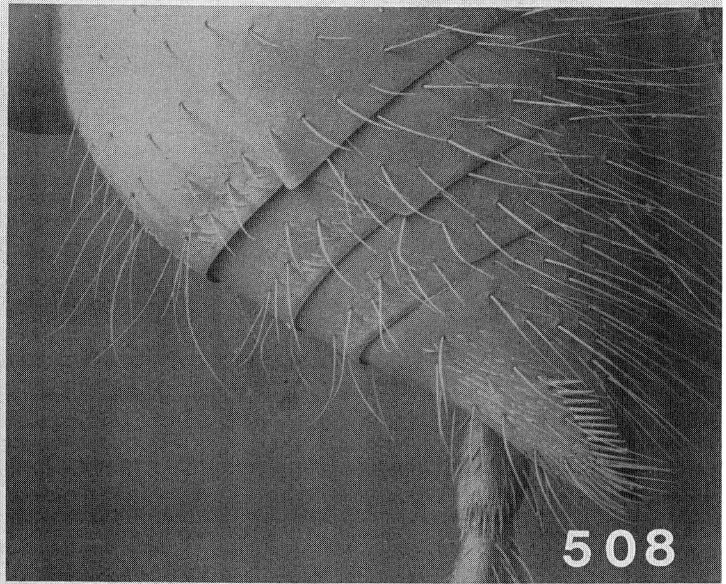
508

509

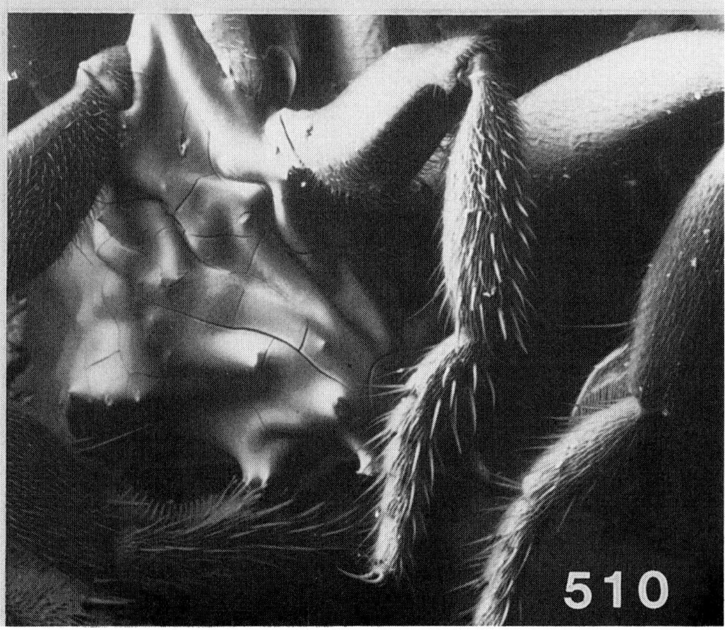
510

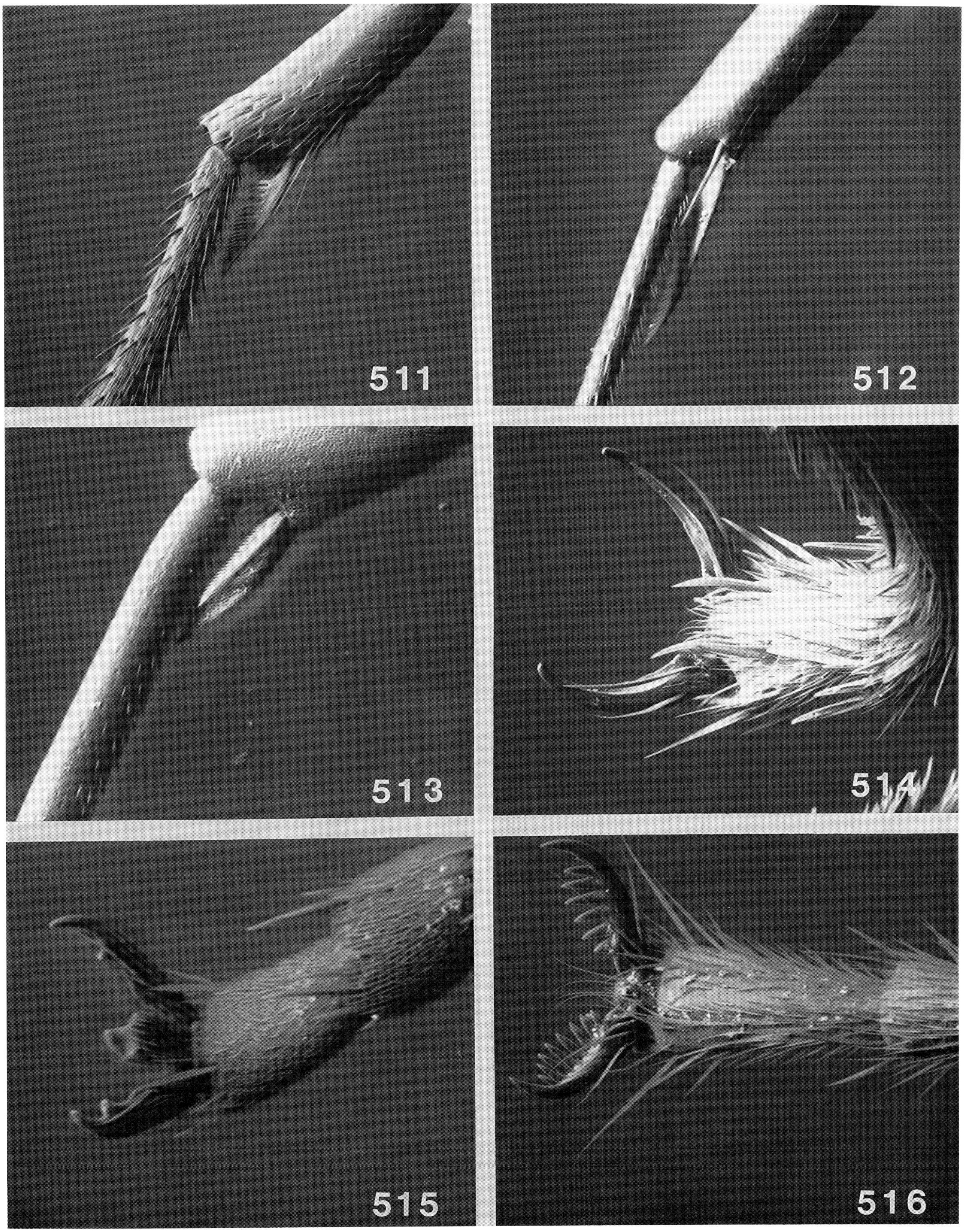
511
512
513
514
515
516

Subfamily PSEUDOMYRMECINAE

Diagnosis of Worker (Figs. 517–522)

Ants with the following combination of characters together.

1 Median portion of clypeus not conspicuously extended backwards between the frontal carinae, its posteromedian margin more or less straight.
2 Antennal sockets well behind anterior margin of head, inclined, the portion of the socket margin and torulus closest to the dorsal midline of the head on a higher level than the portion of the margin most distant from the midline.
3 Frontal carinae not expanded into frontal lobes over the antennal sockets; torular sclerites fully exposed, fused to the frontal carinae on each side, and to some extent covering the condylar bulbs of the scapes.
4 Narrow neck joining condylar bulb of antennal scape to shaft of scape proper straight.
5 Eyes present, relatively large; antenna with 11 or 12 segments.
6 Promesonotal suture always present and flexible, the pronotum capable of movement with respect to the mesonotum.
7 Metapleural gland orifice in lower posterior corner of metapleuron, opening laterally or ventrolaterally, and not concealed by a cuticular flange or flap.
8 Mesonotum usually defined and metanotum present on dorsal alitrunk.
9 Propodeal spiracle situated high on the side and far forward.
10 Metacoxal cavities closed; cuticular annulus around each cavity broad and complete, not interrupted by a suture or gap linking the coxal cavity to the cavity in which the petiole articulates.
11 Propodeal lobes present.
12 Waist of 2 segments, the petiole and postpetiole (= abdominal segments 2 and 3).
13 Helcium sternite small, retracted, concealed by the tergite; not visible without dissection.
14 Abdominal stridulatory system present, the stridulitrum situated on the pretergite of abdominal segment 4 (= gastral segment 1), the plectrum posterior on the preceding tergite.
15 Abdominal segment 4 (= gastral segment 1) with sharply defined and differentiated short, narrow prescleritess, which fit tightly within the posterior end of segment 3.
16 Abdominal spiracles 5–7 (= gastral 2–4) concealed by posterior margins of preceding tergites, not visible without distension of the abdomen.
17 Abdominal segments 2–7 (= petiole to gastral segment 4) without tergosternal fusion.
18 Pygidium (tergite of abdominal segment 7, the last visible gastral tergite) large, simple.
19 Sting present, usually large and strongly developed.

Key to World PSEUDOMYRMECINAE (Workers)

1 Antenna with 11 segments. (Neotropical) .... ***Myrcidris***
— Antenna with 12 segments **2**

2 Basal margin of mandible with a tooth close to the articulation. Basitarsus of hind leg lacking a longitudinal sulcus basally on its anterior surface. Eyes elongate, the width two-thirds or less than the length. (Southern Nearctic, Neotropical) .... ***Pseudomyrmex***
— Basal margin of mandible unarmed, without a tooth close to the articulation. Basitarsus of hind leg usually with a distinct longitudinal sulcus basally on its anterior surface. Eyes broader, the width two-thirds or more than the length. (Old World tropics) .... ***Tetraponera***

Synoptic Classification

Subfamily **PSEUDOMYRMECINAE** (= Leptaleinae, = Pseudomyrmidae).

Tribe **Pseudomyrmecini.** Genera: ***Myrcidris*** (Figs. 517, 518), ***Pseudomyrmex*** (Figs. 519, 520) (= *Apedunculata*, = *Clavanoda*, = *Latinoda*, = *Leptalaea* (misspelling), = *Leptalea*, = *Myrmex* (homonym), = *Ornatinoda*, = *Pseudomyrma*, = *Triangulinoda*), ***Tetraponera*** (Figs. 521, 522) (= *Pachysima*, = *Parasima*, = *Sima*, = *Viticicola*).

Distribution

Of the 3 genera which make up this subfamily, *Myrcidris* and *Pseudomyrmex* are entirely New World in distribution, while *Tetraponera* is very widely distributed in the Old World. The first genus is known only from Brazil, but the second occurs in both the Neotropical and Nearctic regions, with an overwhelming majority of species in the former. *Tetraponera* is present in all zoogeographical regions of the Old World but numbers of species are relatively low in the Palaearctic and Australasian regions. In contrast to many other widely distributed subfamilies the Malagasy region has a rich endemic fauna of pseudomyrmecines.

Taxonomic References

Identification of extant species

Myrcidris: Ward (1990). ***Pseudomyrmex:*** Kempf (1958b) [*gracilis*-group, Neotropical]; Kempf (1960a) [*tenuis*-group, Neotropical]; Kempf (1961a) [*tenuis-*, *oculatus-*, *pallens-*, and *latinodus*-groups, Neotropical]; Ward (1989) [*oculatus-* and *subtilissimus*-groups, Neotropical]; Ward (1985) [Nearctic]. ***Tetraponera:*** Wu and Wang (1990) [China].

Other taxonomic references

Pseudomyrmecinae: Snelling (1981); Ward (1990); Hölldobler and Wilson (1990); Baroni Urbani, Bolton, and Ward (1992).

Figures 517–522 PSEUDOMYRMECINAE workers, heads in full-face view and bodies in profile: 517–518, *Myrcidris;* 519–520, *Pseudomyrmex;* 521–522, *Tetraponera.*

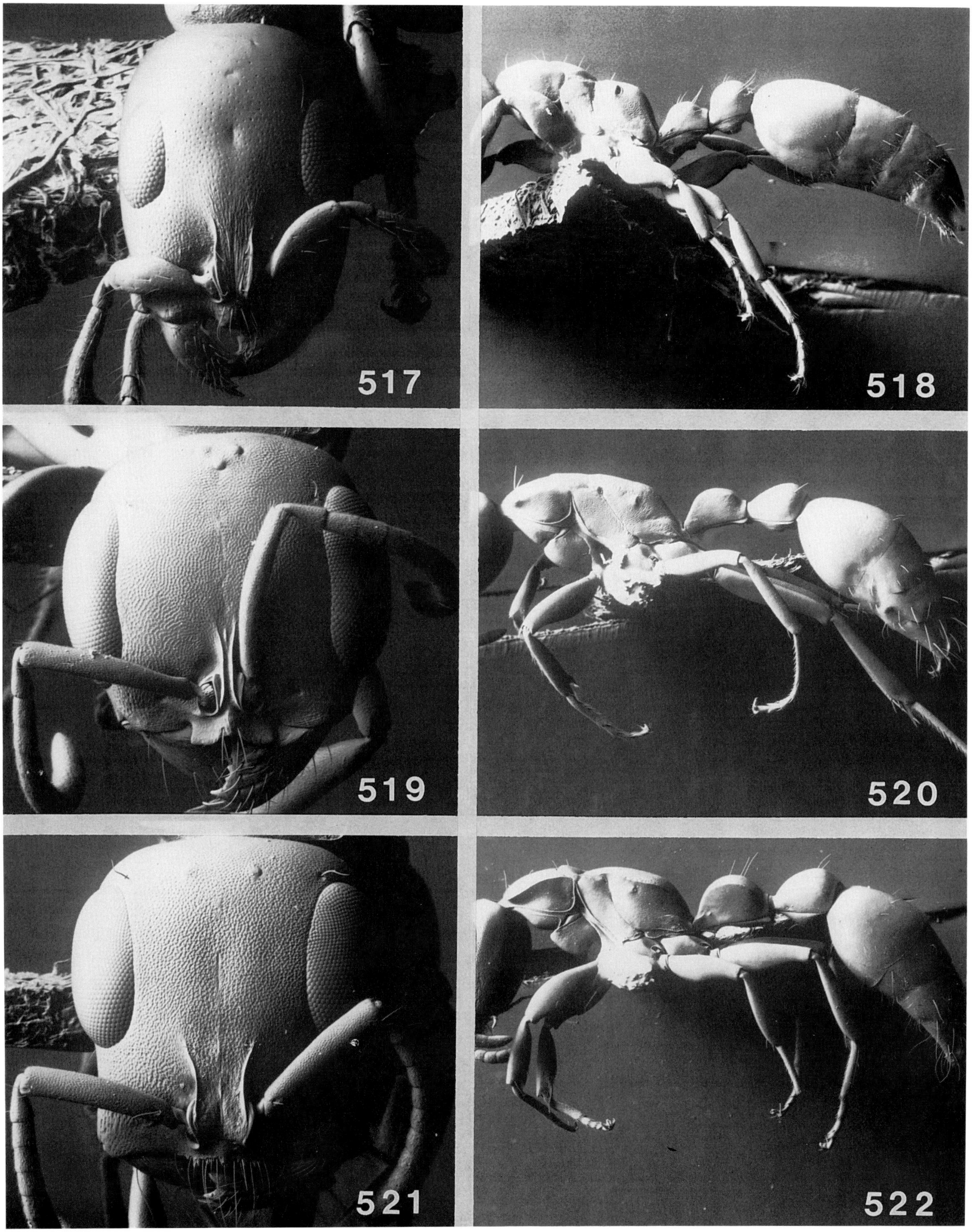
517
518
519
520
521
522

The Extinct Subfamilies

Four subfamilies of ants are known only from fossil representatives. As no extant members exist, the subfamilies are not formally diagnosed here, but to complete the picture of the current constitution of family Formicidae their component genera are listed below. As elsewhere in this survey, names of extinct taxa are prefixed by *. Apart from these there are the names of two fossil genera that have been referred to Formicidae but are almost certainly not ants: ***Calyptites*** and ***Cretacoformica.*** The true taxonomic position of these remains unknown.

Carpenter (1992) provides a list and short diagnoses of ant genera that either contain fossil species or are entirely composed of fossil forms. Unfortunately this list omits a number of genera so care should be taken when consulting it.

Subfamily *ARMANIINAE

Subfamily ***ARMANIINAE.***
Tribe ***Armaniini**. Genera: ***Archaeopone, *Armania, *Armaniella, *Cretopone, *Dolichomyrma, *Petropone, *Poneropterus, *Pseudarmania.***
The arrangement of genera in *Armaniinae and *Sphecomyrminae, below, follows Dlussky and Fedoseeva (1988). The genera in these 2 subfamilies may be oversplit, and Wilson (1987) argues for the synonymy of many genus-rank names. All genera of *Armaniinae are from the Upper Cretaceous.

Subfamily *FORMICIINAE

Subfamily ***FORMICIINAE.***
Tribe ***Formiciini.*** Genus: ***Formicium*** (= **Eoponera,* = **Megapterites*).
The taxonomy of this Eocene subfamily follows Lutz (1986, 1990).

Subfamily *SPHECOMYRMINAE

Subfamily ***SPHECOMYRMINAE.***
Tribe ***Sphecomyrmini.*** Genera: ***Baikuris, *Cretomyrma, *Dlusskyidris*** [New name for **Palaeomyrmex* Dlussky (1975: 118), junior homonym of **Palaeomyrmex* Heer (1865: 91)], ***Sphecomyrma.***
See note under *Armaniinae. All genera of *Sphecomyrminae are from the Cretaceous.

SUBFAMILY *PALEOSMINTHURINAE

Subfamily ***PALEOSMINTHURINAE.***
Tribe ***Paleosminthurini.*** Genus: ***Paleosminthurus.***
The single Miocene genus included here was described by Pierce and Gibron (1962) as a new fossil family (*Paleosminthuridae) in the order Collembola. Najt (1987) recognised that it was an ant, but placed it as Formicidae *incertae sedis.* Until the situation is properly reviewed, and considering that a family-group name already exists, it seems reasonable to treat the group as an extinct subfamily of Formicidae.

References to Faunistic Studies

Glossary of Morphological Terms

Bibliography

Index and Checklist

References to Faunistic Studies

The references given at the end of each subfamily deal with the species of stated genus-rank taxa. Those presented below include faunal works whose identification keys cover the species of all genera that occur in a stated country or zone. References to faunal papers covering extremely small areas are omitted, and the references are arranged by country or zone in alphabetical order.

Country or zone	References
Balkan States	Agosti and Collingwood (1987)
Belgium	Boven (1977)
Bulgaria	Atanassov and Dlussky (1992)
Canada	Francoeur (1977, 1979) [Quebec]
Chile	Snelling and Hunt (1976)
Corsica	Casevitz-Weulersse (1990a,b)
Cuba	Alayo (1974)
Fennoscandia and Denmark	Collingwood (1979)
Fiji Islands	Mann (1921) [out of date]
France	Bernard (1967)
Germany	Gösswald (1985)
Great Britain	Bolton and Collingwood (1975)
Iberian Peninsula	Collingwood (1978)
India	Bingham (1903) [out of date]
Italy	Baroni Urbani (1964a,b, 1969a, 1971)
Japan	Morisita et al. (1989, 1991, 1992); Onoyama (1980)
Kirgizia	Tarbinsky (1976)
Malta	Baroni Urbani (1968); Schembri and Collingwood (1981)
Melanesia	Mann (1919, 1921) [out of date]; Wilson (1958a,b, 1959a,b)
New Zealand	Brown (1958c)
North America	Creighton (1950); Gregg (1963) [Colorado]; G. C. Wheeler and J. Wheeler (1986) [Nevada]
Palaearctic region	Emery (1908a–e, 1909a–d, 1910a) [out of date]
Polynesia	Wilson and Taylor (1967)
Portugal	*See* Iberian Peninsula
Saudi Arabia	Collingwood (1985)
Solomon Islands	Mann (1919) [out of date]
South Africa	Arnold (1915, 1916, 1917, 1920, 1922, 1924, 1926) [out of date]
Spain	*See* Iberian Peninsula
Sweden	Nilsson and Douwes (1987)
Switzerland	Kutter (1977)
Turkmenistan	Dlussky, Soyunov, and Zabelin (1990)
U.S.S.R. (former)	Arnol'di and Dlussky (1978) [former European U.S.S.R.]; Kupyanskaya (1990) [far eastern Russia]
West Palaearctic region	Bernard (1967)
Zaire	W. M. Wheeler (1922) [partial, out of date]

Glossary of Morphological Terms

The glossary, illustrated by Figures 523–531, deals with morphological terms encountered in this survey. Some terms are merely defined but others are discussed in detail for the benefit of first-time users and others not well acquainted with ant morphology. The list is by no means exhaustive, as many specialized terms encountered in males and queens are not covered, nor are the extensive vocabularies applied to forms of sculpture, pilosity, internal anatomy, and morphometrics utilized below genus level. Terms are listed in alphabetical order, and acceptable alternatives are given in parentheses; for instance, alitrunk (= mesosoma). Abbreviations used in the figures are given here in *italic*. For a more general overview of hymenopterous morphology see Gauld and Bolton (1988).

Abdomen The classical third tagma of the insect body. The abdomen in worker ants consists of seven visible segments (Fig. 530; *A1–7*) and each bears a spiracle (Fig. 530; sp), which may be exposed or concealed. The first abdominal segment is the *propodeum* (Fig. 529, 530; *ppd* = *A1*), represented only by its tergite (the sternite has been lost) and immovably fused to the thorax. The tagma formed by the fusion of thorax plus propodeum is termed the *alitrunk* (= mesosoma) (Figs. 528, 529; *al* = *mes*). The second abdominal segment, the *petiole* (Figs. 528, 530; *pt, A2*), is always specialized. It is usually reduced in size, always separated from the preceding propodeum by a complex narrow articulation, and usually separated from the following abdominal segment by at least a constriction. In the vast majority of ants the petiole is distinctly isolated both anteriorly and posteriorly. Abdominal segments 2 to the apex are sometimes collectively called the *metasoma* (Fig. 530; *mt*), but this term, useful elsewhere in the Hymenoptera, has little to recommend it in the ants. Abdominal segment 3 is termed the first gastral segment when it is full-sized and broadly articulated to the following segment (Fig. 530; *G1* = *A3*), but when reduced and isolated it is called the *postpetiole* (Fig. 528; *ppt, A3*). Confusingly, it is sometimes also called the postpetiole when full-sized. Abdominal segment 3 articulates with the preceding petiole by means of the *helcium* (Figs. 528, 530, 531; *he*). The petiole alone, or the petiole plus postpetiole together, when the latter is also reduced and separated, may be termed the *waist* (Figs. 528, 530; *w*). [An older term, *pedicel,* should be abandoned, as it is used for a different body part elsewhere throughout the Hymenoptera.] Abdominal segment 4 is the first gastral segment when the waist consists of petiole plus postpetiole (Fig. 528; *G1* = *A4*), but it is the second gastral segment when the waist consists of petiole alone (Fig. 530; *G2* = *A4*). Abdominal segments 3 or 4 through to 7 are collectively called the *gaster* (Figs. 528, 530; *ga*), the enlarged apparent "abdomen" that comprises the terminal part of the body. In referring to parts of the gaster the term *gastral* is preferred, as this leaves the form *gastric* free for use in connection with the intestine.

Each abdominal segment behind the first consists of a pair of sclerites, a dorsal *tergite* (Figs. 528, 530; *tr*) and a ventral *sternite* (Figs. 528, 530; *st*). These may all be similar, or some may be specialized by fusion, reduction, or division into anterior and posterior portions (see **Presclerite**). Tergites and sternites may be referred to as abdominal or gastral (e.g., abdominal tergite 4 = second gastral tergite when the waist is of a single segment). In workers the last visible abdominal tergite, that of segment 7, is the *pygidium,* and the last visible sternite is the *hypopygium* (Fig. 530; *py, hy*).

The terminology of the ant abdomen may at first seem confusing. This is because two different systems are superimposed and in places are not strictly compatible:

(1) A terminology based strictly on morphology, which numbers the abdominal segments as 1 to 7, from front to back (Fig. 530). This has the advantage of indicating homologous segments between different ant taxa, regardless of the specializations of individual segments or groups of segments.

(2) A more utilitarian terminology based on observed subdivisions of the abdominal segments, which names various specialized segments and groups of segments (Figs. 528 and 529). The advantage here is that the subdivisions and specializations are generally easily visible.

See also **Alitrunk, Helcium, Petiole, Postpetiole, Propodeum, Sternite, Tergite, Waist.**

Acidopore The orifice of the formic acid projecting system peculiar to, and diagnostic of, the ant subfamily Formicinae. It is formed from the apex of the hypopygium and is usually plainly visible, appearing as a short nozzle, generally with a fringe of short setae at its apex (Fig. 160). In most formicines the acidopore is always exposed, but in some it may be concealed by the posterior margin of the pygidium when not in use. In such groups the acidopore usually lacks a nozzle and takes the form of a semicircular to circular emargination of the apical margin of the hypopygium.

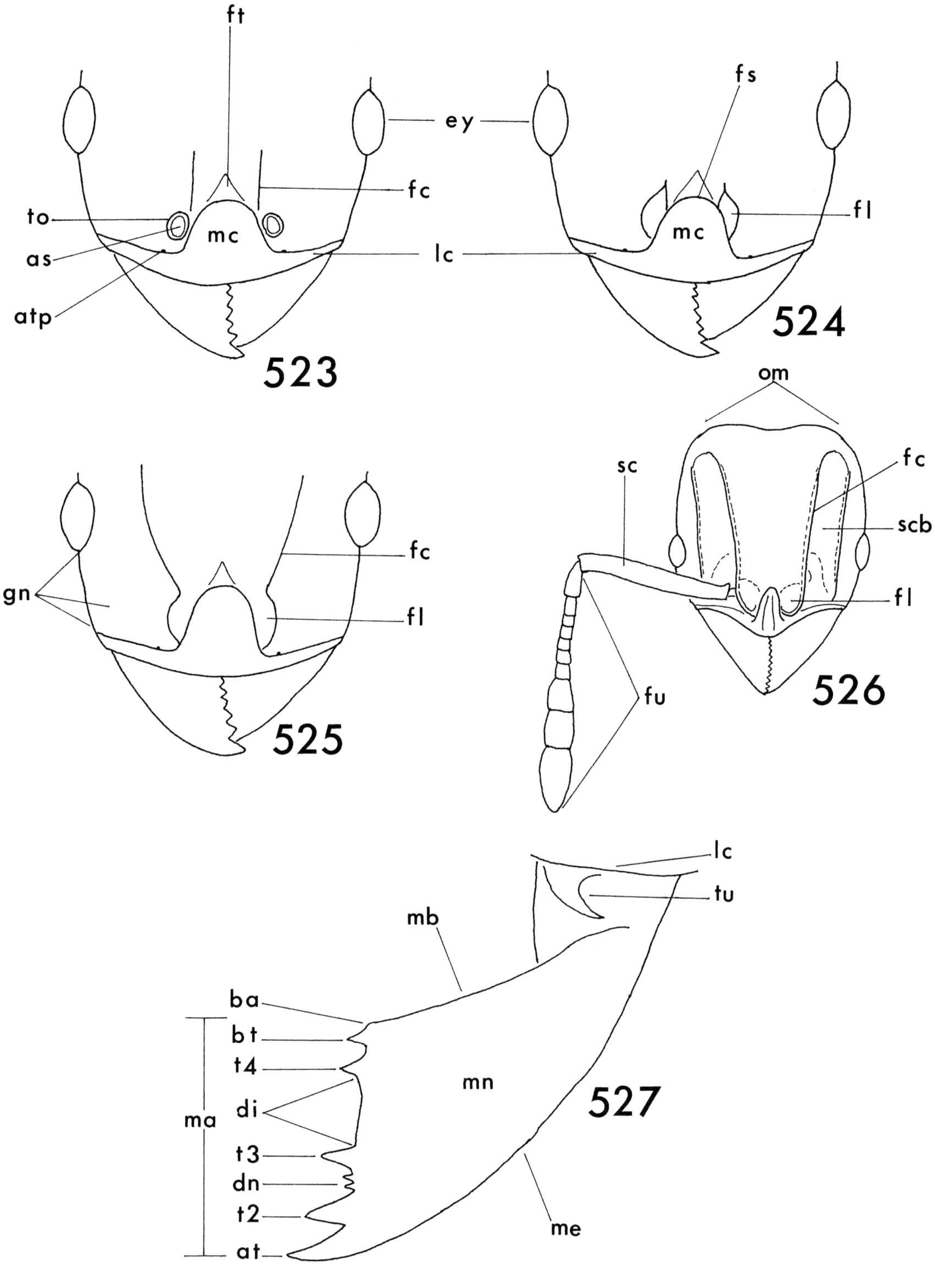

as	antennal socket
at	apical tooth of mandible
atp	anterior tentorial pit
ba	basal angle of mandible
bt	basal tooth of mandible
di	diastema
dn	denticle
ey	eye
fc	frontal carina
fl	frontal lobe
fs	fronto-clypeal suture (= posterior clypeal margin)
ft	frontal triangle
fu	funiculus of antenna
gn	gena
lc	lateral portion of clypeus
ma	apical (masticatory) margin of mandible
mb	basal margin of mandible
mc	median portion of clypeus
me	external margin of mandible
mn	mandible
om	occipital margin of head
sc	scape of antenna
scb	antennal scrobe
t	tooth number
to	torulus
tu	trulleum

Figures 523–527 Morphological features of ants. Drawings are composite, not based on any particular species; sculpture and pilosity omitted. Figs. 523–525, anterior halves of head in full-face view, antennae omitted. Fig. 526, whole head in full-face view, left antenna omitted. Fig. 527, fully opened triangular left mandible.

Aliform Shaped like a wing, approximately wing-like.

Alitrunk (= mesosoma) The second visible tagma of an ant's body, following the head. Morphologically the alitrunk consists of the three segments of the true thorax (pro-, meso-, and metathorax) to which is fused the propodeum, the tergite of the first abdominal segment, to form a single unit (Figs. 528, 529; *al* = *mes*). See also **Abdomen, Propodeum, Thorax.**

Anepisternum See **Pleurite.**

Annulus (pl. annuli) A simple, generally nonsegmental, ring of cuticle.

Antenna (pl. antennae) The antenna in ants consists of an elongate basal segment, the *scape,* followed distally by 3–11 smaller segments which together constitute the *funiculus* (= flagellum) (Fig. 526; *sc, fu*), giving a total antennal segment count (= antennomere count) of 4 to 12. The scape articulates with the head in the *antennal socket* (= antennal insertion) (Figs. 523, 528; *as*), a foramen located behind the clypeus. The antennal socket itself is encircled by a narrow annular sclerite, the *torulus* (Fig. 523; *to*), and may be overhung and concealed by the *frontal lobe* (Figs. 524, 525; *fl*). At the base of the scape is a ball-like *condylar bulb* (= articulatory bulb), the part which actually articulates within the socket. Just distal of the condylar bulb is a short constriction or neck, which may be straight or curved, beyond which the scape shaft proper commences. The funicular segments may be filiform or the apical 1–4 may be enlarged to form a *club.* See also **Frontal carinae, Torulus.**

Antennal scrobe A groove, impression, or excavation in the side of the head, which runs above or below the eye, to accommodate at least the antennal scape, but often the entire antenna, when the latter is folded back. Antennal scrobes vary in development from simple broad shallow grooves to extensive deep trenches (Figs. 526, 528; *scb*). Antennal scrobes are absent from most ant genera.

Antennal socket/insertion See **Antenna.**

Antennomere See **Antenna.**

Anterior tentorial pits A pair of pits or impressions located anteriorly on the dorsal surface of the head, at or very close to the posterior clypeal margin (Fig. 523; *atp*). The pits indicate the points of attachment of the anterior arms of the internal skeleton (tentorium) of the head to the head capsule. The termination of the posterior arms of the tentorium are marked by a pair of *posterior tentorial pits,* which are located close to the occipital foramen.

Apical margin/tooth (of mandible) See **Mandibles.**

Apophyseal lines Externally visible lines marking the internal track of cuticular processes for muscle attachment.

Basal angle/lamella/margin/tooth (of mandible) See **Mandibles.**

Basitarsal sulcus A longitudinal groove in the surface of the first (basal) tarsal segment of the leg.

Basitarsus (pl. basitarsi) The first, basal, of the five tarsal segments of the leg; the tarsal segment that articulates with the tibia.

Buccal cavity The anteroventral cavity of the head which contains the labium and maxillae.

Bulla (pl. bullae) See **Pleurite.**

Calyx See **Proventriculus.**

Carina (pl. carinae) A ridge or low, keel-like crest.

Carinula (pl. carinulae) Diminutive form of carina.

Cephalic Pertaining to the head.

CI (Cephalic Index) See **Standard measurements.**

Clavate/claviform (antenna) With the apical 1–4 funicular segments enlarged and forming a club (Fig. 526).

Claw See **Pretarsal claw.**

Club (antennal) See **Antenna.**

Clypeus Anterior sclerite of the dorsal head, bounded posteriorly by the fronto-clypeal suture (Fig. 524; *fs*), which is also very commonly called the posterior clypeal margin or border. The anterior clypeal margin usually forms the anterior margin of the head in full-face view (but a projection of the labrum may be anterior to the clypeus in some taxa). The body of the clypeus consists of a pair of lateral portions, or narrow bands of cuticle, on each side of a shield-like median portion (Figs. 523, 524; *lc, mc*). The median portion of the clypeus may be equipped with one or more longitudinal carinae, or may be variously specialized in shape. Posteriorly the median portion of the clypeus may end in front of the antennal sockets/frontal carinae or lobes, or may project backwards between them. In some taxa the clypeus is very reduced and extremely narrow from front to back.

Condylar bulb See **Antenna.**

Coxa (pl. coxae) The first, most basal, segment of a leg; the leg segment that articulates with the thorax (Fig. 529; *c1–3*).

Declivity (of propodeum) See **Propodeum.**

Dentate/denticle/denticulate See **Mandibles.**

Diastema (pl. diastemata) See **Mandibles.**

Dimorphic Occurring in two morphologically distinct forms; in the sense of the keys presented here, ants with two morphologically differentiated castes of worker.

Edentate See **Mandibles.**

Elongate-triangular See **Mandibles.**

Emarginate Having a notch, impression, or indentation in a margin, border, or edge.

Endophragmal pit A pit in the lateral alitrunk wall which is an external indication of the position of attachment of part of the endoskeleton.

Epinotum An archaic name for the propodeum, used only by myrmecologists. Propodeum is the recommended term, because it is universally used elsewhere in hymenopterous morphology, and abandoning *epinotum* in favor of *propodeum* brings ant morphological nomenclature into line with the vast majority of the order.

External margin (of mandible) See **Mandibles.**

Falcate (mandible) See **Mandibles.**

Femur The third segment of any leg, counting from the basal coxal segment that articulates with the alitrunk. The femur is generally the

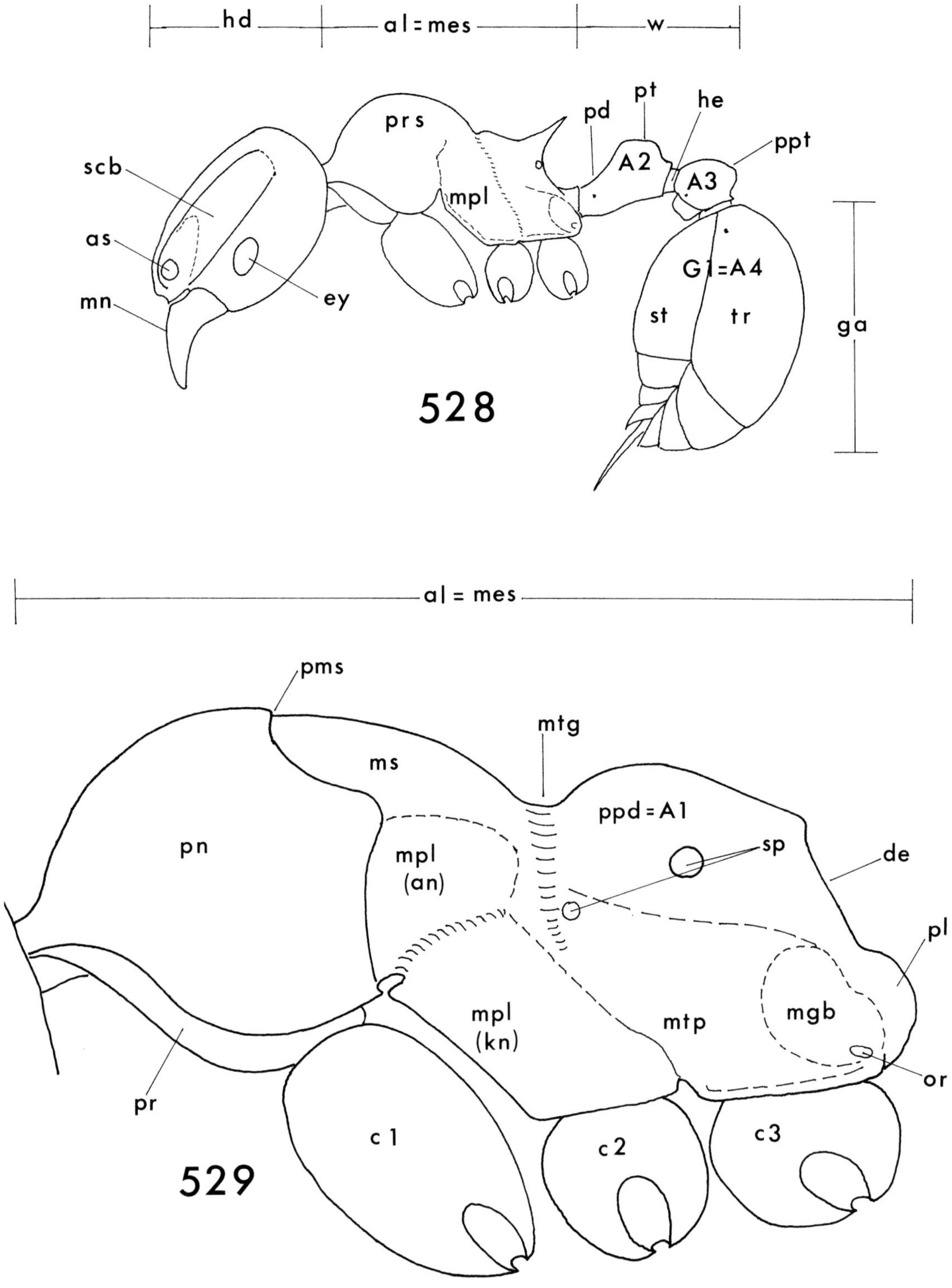

A	abdominal segment number
al	alitrunk
an	anepisternum
as	antennal socket
c	coxa number
de	declivity of propodeum
ey	eye
G	gastral segment number
ga	gaster
hd	head
he	helcium
kn	katepisternum
mes	mesosoma
mgb	metapleural gland bulla
mn	mandible
mpl	mesopleuron
ms	mesonotum
mtg	metanotal groove
mtp	metapleuron
or	orifice of metapleural gland
pd	peduncle of petiole
pl	propodeal lobe
pms	promesonotal suture
pn	pronotum
ppd	propodeum
ppt	postpetiole
pr	propleuron
prs	promesonotum
pt	petiole
scb	antennal scrobe
sp	spiracle
st	sternite
tr	tergite
w	waist

Figures 528–529 Morphological features of ants. Drawings are composite, not based on any particular species; sculpture and pilosity omitted. Fig. 528, whole ant in profile, legs below coxae omitted. Fig. 529, alitrunk (= mesosoma) in profile, legs below coxae omitted.

longest leg segment and is separated from the coxa only by a small segment, the *trochanter.*

Fenestra (pl. fenestrae) In the sense of the keys presented here, a translucent cuticular thin-spot.

Filiform (antenna) With the antennal funiculus thread-like, the segments all of approximately the same size. The contrasting antenual shape is club-like, with the apical segments of the antenna disproportionately enlarged.

Flagellum (= funiculus) See **Antenna.**

Foliaceous (outgrowths) Small to large roughly leaf-like cuticular projections.

Foramen (pl. foramena) An opening or perforation in a sclerite.

Fovea A depression or impressed pit.

Foveola Diminutive of the above; a small pit or depression.

Frontal carinae (sing. frontal carina) A pair of longitudinal ridges on the head, located dorsally behind the clypeus and between the antennal sockets. They are very variable in length and strength of development, frequently being short and simple (Fig. 523; *fc*) but sometimes extending back to the occipital margin of the head (Fig 525; *fc*). In some groups the frontal carinae are vestigial or absent, but elsewhere they may be strongly developed or form the dorsal margins of extensive antennal scrobes (Fig. 526; *fc, scb*). Commonly the frontal carinae anteriorly are expanded into projecting lobate extensions, the *frontal lobes* (Figs. 524–526; *fl*), which partially or entirely cover and conceal the antennal sockets. Frontal lobes may be the only expression of the frontal carinae in some groups. Sometimes the portion of the torulus closest to the cephalic midline is raised and expanded into a small, laterally projecting lobe, which may extend beyond the lateral margin of the frontal carina or lobe.

Frontal lobes See **Frontal carinae.**

Frontal triangle A small triangular patch of cuticle located mediodorsally on the head immediately behind the clypeus and approximately between the antennal sockets or anterior parts of the frontal carinae (Figs. 523–525; *ft*). Not apparent in many ant taxa.

Fronto-clypeal suture The suture forming the posterior margin or boundary of the clypeus (Fig. 524; *fs*); frequently referred to as the posterior clypeal margin.

Full-face view Orientation of the head in which the midpoint of the anterior clypeal margin, the midpoint of the occipital margin, and the midpoints of the sides are in focus at the same time (as Fig. 526).

Funiculus (= flagellum) See **Antenna.**

Gaster Morphologically, abdominal segments 3–7 when the waist is of a single segment (the petiole) (Fig. 530; *ga, G1–5*), or abdominal segments 4–7 when the waist is of two segments (petiole plus postpetiole) (Fig. 528; *ga*); functionally, the terminal, enlarged tagma of the body. See also **Abdomen, Waist.**

Gena (pl. genae) Area of front of head bounded in front by the posterior margin of the clypeus, behind by the anterior margin of the eye, and medially by the antennal socket (Fig. 525; *gn*). The gena thus includes part of the cephalic dorsum and the side of the head capsule between the eye and the clypeus.

Geniculate Bent like a knee-joint.

Girdling constriction A constriction or sudden and marked narrowing of an abdominal segment, which runs around the entire circumference of the segment (Fig. 530; *gc*). For convenience it is usually stated in keys that girdling constrictions are present between two segments. This is not strictly true as the constriction morphologically really represents the junction between the presclerites (see there) and postsclerites of the more posterior segment. The greater parts of these presclerites are usually inserted in the posterior end of the preceding segment and are invisible, leaving only the constriction visible externally (Fig. 531; *gc*).

Guard setae (= guard hairs) Specialized setae that traverse and protect the orifice of the metapleural gland (Fig. 529; *or*). See also **Pleurite.**

Gula Some authors have incorrectly used this term when referring to the ventral surface of the head capsule in ants. Morphologically, the gula is a separate medioventral sclerite of the head which is bounded anteriorly by the posterior tentorial pits. In ants the posterior tentorial pits are located near the occipital foramen and no gula is present.

Helcium The very reduced and specialized presclerites of abdominal segment 3, which form a complex articulation within the posterior foramen of the petiole (= abdominal segment 2) (Figs. 528, 530; *he*). In general the helcium is mostly or entirely concealed within the posterior orifice of the petiole, but in certain groups it is partly visible. Disarticulation of abdominal segments 2 and 3 is necessary to examine the structure and its variation in detail (Fig. 531; *he*). See also **Abdomen.**

Humeral angles (= humeri) The anterolateral dorsal angles of the pronotum.

Hypopygium The sternite of abdominal segment 7; the terminal visible gastral sternite (Fig. 530; *hy*).

Hypostoma The anteroventral region of the head; the area of cuticle immediately behind the buccal cavity and forming its posterior margin.

Hypostomal teeth One or more pairs of triangular or rounded teeth that project forward from the anterior margin of the hypostoma.

Intercalary (teeth) See **Mandibles.**

Katepisternum See **Pleurite.**

Labial palps (alternatively, labial palpus; pl. labial palpi) A pair of sensory palps, with a maximum of 4 segments, that arise anterolaterally on the labium. In ventral view the labium is longitudinal and situated centrally in the buccal cavity, flanked by a maxilla on each side. See also **Palp Formula.**

Labrum Mouthpart sclerite that hinges on the anterior margin of the clypeus and usually folds back and down over the apices of the maxillae and labium when the mouthparts are not in use. In most ants the labrum is a bilobed plate that is invisible in dorsal view, but in some taxa it projects forward from the anterior clypeal margin even when

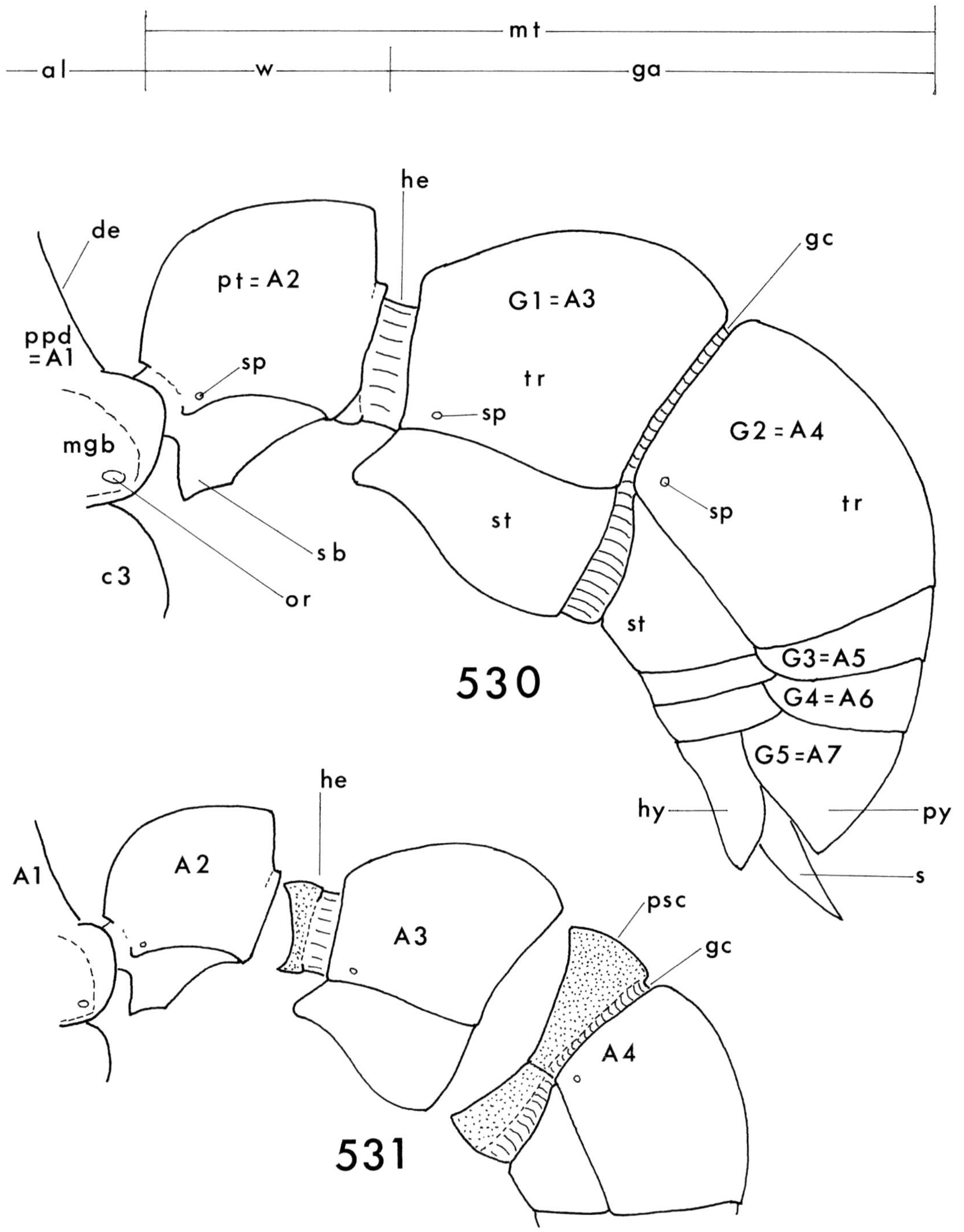

A	abdominal segment number
al	alitrunk
c	coxa number
de	declivity of propodeum
G	gastral segment number
ga	gaster
gc	girdling constriction
he	helcium
hy	hypopygium
mgb	metapleural gland bulla
mt	metasoma
or	orifice of metapleural gland
ppd	propodeum
psc	presclerite
pt	petiole
py	pygidium
s	sting
sb	subpetiolar process
sp	spiracle
st	sternite
tr	tergite
w	waist

Figures 530–531 Morphological features of ants. Drawings are composite, not based on any particular species; sculpture and pilosity omitted. Fig. 530, waist and gaster in profile. Fig. 531, waist and gaster in profile with abdominal segments 2–4 disarticulated to show concealed portions of presclerites (stippled).

the mouthparts are at rest. Occasionally it is modified into one or more long, prominent labral lobes.

Leg segments Each leg consists of a basal *coxa* that articulates with the alitrunk, followed in order by a small *trochanter,* a long and generally stout *femur,* a *tibia,* and a *tarsus,* the last consisting of five small segments and terminating apically in a pair of claws. The prefixes *pro-, meso-,* and *meta-,* applied to any of these terms, indicate that segment on the leg of a particular thoracic segment. For example, the metatibia is the tibial segment of the metathoracic (hind) leg.

Linear (mandibles): See **Mandibles.**

Mandibles The appendages with which ants manipulate their environment. They are very variable in shape, size, and dentition, and extremely important in ant taxonomy.

Margins. In full-face view, with the mandibles closed, the inner margin or border (closest to an anterior extension of the midline of the head) of each mandibular blade is the *apical margin* (= masticatory margin) (Fig. 527; *ma*), and is usually armed with teeth. Proximally, close to the anterior margin of the clypeus, the apical margin usually passes through a *basal angle* (Fig. 527; *ba*) into a transverse or oblique *basal margin* (Fig. 527; *mb*). The two margins may join through a broad or narrow curve, or meet in an angle or tooth. When the mandibles are narrow or linear, the distinction between apical and basal margins may be lost by obliteration of the basal angle. The *external margin* (= lateral margin) (Fig. 527; *me*) of each mandible forms its outer border in full-face view and may be straight, sinuate, or convex.

Shape. In the vast majority of ants the mandibular margins form a *triangular* or *subtriangular* shape in full-face view (e.g., Figs. 7, 20, 136, 308, 452), but may be drawn out anteriorly while retaining the basic triangular shape and become *elongate-triangular* (e.g., Figs. 174, 246). In several discrete lineages the mandible has become *linear* (e.g., Figs. 146, 172, 202, 226, 426, 459, 470); the blade is long and narrow and the apical and external margins are approximately parallel or taper very gradually to the apex; the whole blade may be straight or curved. Linear mandibles may evolve in one of three ways:

(1) The base of the mandible narrows and the basal angle is obliterated so that the apical and basal margins form a single margin.

(2) The apical margin is elongated and the basal margin contracted.

(3) The basal margin is elongated and the apical margin reduced.

Extremely curved mandibles, usually quite short and with few or no teeth on the apical margin, are termed *falcate* (e.g., Figs. 106, 324, 420).

Dentition. The apical margin of each mandible is usually armed with a series of teeth (Fig. 527; *at, t2–t4, bt*) or denticles (short or very reduced acute teeth) (Fig. 527; *dn*) or both, which generally run the length of the apical margin. If teeth alone are present, or a combination of teeth and denticles, the mandible is *dentate.* If only tiny denticles occur the mandible is *denticulate,* and if the margin lacks armament it is *edentate.* A natural gap in a row of teeth (as opposed to a site where teeth have broken off or been worn down) is a *diastema* (pl. diastemata) (Fig. 527; *di*) and an elongate mandible with an uninterrupted series of teeth may be described as *serially dentate.* Teeth are usually sharp and triangular in shape but may be rounded *(crenulate);* long, narrow, and spine-like (*spiniform*), or peg-like. Reduced teeth or denticles that occur between full-sized teeth are *intercalary.* In general the first, distalmost, or *apical tooth* (Fig. 527; *at*), the one farthest away from the anterior clypeal margin, is the largest on the apical (masticatory) margin, although in some taxa median or basal teeth may be the largest. The tooth at or nearest to the basal angle is the *basal tooth* (Fig. 527; *bt*). The tooth immediately behind the apical (Fig. 527; *t2*) may be termed the *preapical* (= subapical), though this term may also be applied more generally to include several teeth that are behind the apical but distal of the mandibular midlength. Similarly, the tooth immediately preceding the basal (Fig. 527; *t4*) may be termed the *prebasal* (= subbasal). In a few taxa teeth may occur on the basal margin of the mandible (Fig. 527; *mb*), but in most this margin is unarmed. Many dacetonine myrmicines have a *basal lamella,* a thin strip or plate of cuticle, on the apical margin proximal to any teeth that may be present.

Marginate Having a sharply defined rim, edge, or margin separating one face of a sclerite, segment, or tagma from another.

Masticatory margin/border (of mandible) See **Mandibles.**

Maxillary palps (alternatively, maxillary palpus; pl. maxillary palpi) The segmented sensory palps of the maxillae. Each palp may have at most 6 segments but these are variously reduced in number in different ant groups; only very rarely are maxillary palps absent. In ventral view the maxillae are situated in the buccal cavity, one on each side of the central labium (which itself possesses a pair of palps), and the palps are articulated to the maxilla anterolaterally on each side. See also **Palp Formula.**

Mesad Medially, toward the middle, towards the midline.

Mesonotum See **Tergite.**

Mesopleuron See **Pleurite.**

Mesosoma See **Alitrunk.**

Mesothoracic spiracle See **Spiracle.**

Mesothorax See **Thorax.**

Metacoxa (pl. metacoxae) The coxa of the metathoracic (= hind, = third) leg (Fig. 529; *c3*). The metacoxae insert posterolaterally in the ventral alitrunk, close to the median emargination in which the petiole articulates. The cavity of this median petiolar articulation may be separated from the cavities in which the metacoxae articulate (metacoxal cavities) by a bar or annulus of cuticle, or the cavities may be confluent. To observe these structures it is necessary to remove the hind legs and mount the ant ventral side uppermost. See also **Leg segments.**

Metanotal groove/metanotum See **Tergite.**

Metapleural gland See **Pleurite.**

Metapleural lobe See **Propodeal lobe.**

Metapleuron (pl. metapleura) See **Pleurite.**

Metasoma See **Abdomen.**

Metasternal process A paired cuticular projection of the posteroventral alitrunk. When present it is located astride the ventral midline, anterior to the apex of the cavity in which the petiole articulates and close to the level of the anterior margins of the metacoxal cavities. To view the process clearly it is necessary to remove the middle and hind legs.

Metathorax See **Thorax.**

Metathoracic spiracle See **Spiracle.**

Metatibia The tibial segment of the metathoracic (= hind, = third) leg. See also **Leg segments.**

Metatibial gland A presumably exocrine gland located ventrally on the metatibia just posterior to the tibial spur in several ant subfamilies. When present it varies considerably in shape and size.

MI (Mandibular Index) See **Standard measurements.**

Monomorphic Occurring in only a single morphological form; in the sense of the keys presented here, having only one form of worker caste.

Node/nodiform See **Petiole.**

Nuchal carina A ridge situated posteriorly on the head that separates the dorsal and lateral surfaces from the occipital surface (Figs. 458, 460).

Occipital corners With the head in full-face view, the rounded to acute posterolateral angles, where the sides of the head curve in to the occipital margin (Fig. 526).

Occipital margin In common usage, the transverse posterior margin of the head in full-face view (Fig. 526; *om*); morphologically the term is incorrect, as the occiput proper usually begins behind this level, but the name is suitable for most purposes.

Ocular Pertaining to the eye.

Ommatidium (pl. ommatidia) A single optical component of the compound eye.

Palp Formula (PF) A standardized way of indicating the number of segments in the maxillary and labial palps. The number of maxillary palp segments is given first, the number of labial palp segments second; thus PF 6,4 indicates that the maxillary palp has six segments, the labial four.

Pectinate Comb-like.

Pedicel An old term used in ants for the isolated body segments between the alitrunk (= mesosoma) and gaster, namely the petiole or petiole plus postpetiole. The use of *pedicel* is not recommended as it is employed elsewhere throughout the Hymenoptera as the name for the first funicular segment of the antenna. Use of the term *waist* is recommended in referring to these isolated segments.

Peduncle (of petiole) The relatively narrow anterior section of the petiole which begins immediately behind the propodeal-petiolar articulation and runs back to the petiolar node or scale (Fig. 528; *pd*). It is very variable in length and thickness but when present in any form the petiole is termed *pedunculate.* When the peduncle is absent, so that the node or scale of the petiole immediately follows the articulation with the propodeum, the petiole is termed *sessile* (Fig. 530). If an extremely short peduncle occurs the petiole is termed *subsessile.*

Petiole Morphologically, the second abdominal segment; the segment immediately following the alitrunk (= mesosoma), which is usually reduced and always isolated (Figs. 528, 530; *pt* = *A2*). Generally the petiole takes the form of a node *(nodiform)* or of a scale *(squamiform)* of varying shape and size, but in some taxa it may be very reduced, represented by only a narrow, subcylindrical segment that may be overhung and concealed by the gaster. The petiole bears the second abdominal spiracle and usually consists of a distinct tergite and sternite. The former may have differentiated laterotergites low down on the side. In some groups the petiolar tergite and sternite have fused together. See also **Abdomen, Peduncle.**

Plectrum See **Stridulatory system.**

Pleurite/pleuron The lateral sclerites of the thorax proper, excluding the propodeum (Fig. 529; *ppd* = *A1*), which is morphologically the first abdominal tergite. The *propleuron* (pleuron of the prothorax) (Fig. 529; *pr*) is relatively small in ants and is mostly or entirely overlapped and concealed by the lateral part of the pronotum (Fig. 529; *pn*) when viewed in profile, but can be seen clearly in ventral view. The *mesopleuron* (pleuron of the mesothorax) (Figs. 528, 529; *mpl*) is the largest pleurite. It may consist of a single sclerite running almost the entire height of the mesothorax (Fig. 528) or may be divided by a transverse groove into an upper *anepisternum* and a lower *katepisternum* (Fig. 529; *an, kn*). The *metapleuron* (pleuron of the metathorax) (Fig. 529; *mtp*) is located posteriorly on the side of the alitrunk, below the level of the propodeum (Fig. 529; *ppd* = *A1*). The metapleuron bears, in most groups of ants, the *metapleural gland.* This is an exocrine gland whose orifice (Figs. 529, 530; *or*) is usually situated in the posteroventral corner of the side of the alitrunk, above the level of the metacoxa (Fig. 529; *c3*) and below the level of the propodeal spiracle (Fig. 529; *sp* on *ppd* = *A1*). The swollen *bulla* (Fig. 529; *mgb*) of the metapleural gland is often more conspicuous than the gland's orifice, taking the form of a shallow blister or convexity on the metapleuron and sometimes reaching almost to the propodeal spiracle. The orifice of the metapleural gland may be a simple pore, or may be protected by cuticular flanges or other outgrowths, or by guard setae traversing the orifice. See also **Sternite, Tergite.**

Polymorphic Occurring in more than two morphologically distinct forms; in the sense of the keys presented here, having more than two different forms of the worker caste.

Posterior tentorial pits See **Anterior tentorial pits.**

Postpetiole Morphologically, the third abdominal segment (Fig. 528; *ppt* = *A3*). In strict usage the term *postpetiole* should only be applied when the third abdominal segment is reduced and separated from the petiole in front and the fourth abdominal segment behind. See also **Abdomen.**

Postsclerite/poststernite/posttergite See **Presclerite.**

Preapical/prebasal teeth See **Mandibles.**

Presclerite A distinctly differentiated anterior section of an abdominal sclerite, separated from the remainder of the sclerite by a ridge, constriction, or both (Fig. 531; psc). In the ant abdomen it is usual for the posterior portion of each sclerite to overlap the anterior portion of the following segment. The overlapped area usually lacks sculpture and pilosity, but the absence of these features alone does not constitute a presclerite. Presclerites derived from tergites are termed *pretergites,* those from sternites, *presternites.* The remainder of each sclerite, posterior to these developments, is the *postsclerite* and may be termed *post-*

tergite and *poststernite,* repectively. The presclerites of abdominal segment 3 form a very specialized articulation with the posterior end of abdominal segment 2, termed the *helcium* (Figs. 528, 530, 531; *he*).

Presternite See **Presclerite.**

Pretarsal claws A pair of claws on the pretarsal (= apical or terminal) tarsal segment of the leg. Usually the inner curvature of each claw is a simple smooth concave surface, but in some taxa one or more preapical teeth may be present, or the claw may be pectinate (Figs. 514-516). See also **Leg segments, Tarsus.**

Pretergite See **Presclerite.**

Profile Orientation of part of the body (usually the alitrunk) in side (lateral) view so that the anterior, posterior, dorsal, and ventral outlines are in focus at the same time (Figs. 528, 529).

Promesonotal suture The transverse suture across the dorsal alitrunk (Fig. 529; *pms,* shown in profile) that separates the pronotum *(pn)* from the mesonotum *(ms).* In some groups of ants the promesonotal suture is well developed and flexible. The pronotum slightly overlaps the mesonotum and the two sclerites are linked by intersegmental membrane so that they are capable of movement relative to each other. Elsewhere, and very commonly, the suture is reduced from this condition. Initially in the sequence of reduction the suture is present and distinct but inflexible, as the posterior pronotal margin has fused to the anterior mesonotal margin. Beyond this fused condition the suture shows a gradual morphoclinal reduction in size and degree of definition, eventually becoming nothing more than a faint line or impression across the dorsum, or often disappearing altogether. When fusion and obliteration of the suture is advanced, and there is little or no sign of separation of the original pronotum and mesonotum, the resulting sclerite is called the *promesonotum* (Fig. 528; *prs*).

Promesonotum See **Promesonotal suture.**

Pronotum See **Tergite.**

Propodeal lobe (= metapleural lobe, = inferior propodeal plate) See **Propodeum.**

Propodeal spiracle See **Propodeum.**

Propodeum Morphologically, the tergite of the first abdominal segment, fused to the thorax and forming most of the posterior section of the alitrunk (= mesosoma) (Fig. 529; *ppd = A1*). An older term for this sclerite, *epinotum* (see there), should not be used. The propodeal dorsum is usually unspecialized but frequently terminates posteriorly in a pair of teeth or spines. Confusingly, the propodeal dorsum is sometimes referred to as its base or basal surface. The dorsum is not basal to anything and the term should be abandoned. The sloping posterior surface is the *propodeal declivity* (Fig. 529; *de*), and may bear a number of specializations. Most common of these is the development of a pair of *propodeal lobes* (= metapleural lobes, = inferior propodeal plates) (Fig. 529; *pl*). When present they are situated at the base of the declivity, one on each side of the propodeal-petiolar articulation. These lobes, which vary considerably in shape and size, are frequently termed *metapleural lobes,* but this name should be abandoned as they are formed from the propodeum and not the metapleuron. The side of the propodeum bears the *propodeal spiracle* (Fig. 529; *sp* on *ppd = A1*), morphologically the first abdominal spiracle. Its shape, size, and location are variable and of considerable taxonomic value.

Prothorax See **Thorax.**

Proventriculus A muscular pump located in the intestine between the crop and the midgut. In all groups the proventriculus has a basal *bulb,* but in some the bulb is surmounted by a ring of four sclerotized *sepals,* collectively termed the *calyx.*

Psammophore A basket-like series of long and usually stout, curved setae arising on the ventral surfaces of the head and mandibles in deserticolous ants, used for carrying sand grains (Fig. 311).

Pubescence Small to minute hair-like cuticular projections which are not socketed basally. See also **Seta.**

Pygidium The tergite of abdominal segment 7; the terminal visible gastral tergite (Fig. 530; *py*).

Scale See **Petiole.**

Scape See **Antenna.**

Sclerite Functionally, a general term for any single plate of the exoskeleton (e.g., pronotal sclerite, abdominal sclerites); more specifically, an integumental plate in which the protein sclerotin has been deposited. In the case of ants the latter applies to all parts of the exoskeleton.

Scrobe (antennal) See **Antennal scrobe.**

Sepals See **Proventriculus.**

Serially dentate See **Mandibles.**

Sessile (petiole) See **Peduncle.**

Seta (pl. setae) Any stout hair that is socketed basally. Generally, as here, the terms *seta* and *hair* are interchangeable, but care must be taken to differentiate between setae and pubescence, as the latter may also sometimes be referred to as hairs.

Spiniform (teeth) See **Mandibles.**

Spiracle An orifice of the tracheal system by which gases enter and leave the body. Ants have 9 or 10 spiracles on each side of the body. The prothoracic spiracles have been lost, so the first opening occurs on the mesothorax. This spiracle is situated forward and quite high on the side of the segment, and is usually concealed by a backward-projecting lobe of the pronotum (Fig. 529). Metathoracic spiracles (Fig. 529; *sp* on *mtp*) may be dorsal (especially when the metathorax forms part of the dorsal alitrunk), lateral, concealed by a small backward-projecting lobe of the mesopleuron, or absent. The propodeal (first abdominal) spiracle is usually the largest on the body (Fig. 529; *sp* on *ppd = A1*). Spiracles are always visible on abdominal segments 2–4 (Figs. 528, 530; *sp*), but those on abdominal segments 5–7 are frequently concealed beneath the posterior margins of the preceding tergites. The spiracle of abdominal segment 8 is always hidden. The sclerite to which it belongs is internal and forms part of the sting apparatus (the spiracular plate).

Spongiform (tissue) Specialized sponge-like external cuticular tissue, distributed mainly about the waist segments in some groups of ants (e.g., Figs. 243, 399).

Spur See **Tibial spur.**

Squamiform In the form of a scale.

Standard measurements A series of external measurements and ratios of the body used in ant taxonomy. Different groups of ants may require different combinations of standard measurements. Any recent survey or revision of a group or genus will list and define the standards used. Some very commonly encountered standards include *head length* and *head width* (HL and HW), *mandible length* (ML), *scape length* (SL), *pronotal width* (PW), and *alitrunk length* (AL). Ratios of these, termed *indices,* serve to indicate relative dimensions. Commonly encountered indices include *mandibular index* (MI) = ML × 100 divided by HL; *cephalic index* (CI) = HW × 100 divided by HL; *scape index* = SL × 100 divided by HW. Multiplication by 100 is not essential but it serves to express the ratio as a whole number.

Sternite (= sternum, = sternal plate) The lower sclerite of a segment (the tergite is the upper; pleurites are the laterals on the alitrunk). The sternite may be a simple flat or curved plate, or may be specialized or subdivided on some segments. In the hymenopterous prothorax the sternite is very small. Sternites of the mesothorax and metathorax are reduced and internal, the ventral surface being made up of extensions of the pleurites to the mid-ventral line. The sternite of the propodeum (= abdominal segment 1) has been lost, but those of the remaining abdominal segments are usually distinct (Fig. 530; *st*), although the margins of some may be difficult to discern because of fusion to the tergite.

Stridulatory system A sound-producing system present in a number of ant subfamilies. The system consists of a *plectrum* (= stridulatory file), located on the posterior margin of the third abdominal segment (usually, but not always, on the tergite), and a finely grooved *stridulitrum* or sounding board on the anterior portion of the fourth abdominal segment.

Stridulitrum See **Stridulatory system.**

Subpetiolar process An anteroventral projection on the petiole or its peduncle; sometimes absent but when present very variable in shape and size (Fig. 530; *sb*).

Subsessile (petiole) See **Peduncle.**

Subtriangular (mandible) See **Mandibles.**

Suture Line of junction between sclerites.

Tagma (pl. tagmata) Unit of body; part or section of body separated from other body units.

Tarsal claws See **Pretarsal claws.**

Tarsus (pl. tarsi) Collective term for the five small apical segments of any leg. The first tarsal segment (first tarsomere) articulates with the tibia and is termed the *basitarsus.* The next three tarsomeres are not individually named but the apical (terminal) tarsomere is the *pretarsus* and bears a pair of *pretarsal claws.* See also **Leg segments.**

Teeth (mandibular) See **Mandibles.**

Tentorial pits (anterior) See **Anterior tentorial pits.**

Tergite (= tergum, = tergal plate) The upper sclerite of a segment (the sternite is the lower; the pleurites the laterals on the alitrunk). The tergite may be a simple flat or curved plate, or may be specialized or subdivided on some segments. The tergite of the prothorax is composed entirely of the *pronotum* (Fig. 529; *pn*); this sclerite extends across the dorsum and down the sides of the segment, mostly concealing the propleuron. The *mesonotum* (Fig. 529; *ms*), tergite of the mesothorax, may be separated from the pronotum by the *promesonotal suture* (Fig. 529; *pms*), or may be fused to it to form a single sclerite, the *promesonotum* (Fig. 528; *prs*). The *metanotum,* tergite of the metathorax, may be present on the dorsum, or reduced, or obliterated. The mesonotum and propodeum are often separated by the *metanotal groove* (Fig. 529; *mtg*), a transverse groove or impression representing the last vestige of the metanotum on the dorsal alitrunk. The propodeum (see there) is the tergite of the first abdominal segment. The remaining abdominal segments (2–7) have tergites that are usually simple but that may be subdivided or otherwise specialized (Fig. 530; *tr*).

Tergosternal fusion A condition of the abdominal segments where the tergite and sternite of a single segment fuse together so that they are not capable of movement relative to each other. This may occur in some or all abdominal segments from 2 (petiole) to 4. To investigate this feature properly it is best to disarticulate the specimen, then macerate the appropriate segments in sodium hydroxide to remove soft tissues. This is essential in dried specimens, though good results can be obtained by dissection of fresh material without maceration.

Thorax The second classical body tagma in the insects. In ants, and other Hymenoptera, the apparent thorax consists of the usual three body segments of the true thorax (pro-, meso-, and metathorax) to which the tergite of the first abdominal segment (the propodeum) is immovably fused (Fig. 529). This modification means that the system "true thorax plus propodeum" cannot strictly be called the thorax, as it is not homologous with the term as used otherwise throughout the Insecta. Several names have been proposed for true thorax plus propodeum, of which two, *alitrunk* and *mesosoma,* are currently in common use (Figs. 528, 529; *al* = *mes*). Both names are somewhat misleading as far as ants are concerned, but either is better than *thorax,* which is morphologically incorrect for describing the tagma "thorax plus propodeum." See also **Alitrunk.**

Tibia (pl. tibiae) The fourth segment of any leg, counting from the basal coxa that articulates with the alitrunk. See also **Leg segments.**

Tibial spur A socketed spur located at the apex of each tibia. The forelegs have a single pectinate tibial spur, modified into an antennal cleaning device, the *strigil.* The mesothoracic (middle) and metathoracic (hind) tibiae, also termed mesotibiae and metatibiae, may each have two, one, or no spurs present (Figs. 511–513). When present the spurs may be pectinate or barbed, or be simple cuticular spikes. If two spurs are present it is usual for one to be larger than the other, and in such cases the larger spur is usually pectinate while the smaller is simple. See also **Leg segments.**

Torulus (= torular sclerite, = antennal sclerite) The small annular sclerite that surrounds the antennal socket (Fig. 523; *to*). The torulus may be horizontal, or the part closest to the midline of the head may be elevated, in some cases to such an extent that the torulus is almost vertical. In the latter condition the highest part of its arc may form a laterally projecting small lobe. The lobe may or may not be covered by the frontal lobes; in some taxa where very narrow frontal lobes occur the torulus lobe projects beyond them. See also **Antenna.**

Triangular (mandibles) See **Mandibles.**

Trulleum A basin-shaped depression near the base of the mandible dorsally, bounded distally by the basal margin of the mandibular blade (Fig. 527; *tu*).

Tuberculiform Having the form or appearance or a tubercle.

Tubercle A small, rounded prominence or protuberance.

Tuberculate Bearing one or more tubercles.

Tumulus (pl. tumuli) A prominent, small, mound-like or rounded hill-like to subconical, but not acutely pointed, cuticular excrescence.

Waist Collective term for the one or two separated abdominal segments that occur between the alitrunk (= mesosoma) and gaster (Figs. 528, 530; *w*). When only the petiole (abdominal segment 2) is isolated the waist is said to be one-segmented, but in those subfamilies where the postpetiole (abdominal segment 3) is also separated the waist is said to be two-segmented (of petiole plus postpetiole). See also **Abdomen, Gaster, Pedicel.**

Xenobiosis A lifeway in which one eusocial species nests within the walls or in the nest chambers of another eusocial species. The *xenobiotic* species (xenobiont) not only nests within the nest of its host but also shares its trails and solicits and receives food from the host species.

Bibliography

Agosti, D. 1990a. Review and reclassification of *Cataglyphis. Journal of Natural History* **24:** 1457–1505.

———1990b. What makes the Formicini the Formicini? *Actes des Colloques Insectes Sociaux* **6:** 295–303.

———1991. Revision of the Oriental ant genus *Cladomyrma,* with an outline of the higher classification of the Formicinae. *Systematic Entomology* **16:** 293–310.

———1992. Revision of the ant genus *Myrmoteras* in the Malay Archipelago. *Revue Suisse de Zoologie* **99:** 405–429.

Agosti, D., and C. A. Collingwood. 1987. A provisional list of the Balkan ants with a key to the worker caste. 2. Key to the worker caste, including the European species without the Iberian. *Mitteilungen der Schweizerischen Entomologischen Gesellschaft* **60:** 261–293.

Alayo, P. D. 1974. Introducción al estudio de los Himenópteros de Cuba. Superfamilia Formicoidea. *Academia de Ciencias de Cuba. Instituto de Zoología. Serie Biológica* (Havana), no. **53:** 58 pp.

Arnold, G. 1915. A monograph of the Formicidae of South Africa. Part 1. (Ponerinae; Dorylinae.) *Annals of the South African Museum* **14:** 1–159.

———1916. A monograph of the Formicidae of South Africa. Part 2. (Ponerinae; Dorylinae.) *Annals of the South African Museum* **14:** 159–270.

———1917. A monograph of the Formicidae of South Africa. Part 3. (Myrmicinae.) *Annals of the South African Museum* **14:** 271–402.

———1920. A monograph of the Formicidae of South Africa. Part 4. (Myrmicinae.) *Annals of the South African Museum* **14:** 403–578.

———1922. A monograph of the Formicidae of South Africa. Part 5. (Myrmicinae.) *Annals of the South African Museum* **14:** 579–674.

———1924. A monograph of the Formicidae of South Africa. Part 6. (Camponotinae.) *Annals of the South African Museum* **14:** 675–766.

———1926. A monograph of the Formicidae of South Africa. Appendix. *Annals of the South African Museum* **23:** 191–295.

———1951. The genus *Hagensia* Forel. *Journal of the Entomological Society of Southern Africa* **14:** 53–56.

———1952a. New species of African Hymenoptera. No. 10. *Occasional Papers of the National Museum of Southern Rhodesia* **2** (no. 17): 460–493.

Arnol'di, K. V. 1970. Obzor murav'ev roda *Myrmica* evropeiskoi chasti SSSR. *Zoologicheskii Zhurnal* **49:** 1829–1844.

———1975. Obzor vidov roda *Stenamma* soyuza SSR i opisanie novykh vidov. *Zoologicheskii Zhurnal* **54:** 1819–1829.

———1976a. Murav'i roda *Myrmica* Latr. srednei azii i yuzhnogo Kazakhstana. *Zoologicheskii Zhurnal* **55:** 547–558.

———1976b. Obzor roda *Aphaenogaster* SSSR. *Zoologicheskii Zhurnal* **55:** 1019–1026.

———1977. Obzor murav'ev-zhnetsov roda *Messor* fauny SSSR. *Zoologicheskii Zhurnal* **56:** 1637–1648.

Arnol'di, K. V., and G. M. Dlussky. 1978. Formicoidea. In G. S. Medvedev, ed., *Opredelitel' Nasekomykh Evropeiskoi Chasti SSSR.* Vol. **3:** *Pereponchatokrylye, pervaya chast',* 519–556. Leningrad.

Ashmead, W. H. 1905. A skeleton of a new arrangement of the families, subfamilies, tribes and genera of the ants, or the superfamily Formicoidea. *Canadian Entomologist* **37:** 381–384.

Atanassov, N., and G. M. Dlussky. 1992. *Fauna na Bulgariya.* Vol. **22:** *Hymenoptera, Formicidae.* Sofia. 310 pp.

Baltazar, C. R. 1966. A catalogue of the Philippine Hymenoptera. (With a bibliography, 1758–1963.) *Pacific Insects Monograph* **8:** 1–488.

Baroni Urbani, C. 1964a. Studi sulla mirmecofauna d'Italia. 2. Formiche di Sicilia. *Atti della Accademia Gioenia di Scienze Naturali in Catania* **16:** 25–66.

———1964b. Studi sulla mirmecofauna d'Italia. 3. Formiche dell'Italia Appenninica. *Memorie del Museo Civico di Storia Naturale di Verona* **12:** 149–172.

———1968. Studi sulla mirmecofauna d'Italia. 4. La fauna mirmecologica delle isole maltese ed il suo significato ecologico e biogeografica. *Annali del Museo Civico di Storia Naturale di Genova* **77:** 408–559.

———1969a. Studi sulla mirmecofauna d'Italia. 8. L'Isola di Giannutri ed alcuni scogli minori dell'arcipelago Toscano. *Atti della Società Toscana di Scienze Naturali. Memorie (B)* **75** (1968): 325–338.

———1969b. Gli *Strongylognathus* del gruppo *huberi* nell'Europa occidentale: saggio di una revisione basata sulla casta operaia. *Bollettino della Società Entomologica Italiana* **99–101:** 132–168.

———1971. Studien zur Ameisenfauna Italiens. 11. Die Ameisen des Toskanischen Archipels. Betrachtungen zur Herkunft der Inselfaunen. *Revue Suisse de Zoologie* **78:** 1037- 1067.

———1975a. Primi reperti del genere *Calyptomyrmex* Emery nel subcontinente Indiano. *Entomologica Basiliensia* **1:** 395–411.

———1975b. Contributo alla conoscenza dei generi *Belonopelta* Mayr e *Leiopelta* gen. n. *Mitteilungen der Schweizerischen Entomologischen Gesellschaft* **48:** 295–310.

———1977a. Materiali per una revisione della sottofamiglia Leptanillinae Emery. *Entomologica Basiliensia* **2:** 427–488.

———1977b. Les espèces européennes du genre *Proceratium* Roger. *Mitteilungen der Schweizerischen Entomologischen Gesellschaft* **50:** 91–93.

———1978a. Contributo alla conoscenza del genere *Amblyopone* Erichson. *Mitteilungen der Schweizerischen Entomologischen Gesellschaft* **51:** 39–51.

———1978b. Materiali per una revisione dei *Leptothorax* neotropicali al sottogenere *Macromischa* Roger. *Entomologica Basiliensia* **3:** 395–618.

Baroni Urbani, C., B. Bolton, and P. S. Ward. 1992. The internal phylogeny of ants. *Systematic Entomology* **17:** 301–329.

Beattie, A. J. 1985. *The evolutionary ecology of ant-plant mutualisms:* Cambridge. 182 pp.

Belshaw, R., and B. Bolton. 1994. A new myrmicine ant genus from cocoa leaf litter in Ghana. *Journal of Natural History* (currently in press).

Bernard, F. 1950. Notes sur les fourmis de France. 2. Peuplement des montagnes méridionales. *Annales de la Société Entomologique de France* **115** (1946): 1–36.

———1955. Fourmis moissonneuses nouvelles ou peu connues des montagnes d'Algérie et revision des *Messor* du groupe *structor* (Latr.). *Bulletin de la Société d'Histoire Naturelle de l'Afrique du Nord* **45** (1954): 345–365.

———1956a. Révision des *Leptothorax* d'Europe occidentale, bassée sur la biométrie et les genitales males. *Bulletin de la Société Zoologique de France* **81:** 151–165.

———1956b. Révision des fourmis paléarctiques du genre *Cardiocondyla* Emery. *Bulletin de la Société d'Histoire Naturelle de l'Afrique du Nord* **47:** 299–306.

———1967. *Faune de l'Europe et du Bassin Méditerranéen.* Vol. **3:** *Les fourmis d'Europe occidentale et septentrionale.* Paris, 1968. 411 pp.

———1978. Révision des *Diplorhoptrum* de France, fourmis plus différenciées par l'écologie que par leurs formes. *Annales de la Société Entomologique de France* (n.s.) **13** (1977): 543–577.

———1980. *Messor carthaginensis* n. sp. de Tunis, et révision des *Messor* du groupe *barbara. Bulletin de la Société Entomologique de France* **84** (1979): 265–269.

Bingham, C. T. 1903. *The fauna of British India, including Ceylon and Burma. Hymenoptera.* Vol. **2:** *Ants and Cuckoo-Wasps.* London. 506 pp.

Blum, M. S. 1985. Poisonous ants and their venoms. In A. T. Tu, ed., *Handbook of Natural Toxins.* Vol. **2:** *Insect Poisons, Allergens, and Other Invertebrate Venoms,* 225–242. New York. 732 pp.

Bolton, B. 1972. Two new species of the ant genus *Epitritus* from Ghana, with a key to the world species. *Entomologist's Monthly Magazine* **107** (1971): 205–208.

———1973. The ant genus *Polyrhachis* F. Smith in the Ethiopian region. *Bulletin of the British Museum (Natural History)* (Entomology) **28:** 283–369.

———1974a. A revision of the Palaeotropical arboreal ant genus *Cataulacus* F. Smith. *Bulletin of the British Museum (Natural History)* (Entomology) **30:** 1–105.

———1974b. A revision of the ponerine ant genus *Plectroctena* F. Smith. *Bulletin of the British Museum (Natural History)* (Entomology) **30:** 309–338.

———1975a. A revision of the ant genus *Leptogenys* Roger in the Ethiopian region, with a review of the Malagasy species. *Bulletin of the British Museum (Natural History)* (Entomology) **31:** 235–305.

———1975b. A revision of the African ponerine ant genus *Psalidomyrmex* André. *Bulletin of the British Museum (Natural History)* (Entomology) **32:** 1–16.

———1975c. The *sexspinosa*-group of the ant genus *Polyrhachis* F. Smith. *Journal of Entomology* (ser. B) **44:** 1–14.

———1976. The ant tribe Tetramoriini. Constituent genera, review of smaller genera and revision of *Triglyphothrix* Forel. *Bulletin of the British Museum (Natural History)* (Entomology) **34:** 281–379.

———1977. The ant tribe Tetramoriini. The genus *Tetramorium* Mayr in the Oriental and Indo-Australian regions, and in Australia. *Bulletin of the British Museum (Natural History)* (Entomology) **36:** 67–151.

———1979. The ant tribe Tetramoriini. The genus *Tetramorium* Mayr in the Malagasy region and in the New World. *Bulletin of the British Museum (Natural History)* (Entomology) **38:** 129–181.

———1980. The ant tribe Tetramoriini. The genus *Tetramorium* Mayr in the Ethiopian zoogeographical region. *Bulletin of the British Museum (Natural History)* (Entomology) **40:** 193–384.

———1981a. A revision of the ant genera *Meranoplus* F. Smith, *Dicroaspis* Emery and *Calyptomyrmex* Emery in the Ethiopian zoogeographical region. *Bulletin of the British Museum (Natural History)* (Entomology) **42:** 43–81.

———1981b. A revision of six minor genera of Myrmicinae in the Ethiopian zoogeographical region. *Bulletin of the British Museum (Natural History)* (Entomology) **43:** 245–307.

———1982. Afrotropical species of the myrmicine ant genera *Cardiocondyla, Leptothorax, Melissotarsus, Messor* and *Cataulacus. Bulletin of the British Museum (Natural History)* (Entomology) **45:** 307–370.

———1983. The Afrotropical dacetine ants. *Bulletin of the British Museum (Natural History)* (Entomology) **46:** 267–416.

———1984. Diagnosis and relationships of the myrmicine ant genus *Ishakidris* gen. n. *Systematic Entomology* **9:** 373–382.

———1986. A taxonomic and biological review of the tetramoriine ant genus *Rhoptromyrmex. Systematic Entomology* **11:** 1–17.

———1987. A review of the *Solenopsis* genus-group and revision of Afrotropical *Monomorium* Mayr. *Bulletin of the British Museum (Natural History)* (Entomology) **54:** 263–452.

———1988a. A new socially parasitic *Myrmica,* with a reassessment of the genus. *Systematic Entomology* **13:** 1–11.

———1988b. A review of *Paratopula* Wheeler, a forgotten genus of myrmicine ants. *Entomologist's Monthly Magazine* **124:** 125–143.

———1988c. *Secostruma,* a new subterranean tetramoriine ant genus. *Systematic Entomology* **13:** 263–270.

———1990a. Abdominal characters and status of the cerapachyine ants. *Journal of Natural History* **24:** 53–68.

———1990b. The higher classification of the ant subfamily Leptanillinae. *Systematic Entomology* **15:** 267–282.

———1990c. Army ants reassessed: the phylogeny and classification of the doryline section. *Journal of Natural History* **24:** 1339–1364.

———1991. New myrmicine ant genera from the Oriental region. *Systematic Entomology* **16:** 1–13.

———1992. A review of the ant genus *Recurvidris,* a new name for *Trigonogaster* Forel. *Psyche* **99:** 35–48.

Bolton, B., and R. Belshaw. 1993. Taxonomy and biology of the supposedly lestobiotic ant genus *Paedalgus. Systematic Entomology* **18:** 181–189.

Bolton, B., and C. A. Collingwood. 1975. Hymenoptera: Formicidae. *Handbooks for the Identification of British Insects.* Vol. **4,** part 3(c). London. 34 pp.

Bolton, B., and A. C. Marsh. 1989. The Afrotropical thermophilic ant genus *Ocymyrmex. Journal of Natural History* **23:** 1267–1308.

Borgmeier, T. 1955. Die Wanderameisen der Neotropischen Region. *Studia Entomologica* **3:** 1–720.

———1959. Revision der Gattung *Atta* Fabricius. *Studia Entomologica* (N.S.) **2:** 321–390.

Boven, J. K. A. van. 1977. De Mierenfauna van Belgie. *Acta Zoologica et Pathologica Antverpiensia* **67:** 1–191.

Brandão, C. R. F. 1990. Systematic revision of the Neotropical ant genus *Megalomyrmex* Forel, with the description of thirteen new species. *Arquivos de Zoologia* **31:** 411–481.

———1991. Adendos ao catálogo abreviado das formigas da regiao Neotropical. *Revista Brasileira de Entomologia* **35:** 319–412.

Brown, W. L., Jr. 1948. A preliminary generic revision of the higher Dacetini. *Transactions of the American Entomological Society* **74:** 101–129.

———1952a. Revision of the ant genus *Serrastruma. Bulletin of the Museum of Comparative Zoology at Harvard College* **107:** 67–86.

———1952b. The dacetine ant genus *Mesostruma* Brown. *Transactions of the Royal Society of South Australia* **75:** 9- 13.

———1953a. Revisionary studies in the ant tribe Dacetini. *American Midland Naturalist* **50:** 1–137.

———1953b. Revisionary notes on the ant genus *Myrmecia* of Australia. *Bulletin of the Museum of Comparative Zoology at Harvard College* **111:** 1–35.

———1953c. A revision of the dacetine ant genus *Orectognathus. Memoirs of the Queensland Museum* **13:** 84–104.

———1954a. Remarks on the internal phylogeny and subfamily classification of the family Formicidae. *Insectes Sociaux* **1:** 21–31.

———1954b. The ant genus *Strumigenys* Fred. Smith in the Ethiopian and Malagasy regions. *Bulletin of the Museum of Comparative Zoology at Harvard College* **112:** 3–34.

———1955. A revision of the Australian ant genus *Notoncus* Emery, with notes on the other genera of Melophorini. *Bulletin of the Museum of Comparative Zoology at Harvard College* **113:** 471–494.

———1958a. A supplement to the revisions of the dacetine ant genera *Orectognathus* and *Arnoldidris,* with keys to the species. *Psyche* **64** (1957): 17–29.

———1958b. Contributions toward a reclassification of the Formicidae. 2. Tribe Ectatommini. *Bulletin of the Museum of Comparative Zoology at Harvard College* **118:** 175–362.

———1958c. A review of the ants of New Zealand. *Acta Hymenopterologica* **1:** 1–50.

———1959. A revision of the dacetine ant genus *Neostruma. Breviora* **107:** 1–13.

———1960. Contributions toward a reclassification of the Formicidae. 3. Tribe Amblyoponini. *Bulletin of the Museum of Comparative Zoology at Harvard College* **122:** 145- 230.

———1962. The Neotropical species of the ant genus *Strumigenys* Fr. Smith: synopsis and key to species. *Psyche* **69:** 238–267.

———1963. Characters and synonymies among the genera of ants. Part 3. Some members of the tribe Ponerini. *Breviora* **190:** 1–10.

———1964a. The ant genus *Smithistruma:* a first supplement to the world revision. *Transactions of the American Entomological Society* **89:** 183–200.

———1964b. Genus *Rhoptromyrmex,* revision of, and key to species. *Pilot Register of Zoology,* cards 11–19.

———1965. Contributions toward a reclassification of the Formicidae. 4. Tribe Typhlomyrmecini. *Psyche* **72:** 65–78.

———1967. Studies on North American ants. 2. *Myrmecina. Entomological News* **78:** 233–240.

———1972. *Asketogenys acubecca,* a new genus and species of dacetine ants from Malaya. *Psyche* **79:** 23–26.

———1973. A comparison of the Hylean and Congo-West African rain forest ant faunas. In B. J. Meggers, E. S. Ayensu, and W. D. Duckworth, eds., *Tropical Forest Ecosystems in Africa and South America: A Comparative Review,* 161–185 Washington, D.C.

———1974a. A supplement to the revision of the ant genus *Basiceros. Journal of the New York Entomological Society* **82:** 131–140.

———1974b. *Concoctio* genus nov.; *C. concenta* species nov. *Pilot Register of Zoology,* cards 29–30.

———1974c. *Dolioponera* genus nov.; *D. fustigera* species nov. *Pilot Register of Zoology,* cards 31–32.

———1975. Contributions toward a reclassification of the Formicidae. 5. Ponerinae, tribes Platythyreini, Cerapachyini, Cylindromyrmecini, Acanthostichini, and Aenictogitini. *Search Agriculture* **5.** Entomology (Ithaca), **15:** 1–115.

———1976a. Contributions toward a reclassification of the Formicidae. Part 6. Ponerinae, tribe Ponerini, subtribe Odontomachiti. Section A. Introduction, subtribal characters, genus *Odontomachus. Studia Entomologica* (n.s.) **19:** 67–171.

———1976b. *Cladarogenys* genus nov.; *C. lasia* species nov. *Pilot Register of Zoology,* cards 33–34.

———1978a. An aberrant new genus of myrmicine ant from Madagascar. *Psyche* **84** (1977): 218–224.

———1978b. Contributions toward a reclassification of the Formicidae. Part 6. Ponerinae, tribe Ponerini, subtribe Odontomachiti. Section B. Genus *Anochetus* and bibliography. *Studia Entomologica* (n.s.) **20:** 549–652.

———1980a. *Protalaridris* genus nov.;*P. armata* species nov. *Pilot Register of Zoology,* cards 36–37.

———1980b. A remarkable new species of *Proceratium,* with dietary and other notes on the genus. *Psyche* **86** (1979): 337–346.

———1986. *Indomyrma dasypyx,* new genus and species, a myrmicine ant from peninsular India. *Israel Journal of Entomology* **19** (1985): 37–49.

Brown, W. L., Jr., and R. G. Boisvert. 1979. The dacetine ant genus *Pentastruma. Psyche* **85** (1978): 201–207.

Brown, W. L., Jr., W. H. Gotwald, and J. Lévieux. 1971. A new genus of ponerine ants from West Africa, with ecological notes. *Psyche* **77** (1970): 259–275.

Brown, W. L., Jr., and W. W. Kempf. 1960. A world revision of the ant tribe Basicerotini. *Studia Entomologica* (n.s.) **3:** 161–250.

———1968. *Tatuidris,* a remarkable new genus of Formicidae. *Psyche* **74** (1967): 183–190.

———1969. A revision of the Neotropical dacetine ant genus *Acanthognathus. Psyche* **76:** 87–109.

Buren, W. F. 1968a. Some fundamental taxonomic problems in *Formica. Journal of the Georgia Entomological Society* **3:** 25- 40.

———1968b. A review of the species of *Crematogaster, sensu stricto,* in North America. 2. Descriptions of new species. *Journal of the Georgia Entomological Society* **3:** 91- 121.

Burnham, L. 1979. Survey of social insects in the fosil record. *Psyche* **85** (1978): 85–134.

Buschinger, A. 1989. Evolution, speciation and inbreeding in the parasitic ant genus *Epimyrma. Journal of Evolutionary Biology* **2:** 265–283.

Buschinger, A., W. Erhardt, K. Fischer, and J. Ofer. 1988. The slave-making ant genus *Chalepoxenus.* 1. Review of literature, range, slave species. *Zoologische Jahrbücher. Abteilung für Systematik und Geographie der Tiere* **115:** 383- 401.

Buschinger, A., K. Fischer, H.-P. Guthy, K. Jessen, and U. Winter. 1987. Biosystematic revision of *Epimyrma kraussei, E. vandeli,* and *E. foreli. Psyche* **93** (1986): 253–276.

Carpenter, F. M. 1992. In R. L. Kaesler, ed., *Treatise on Invertebrate Paleontology.* Part R. Arthropoda **4.** Vol. 4: Superclass Hexapoda: 279–655. Boulder. [Formicidae, pp.490- 495.]

Casevitz-Weulersse, J. 1990a. Etude systématique de la myrmécofaune corse. (Première partie.) *Bulletin du Muséum National d'Histoire Naturelle* (4) **12** (A): 135–163.

———1990b. Etude systématique de la myrmécofaune corse. (Deuxième partie.) *Bulletin du Muséum National d'Histoire Naturelle* (4) **12** (A): 415–442.

Chapman, J. W., and S. R. Capco. 1951. Check list of the ants of Asia. *Monographs of the Institute of Science and Technology, Manila* **1:** 327 pp.

Clark, J. 1930. The Australian ants of the genus *Dolichoderus,* subgenus *Hypoclinea* Mayr. *Australian Zoologist* **6:** 252–268.

———1936. A revision of Australian species of *Rhytidoponera* Mayr. *Memoirs of the National Museum, Victoria* **9:** 14–88.

———1951. *The Formicidae of Australia.* Vol. **1:** *Subfamily Myrmeciinae.* CSIRO, Melbourne. 230 pp.

Cole, A. C., Jr. 1949. A study of the genus *Gesomyrmex* Mayr, and a description of a species new to the genus. *Annals of the Entomological Society of America* **42:** 71–76.

———1968. *Pogonomyrmex Harvester Ants. A Study of the Genus in North America.* Knoxville. 222 pp.

Collingwood, C. A. 1978. A provisional list of Iberian Formicidae with a key to the worker caste. *Eos. Revista Española de Entomología* **52** (1976): 65–95.

———1979. The Formicidae of Fennoscandia and Denmark. *Fauna Entomologica Scandinavica* **8:** 1–174.

———1982. Himalayan ants of the genus *Lasius. Systmatic Entomology* **7:** 283–296.

———1985. Hymenoptera: Fam. Formicidae of Saudi Arabia. *Fauna of Saudi Arabia* **7:** 230–302.

Creighton, W. S. 1930. The New World species of the genus *Solenopsis. Proceedings of the American Academy of Arts and Sciences* **66:** 39–151.

———1950. The ants of North America. *Bulletin of the Museum of Comparative Zoology at Harvard College* **104:** 1–585.

———1957. A study of the genus *Xenomyrmex. American Museum Novitates* **1843:** 1–14.

Dalla Torre, C. G. de. 1893. *Catalogus Hymenopterorum, hucusque descriptorum systematicus et synonymicus* **7.** Lipsiae [Leipzig]. 289 pp.

Deyrup, M., J. Trager, and N. Carlin. 1985. The genus *Odontomachus* in the southeastern United States. *Entomological News* **96:** 188–195.

Diniz, J. L. M. 1990. Revisao sistematica da tribo Stegomyrmicini, com a descriçao de uma nova espécie. *Revista Brasileira de Entomologia* **34:** 277–295.

Dlussky, G. M. 1964. Murav'i podroda *Coptoformica* roda *Formica* SSSR. *Zoologicheskii Zhurnal* **43:** 1026–1040.

———1965. Ants of the genus *Formica* L. of Mongolia and northeast Tibet. *Annales Zoologici* **23:** 15–43.

———1967. Murav'i roda *Formica. Institut Morfologii Zhivotnykh, Akademiya Nauk SSSR.* Moscow. 236 pp.

———1969. Murav'i roda *Proformica* Ruzs. SSSR i sopredel'nykh stran. *Zoologicheskii Zhurnal* **48:** 218–232.

———1975. Semeistvo Formicidae. In A. P. Rasnitsyn ed., *Vysshie Pereponchatokrylye Mezozoya. Trudy Paleontologicheskogo Instituta* **147:** 114–122.

———1981. *Murav'i Pustyn'.* Moscow. 230 pp.

Dlussky, G. M., and E. B. Fedoseeva. 1988. Proiskhozhdenie i rannie etapy evolyutsii murav'ev. In A. G. Ponomarenko, ed., *Melovoi Biotsenoticheskii Krizis i Evolyutsiya Nasekomykh,* 70–144. Moscow. 228 pp.

Dlussky, G. M., and B. Pisarski. 1971. Rewizja polskich gatunków mrówek z rodzaju *Formica* L. *Fragmenta Faunistica* **16:** 145- 224.

Dlussky, G. M., and O. S. Soyunov. 1988. Murav'i roda *Temnothorax* Mayr SSSR. *Izvestiya Akademii Nauk Turkmenskoi SSR.* Seriya Biologicheskikh Nauk **1988** (no. 4): 29–37.

Dlussky, G. M., O. S. Soyunov, and S. I. Zabelin. 1990. *Muravy Turkmenistana.* Ashkhebad, Ylym. 273 pp.

Donisthorpe, H. 1941. *Lordomyrma niger* sp. n., with key and notes on the genus. *Entomologist's Monthly Magazine* **77:** 36–38.

DuBois, M. B. 1986. A revision of the native New World species of the ant genus *Monomorium* (*minimum* group). *University of Kansas Science Bulletin* **53:** 65–119.

Dumpert, K. 1978. *Das Sozialleben der Ameisen.* Berlin and Hamburg. 298 pp. [English translation: 1981. *The Social Biology of Ants.* Bath. 298 pp.]

———1986. *Camponotus (Karavaievia) texens* sp. n. and *C. (K.) gombaki* sp. n. from Malaysia in comparison with the other *Karavaievia* species. *Psyche* **92** (1985): 557–574.

Emery, C. 1877. Saggio di un ordinamento naturale dei Mirmicidei e considerazioni sulla filogenesi delle formiche. *Bullettino della Società Entomologica Italiana* **9:** 67–83.

———1895. Die Gattung *Dorylus* Fab. und die systematische Eintheilung der Formiciden. *Zoologische Jahrbücher. Abtheilung für Systematik, Geographie und Biologie der Thiere* **8:** 685–778.

———1896. Clef analytique des genres de la famille des formicides. *Annales de la Société Entomologique de Belgique* **40:** 172–189.

———1899. Intorno alle larve di alcune formiche. *Memorie della R. Accademia delle Scienze dell'Istituto di Bologna* (5) **8:** 3–10.

———1901. Notes sur les sous-familles des dorylines et ponérines (famille des formicides). *Annales de la Société Entomologique de Belgique* **45:** 32–54.

———1908a. Beiträge zur Monographie der Formiciden des paläarktischen Faunengebietes. *Deutsche Entomologische Zeitschrift* **1908:** 165–205.

———1908b. Beiträge zur Monographie der Formiciden des paläarktischen Faunengebietes. 3. Die mit *Aphaenogaster* verwandte Gattungengruppe. *Deutsche Entomologische Zeitschrift* **1908:** 305–338.

———1908c. Beiträge zur Monographie der Formiciden des paläarktischen Faunengebietes. 3. *Deutsche Entomologische Zeitschrift* **1908:** 437–465.

———1908d. Beiträge zur Monographie der Formiciden des paläarktischen Faunengebietes. 4. Parasitische und Gast- Myrmicinen mit Ausnahme von *Strongylognathus. Deutsche Entomologische Zeitschrift* **1908:** 549–558.

———1908e. Beiträge zur Monographie der Formiciden des paläarktischen Faunengebietes. 5. *Deutsche Entomologische Zeitschrift* **1908:** 663–686.

———1909a. Beiträge zur Monographie der Formiciden des paläarktischen Faunengebietes. 6. *Deutsche Entomologische Zeitschrift* **1909:** 19–37.

———1909b. Beiträge zur Monographie der Formiciden des paläarktischen Faunengebietes. 7. *Deutsche Entomologische Zeitschrift* **1909:** 179–204.

———1909c. Beiträge zur Monographie der Formiciden des paläarktischen Faunengebietes. 8. *Deutsche Entomologische Zeitschrift* **1909:** 355–376.

———1909d. Beiträge zur Monographie der Formiciden des paläarktischen Faunengebietes. 9. *Deutsche Entomologische Zeitschrift* **1909:** 695–712.

———1910a. Beiträge zur Monographie der Formiciden des paläarktischen Faunengebietes. 10. *Deutsche Entomologische Zeitschrift* **1910:** 127–132.

———1910b. In Hymenoptera, Fam. Formicidae, subfam. Dorylinae. P. Wytsman, ed., *Genera Insectorum,* fasc. 102. Brussels. 34 pp.

———1911. In Hymenoptera, Fam. Formicidae, subfam. Ponerinae. P. Wytsman, ed., *Genera Insectorum,* fasc. 118. Brussels. 124 pp.

———1912. In Hymenoptera, Fam. Formicidae, subfam. Dolichoderinae. P. Wytsman, ed., Genera Insectorum, fasc. 137. Brussels. 50 pp.

———1914. Intorno alla classificazione dei Myrmicinae. *Rendiconto delle Sessioni della R. Accademia delle Scienze dell'Istituto di Bologna* (n.s.) **18:** 29–41.

———1915. Noms de sous-genres et de genres proposés pour la sous-famille des Myrmicinae. Modifications à la classification de ce groupe. *Bulletin de la Société Entomologique de France* **1915:** 189–192.

———1921. In Hymenoptera, Fam. Formicidae, subfam. Myrmicinae. P. Wytsman, ed., *Genera Insectorum,* fasc. 174A, 1–94. Brussels.

———1922a. In Hymenoptera, Fam. Formicidae, subfam. Myrmicinae. P. Wytsman, ed., *Genera Insectorum,* fasc. 174B, 95–206. Brussels.

———1922b. In Hymenoptera, Fam. Formicidae, subfam. Myrmicinae. P. Wytsman, ed., *Genera Insectorum,* fasc. 174C, 207–397. Brussels.

———1925a. In Hymenoptera, Fam. Formicidae, subfam. Formicinae. P. Wytsman, ed., *Genera Insectorum,* fasc. 183. Brussels. 302 pp.

———1925b. Revision des espèces paléarctiques du genre *Tapinoma. Revue Suisse de Zoologie* **32:** 45–64.

———1925c. Les espèces européennes et orientales du genre *Bothriomyrmex. Bulletin de la Société Vaudoise des Sciences Naturelles* **56:** 5–22.

Ettershank, G. 1966. A generic revision of the world Myrmicinae related to *Solenopsis* and *Pheidologeton. Australian Journal of Zoology* **14:** 73–171.

Forel, A. 1878. Etudes myrmécologiques en 1878 (première partie) avec l'anatomie du gésier des fourmis. *Bulletin de la Société Vaudoise des Sciences Naturelles* **15:** 337–392.

———1885. Etudes myrmécologiques en 1884; avec une description des organes sensoriels des antennes. *Bulletin de la Société Vaudoise des Sciences Naturelles* **20** (1884): 316- 380.

———1893. Sur la classification de la famille des formicides, avec remarques synonymiques. *Annales de la Société Entomologique de Belgique* **37:** 161–167.

———1917. Cadre synoptique actuel de la faune universelle des fourmis. *Bulletin de la Société Vaudoise des Sciences Naturelles* **51:** 229–253.

Fowler, H. G. 1988. Taxa of the Neotropical grass-cutting ants, *Acromyrmex* (*Moellerius*). *Científica, Sao Paulo* **16:** 281–295.

Francoeur, A. 1973. Révision taxonomique des espèces Néarctiques du groupe *fusca,* genre *Formica. Mémoires de la Société Entomologique du Québec* **3:** 1–316.

———1977. Synopsis taxonomique et économique des fourmis du Québec. *Annales de la Société Entomologique de Québec* **22:** 205–212.

———1979. Les fourmis du Québec. 1. Introduction. 2. La famille des Formicidae. 3. La sous-famille des Ponerinae. *Annales de la Société Entomologique de Québec* **24:** 48–64.

Francoeur, A., R. Loiselle, and A. Buschinger. 1985. Biosystématique de la tribu Leptothoracici. 1. Le genre *Formicoxenus* dans la région holarctique. *Naturaliste Canadien* **112:** 343–403.

Gallardo, A. 1916. Las hormigas de la República Argentina. Subfamilia dolicoderinas. *Anales del Museo Nacional de Historia Natural de Buenos Aires* **28:** 1–130.

Gauld, I. D., and B. Bolton. 1988. *The Hymenoptera.* Oxford University Press and British Museum (Natural History). 332 pp.

Gonçalves, C. R. 1961. O gênero *Acromyrmex* no Brasil. *Studia Entomologica* (n.s.) **4:** 113–180.

Gösswald, K. 1985. *Organisation und Leben der Ameisen.* Stuttgart. 355 pp.

———1989. *Die Waldameise.* Vol. **1:** *Biologische Grundlagen, Ökologie und Verhalten.* Wiesbaden. 660 pp.

Gotwald, W. H., Jr. 1982. Army ants. In H. R. Hermann, ed., *Social Insects* **4:** 157–254. New York. 385pp.

Gotwald, W. H., Jr. and W. L. Brown, Jr. 1967. The ant genus *Simopelta. Psyche* **73** (1966): 261–276.

Gregg, R. E. 1959. Key to the species of *Pheidole* in the United States. *Journal of the New York Entomological Society* **66** (1958): 7–48.

———1963. *The Ants of Colorado.* Boulder. 792 pp.

Handlirsch, A. 1906. *Die Fossilen Insekten und die Phylogonie der Rezenten Formen. Ein Handbuch für Paläontologen und Zoologen.* **4:** 481–640. Leipzig.

———1907. *Die Fossilen Insekten und die Phylogenie der Rezenten Formen. Ein Handbuch für Paläontologen und Zoologen.* **6:** 801–960. Leipzig.

Hashmi, A. A. 1973. A revision of the Neotropical ant subgenus *Myrmothrix* of genus *Camponotus. Studia Entomologica* (n.s.) **16:** 1–140.

Heer, O. 1865. *Die Urwelt der Schweiz.* Zürich. 622 pp.

Hermann, H. R., ed. 1979–1982. *Social Insects.* 4 volumes. Vol. **1** (1979): 437 pp.; vol. **2** (1981): 491 pp.; vol. **3** (1982): 468 pp.; vol. **4** (1982): 398 pp. New York.

———ed. 1984. *Defensive Mechanisms in Social Insects.* New York. 259 pp.

Hölldobler, B., and E. O. Wilson. 1990. *The Ants.* Cambridge, Mass. 732 pp.

Hung, A. C. F. 1967. A revision of the ant genus *Polyrhachis* at the subgeneric level. *Transactions of the American Entomological Society* **93:** 395–422.

———1970. A revision of ants of the subgenus *Polyrhachis* Fr. Smith. *Oriental Insects* **4:** 1–36.

Johnson, C. 1988. Species identification in the eastern *Crematogaster. Journal of Entomological Science* **23:** 314–332.

———1989a. Identification and nesting sites of North American species of *Dolichoderus* Lund. *Insecta Mundi* **3:** 1–9.

———1989b. Taxonomy and diagnosis of *Conomyrma insana* (Buckley) and *C. flava* (McCook). *Insecta Mundi* **3:** 179–194.

Jolivet, P. 1986. *Les fourmis et les plantes. Un example de coévolution.* Paris. 254 pp.

Keilbach, R. 1982. Bibliographie und Liste der Arten tierischer Einschlüsse in fossilen Harzen sowie ihrer Aufbewahrungsorte. *Deutsche Entomologische Zeitschrift* (N.F.) **29:** 129–286 (Formicidae, 272–281).

Kempf, W. W. 1951. A taxonomic study on the ant tribe Cephalotini. *Revista de Entomologia* **22:** 1–244.

———1952. A synopsis of the *pinelii*-complex in the genus *Paracryptocerus. Studia Entomologica* **1:** 1–30.

———1958a. New studies of the ant tribe Cephalotini. *Studia Entomologica* (n.s.) **1:** 1–176.

———1958b. Estudos sôbre *Pseudomyrmex.* 2. *Studia Entomologica* (n.s.) **1:** 433–462.

———1959a. A revision of the Neotropical ant genus *Monacis* Roger. *Studia Entomologica* (n.s.) **2:** 225–270.

———1959b. A synopsis of the New World species belonging to the *Nesomyrmex*-group of the ant genus *Leptothorax* Mayr. *Studia Entomologica* (n.s.) **2:** 391–432.

———1959c. Two new species of *Gymnomyrmex* Borgmeier, 1954, from southern Brazil, with remarks on the genus. *Revista Brasileira de Biologia* **19:** 337–344.

———1960a. Estudo sôbre *Pseudomyrmex.* 1. *Revista Brasileira de Entomologia* **9:** 5–32.

———1960b. *Phalacromyrmex,* a new ant genus from southern Brazil. *Revista Brasileira de Biologia* **20:** 89–92.

———1960c. A review of the ant genus *Mycetarotes* Emery. *Revista Brasileira de Biologia* **20:** 277–283.

———1960d. Miscellaneous studies on Neotropical ants. *Studia Entomologica* (n.s.) **3:** 417–466.

———1961a. Estudos sôbre *Pseudomyrmex.* 3. *Studia Entomologica* (n.s.) **4:** 369–408.

———1961b. As formigas do gênero *Pachycondyla* Fr. Smith no Brasil. *Revista Brasileira de Entomologia* **10:** 189–204.

———1962a. Retoques à classificaçao das formigas neotropicas do gênero *Heteroponera. Papeís Avulsos do Departamento de Zoologia* **15:** 29–47.

———1962b. Miscellaneous studies on Neotropical ants. 2. *Studia Entomologica* (n.s.) **5:** 1–38.

———1963a. Additions to the Neotropical ant genus *Rogeria* Emery, with a key to the hitherto recorded South American species. *Revista Brasileira de Biologia* **23:** 189–196.

———1963b. A review of the ant genus *Mycocepurus* Forel, 1893. *Studia Entomologica* (n.s.) **6:** 417–432.

———1964a. A propósito de um estudo sôbre as formigas do gênero *Acanthostichus* Mayr. *Papéis Avulsos do Departamento de Zoologia* **16:** 263–266.

———1964b. A revision of the Neotropical fungus-growing ants of the genus *Cyphomyrmex* Mayr. Part 1. Group of *strigatus* Mayr. *Studia Entomologica* (n.s.) **7:** 1–44.

———1964c. Miscellaneous studies on Neotropical ants. 3. *Studia Entomologica* (n.s.) **7:** 45–71.

———1964d. The ants of the genus *Anochetus* (*Stenomyrmex*) in Brazil. *Studia Entomologica* (n.s.) **7:** 237–246.

———1966. A revision of the Neotropical fungus-growing ants of the genus *Cyphomyrmex* Mayr. Part 2. Group of *rimosus* (Spinola). *Studia Entomologica* (n.s.) **8** (1965): 161–200.

———1967a. New ants from southeastern and central Brazil. *Studia Entomologica* (n.s.) **9** (1966): 121–128.

———1967b. A synopsis of the Neotropical ants of the genus *Centromyrmex* Mayr. *Studia Entomologica* (n.s.) **9** (1966): 401–410.

———1969. Miscellaneous studies on Neotropical ants. 5. *Studia Entomologica* (n.s.) **12:** 273–296.

———1971. A preliminary review of the ponerine ant genus *Dinoponera* Roger. *Studia Entomologica* (n.s.) **14:** 369–394.

———1972a. Catálogo abreviado das formigas da Regiao Neotropical. *Studia Entomologica* (n.s.) **15:** 3–344.

———1972b. A new species of the dolichoderine ant genus *Monacis* Roger, from the Amazon, with further remarks on the genus. *Revista Brasileira de Biologia* **32:** 251–254.

———1973a. A revision of the Neotropical myrmicine ant genus *Hylomyrma* Forel. *Studia Entomologica* (n.s.) **16:** 225- 260.

———1973b. A new *Zacryptocerus* from Brazil, with remarks on the generic classification of the tribe Cephalotini. *Studia Entomologica* (n.s.) **16:** 449–462.

———1974. A review of the Neotropical ant genus *Oxyepoecus* Santschi. *Studia Entomologica* (n.s.) **17:** 471–512.

———1975. A revision of the Neotropical ponerine ant genus *Thaumatomyrmex* Mayr. *Studia Entomologica* (n.s.) **18:** 95–126.

———1976. A new species of *Strumigenys* from the lower Amazon, Brazil. *Studia Entomologica* (n.s.) **19:** 39–44.

Kipyatkov, V. E. 1991. *Mir Obshchestvennykh Nasekomykh.* Leningrad. 406 pp.

Kohout, R. J. 1987. Three new *Polyrhachis sexspinosa*-group species from the Philippines. *Memoirs of the Queensland Museum* **25:** 169–176.

———1988. A new species of *Polyrhachis (Polyrhachis)* from Papua New Guinea with a review of the New Guinean and Australian species. *Memoirs of the Queensland Museum* **25:** 417–427.

———1989. The Australian ants of the *Polyrhachis relucens* species-group. *Memoirs of the Queensland Museum* **27:** 509–516.

———1990. A review of the *Polyrhachis viehmeyeri* species-group. *Memoirs of the Queensland Museum* **28:** 499–508.

Kugler, C. In preparation. Revision of the ant genus *Rogeria,* with descriptions of the sting apparatus.

Kugler, C. and W. L. Brown, Jr. 1982. Revisionary and other studies on the ant genus *Ectatomma,* including the descriptions of two new species. *Search Agriculture* **24:** 1–7.

Kugler, J. 1984. The males of *Cardiocondyla* Emery, with a description of the winged male of *Cardiocondyla wroughtonii* (Forel). *Israel Journal of Entomology* **17** (1983): 1–21.

———1987. The Leptanillinae of Israel and a description of a new species from India. *Israel Journal of Entomology* **20** (1986): 45–57.

Kupyanskaya, A. N. 1980. Murav'i roda *Formica* Linnaeus Sovetskogo dal'nego Vostoka. In P. A. Ler, ed., *Taksonomiya Nasekomykh dal'nego Vostoka,* 95–108. Vladivostok.

———1986. Murav'i gruppy *Myrmica lobicornis* na dal'nem Vostoke. In P. A. Ler and A. N. Kupyanskaya, eds., *Sistematika i Ekologiya Nasekomykh dal'nego Vostoka,* 83–90. Vladivostok.

———1988. Dal'nevostochnyi predstavitel' roda *Liometopum. Vestnyk Zoologii* **1988** (no. 1): 29–34.

———1989. Murav'i podroda *Dendrolasius* Ruzsky, 1912 dal'nego Vostoka SSR. *Entomologicheskoe Obozrenie* **68:** 779–789.

———1990. *Murav'i Dal'nego Vostoka SSR* (1989). Vladivostok. 258 pp.

Kusnezov, N. 1951. El genero *Pogonomyrmex* Mayr. *Acta Zoologica Lilloana* **11:** 227–333.

———1952a. El género *Pheidole* en la Argentina. *Acta Zoologica Lilloana* **12** (1951): 2–88.

———1952b. El género *Camponotus* en la Argentina. *Acta Zoologica Lilloana* **12** (1951): 183–252.

———1962. El género *Acanthostichus* Mayr. *Acta Zoologica Lilloana* **18:** 121–138.

Kutter, H. 1973a. Zur Taxonomie der Gattung *Chalepoxenus. Mitteilungen der Schweizerischen Entomologischen Gesellschaft* **46:** 269–280.

———1973b. Beitrag zur Lösung Taxonomischer Probleme in der Gattung *Epimyrma. Mitteilungen der Schweizerischen Entomologischen Gesellschaft* **46:** 281–289.

———1977. *Insecta Helvetica Fauna.* Vol. **6:** *Hymenoptera, Formicidae.* Zürich. 298 pp.

Lattke, J. E. 1987. Notes on the ant genus *Hypoclinea* Mayr, with descriptions of three new species. *Revista de Biología Tropical* **34** (1986): 259–265.

———1990a. A new genus of myrmicine ants from Venezuela. *Entomologica Scandinavica* **21:** 173–178.

———1990b. Revisión del género *Gnamptogenys* Roger en Venezuela. *Acta Terramaris* **2:** 1–47.

———1991. Studies of Neotropical *Amblyopone* Erichson. *Contributions in Science* **428:** 1–7.

———1992. Revision of the *minuta*-group of the genus *Gnamptogenys. Deutsche Entomologische Zeitschrift* **39:** 123- 129.

Longino, J. T. 1988. Notes on the taxonomy of the Neotropical ant genus *Thaumatomyrmex* Mayr. In J. C. Trager, ed., *Advances in Myrmecology,* New York. 551 pp.

———1989. Taxonomy of the *Cecropia*-inhabiting ants in the *Azteca alfari* species group: evidence for two broadly sympatric species. *Contributions in Science* **412:** 1–16.

———1991. Taxonomy of the *Cecropia*-inhabiting *Azteca* ants. *Journal of Natural History* **25:** 1571–1602.

Lutz, H. 1986. Eine neue Unterfamilie der Formicidae aus dem mitteleozänen Ölschiefer der "Grube Messel" bei Darmstadt (Deutschland, S-Hessen). *Senckenbergiana Lethaea* **67:** 177- 218.

———1990. Systematische und palökologische Untersuchungen an Insekten aus dem Mittel-Eozän der Grube Messel bei Darmstadt. *Courier Forschungsinstitut Senckenberg* **124:** 1- 164.

MacKay, W. P. 1993. A review of the New World ants of the genus *Dolichoderus. Sociobiology* **22:** 1–148.

MacKay, W. P., E. E. MacKay, J. F. P. Dominguez, L. I. V. Sanchez, and P. V. Orozco. 1985. Las hormigas del Estado de Chihuahua, Mexico: el género *Pogonomyrmex. Sociobiology* **1:** 39–54.

Mann, W. M. 1916. The Stanford expedition to Brazil, 1911. John C. Branner, director. The ants of Brazil. *Bulletin of the Museum of Comparative Zoology at Harvard College* **60:** 399–490.

———1919. The ants of the British Solomon Islands. *Bulletin of the Museum of Comparative Zoology at Harvard College* **63:** 273–391.

———1921. The ants of the Fiji Islands. *Bulletin of the Museum of Comparative Zoology at Harvard College* **64:** 401- 499.

McAreavey, J. J. 1947. New species of the genera *Prolasius* Forel and *Melophorus* Lubbock. *Memoirs of the National Museum of Victoria* **15:** 7–27.

———1957. Revision of the genus *Stigmacros* Forel. *Memoirs of the National Museum of Victoria* **21:** 7–64.

Menozzi, C. 1924. Alcune nuove formiche africane. *Annali del Museo Civico di Storia Naturale di Genova* **51:** 220–227.

———1929. Revisione delle formiche del genere *Mystrium* Roger. *Zoologischer Anzeiger* **82:** 518–536.

———1931. Revisione del genere *Epimyrma* Em. e descrizione di una specie inedita di questo genere. *Memorie della Società Entomologica Italiana* **10:** 36–53.

———1939. Formiche dell'Himalaya e del Karakorum raccolte dalla Spedizione Italiana comandata da S.A.R. il Duca di Spoleto (1929). *Atti della Società Italiana di Scienze Naturali* **78:** 285–345.

Moffett, M. W. 1985. Revision of the genus *Myrmoteras. Bulletin of the Museum of Comparative Zoology* **151:** 1–53.

———1986. Revision of the myrmicine genus *Acanthomyrmex. Bulletin of the Museum of Comparative Zoology* **151:** 55–89.

Morisita, M., M. Kubota, K. Onoyama, K. Ogata, M. Terayama, M. Kondoh, and H. T. Imai. 1989. A guide for the identification of Japanese ants. 1. Ponerinae, Cerapachyinae, Pseudomyrmecinae, Dorylinae and Leptanillinae. *Myrmecological Society of Japan:* 42 pp. [In Japanese.]

Morisita, M., M. Kubota, K. Onoyama, K. Ogata, M. Terayama, K. Yamauchi, R. Sonobe, M. Kondoh, and H. T. Imai. 1991. A guide for the identification of Japanese ants. 2. Dolichoderinae and Formicinae. *Myrmecological Society of Japan:* 56 pp. [In Japanese.]

Morisita, M., M. Kubota, K. Onoyama, K. Ogata, M. Terayama, K. Yamauchi, R. Sonobe, S. Yamane, M. Kondoh, and H. T. Imai. 1992. A guide for the identification of Japanese ants. 3. Myrmicinae and supplement to Leptanillinae. *Myrmecological Society of Japan:* 94 pp. [In Japanese.]

Najt, J. 1987. Le collembole fossile *Paleosminthurus juliae* est un hyménoptère. *Revue Francaise d'Entomologie* (n.s.) **9:** 152- 154.

Naves, M. A. 1985. A monograph of the genus *Pheidole* in Florida, USA. *Insecta Mundi* **1:** 53–90.

Nilsson, O. and P. Douwes, 1987. Norrländska myror med bestämningstabell till arbetarna. *Natur i Norr* **6:** 49–90.

Ogata, K. 1982. Taxonomic study of the ant genus *Pheidole* Westwood of Japan, with a description of a new species. *Kontyû* **50:** 189–197.

———1983. The ant genus *Cerapachys* F. Smith of Japan, with description of a new species. *Esakia* **20:** 131–137.

———1987. A generic synopsis of the poneroid complex of the family Formicidae in Japan. Part 1. Subfamilies Ponerinae and Cerapachyinae. *Esakia* **25:** 97–132.

———1991a. Ants of the genus *Myrmecia* Fabricius: a review of the species groups and their phylogenetic relationships. *Systematic Entomology* **16:** 353–381.

———1991b. A generic synopsis of the poneroid complex of the family Formicidae in Japan. Part 2. Subfamily Myrmicinae. *Bulletin of the Institute of Tropical Agriculture, Kyushu University* **14:** 61–149.

Ogata, K., and R. W. Taylor. 1991. Ants of the genus *Myrmecia* Fabricius: a preliminary review and key to the named species. *Journal of Natural History* **25:** 1623–1673.

Onoyama, K. 1980. An introduction to the ant fauna of Japan, with a check list. *Kontyû* **48:** 193–212.

———1989. Notes on the ants of the genus *Hypoponera* in Japan. *Edaphologia* **41:** 1–10.

Passera, L. 1984. *L'Organisation Sociale des Fourmis.* Toulouse. 360 pp. Toulouse.

Perrault, G. H. 1986. *Gymnomyrmex villiersi,* nouvelle espèce de Guyane. *Revue Francaise d'Entomologie* (n.s.) **8:** 1–4.

Petersen, B. 1968. Some novelties in presumed males of Leptanillinae. *Entomologiske Meddelelser* **36:** 577–598.

Pierce, W. D. and J. Gibron, Sr. 1962. Fossil arthropods of California. 24. Some unusual fossil arthropods from the Calico Mountains nodules. *Bulletin of the Southern California Academy of Sciences* **61:** 143–151.

Prins, A. J. 1982. Review of *Anoplolepis* with reference to male genitalia, and notes on *Acropyga. Annals of the South African Museum* **89:** 215–247.

———1983. A new ant genus from southern Africa. *Annals of the South African Museum* **94:** 1–11.

Radchenko, A. G. 1985. Murav'i roda *Strongylognathus* evropeiskoi chasti SSR. *Zoologicheskii Zhurnal* **64:** 1514–1523.

———1989a. Murav'i roda *Chalepoxenus* fauny SSSR. *Vestnik Zoologii* **1989** (no. 2): 37–41.

———1989b. Murav'i roda *Plagiolepis* evropeiskoi chasti SSR. *Zoologicheskii Zhurnal* **68:** 153–156.

———1991. Murav'i roda *Strongylognathus* fauny SSSR. *Zoologicheskii Zhurnal* **70:** 84–90.

Radchenko, A. G. 1992. Murav'i roda *Tetramorium* fauny S.S.S.R. *Zoologicheskii Zhurnal* **71** (8): 39–49.

Radchenko, A. G., and G. R. Arakelian. 1990. Murav'i gruppy *Tetramorium ferox* Ruzky iz Kryma i Kavkaza. *Biologicheskii Zhurnal Armenii* **43:** 371–378.

Robertson, H. G. 1990. Unravelling the *Camponotus fulvopilosus* species complex. In G. K. Veeresh, B. Mallik, and C. A. Viraktamath, eds., *Social Insects and the Environment,* 327–328. *Proceedings of the 11th International Congress of IUSSI, 1990.* New Delhi.

Santschi, F. 1922. Myrmicines, dolichodérines et autres formicides néotropiques. *Bulletin de la Société Vaudoise des Sciences Naturelles* **54:** 345–378.

———1923. Revue des fourmis du genre *Brachymymex* Mayr. *Anales del Museo Nacional de Historia Natural de Buenos Aires* **31:** 650–678.

———1924. Descriptions de nouveaux formicides africains et notes diverses. 2. *Revue Zoologique Africaine* **12:** 195–224.

———1925. Révision des *Myrmicaria* d'Afrique. *Annales de la Société Entomologique de Belgique* **64** (1924): 133–176.

———1929a. Etude sur les *Cataglyphis. Revue Suisse de Zoologie* **36:** 25–70.

———1929b. Fourmis du Maroc, d'Algérie et de Tunisie. *Annales et Bulletin de la Société Entomologique de Belgique* **69:** 138–165.

Schembri, S. P., and C. A. Collingwood. 1981. A revision of the myrmecofauna of the Maltese Islands. *Annali del Museo Civico di Storia Naturale di Genova* **83:** 417–442.

Schneirla, T. C. 1971. *Army Ants. A Study in Social Organisation.* San Francisco. 349 pp.

Seifert, B. 1988a. A revision of the European species of the ant subgenus *Chthonolasius. Entomologische Abhandlungen. Staatliches Museum für Tierkunde Dresden* **51:** 143–180.

———1988b. A taxonomic revision of the *Myrmica* species of Europe, Asia Minor, and Caucasus. *Abhandlungen und Berichte des Naturkundemuseums Görlitz* **62:** 1–75.

———1990. Supplementation to the revision of the European species of the ant subgenus *Chthonolasius. Doriana. Supplemento agli Annali del Museo Civico di Storia Naturale "G. Doria"* **6:** 1–13.

———1992. A taxonomic revision of the Palaearctic members of the ant subgenus *Lasius* s. str. *Abhandlungen und Berichte des Naturkundemuseums Görlitz* **66:** 1–67.

Shattuck, S. O. 1987. An analysis of geographic variation in the *Pogonomyrmex occidentalis* complex. *Psyche* **94:** 159–180.

———1990. Revision of the dolichoderine ant genus *Turneria. Systematic Entomology* **15:** 101–117.

———1991. Revision of the dolichoderine ant genus *Axinidris. Systematic Entomology* **16:** 105–120.

———1992a. Review of the dolichoderine ant genus *Iridomyrmex* Mayr with descriptions of three new genera. *Journal of the Australian Entomological Society* **31:** 13–18.

———1992b. Higher classification of the ant subfamilies Aneuretinae, Dolichoderinae and Formicinae. *Sysytematic Entomology* **17:** 199–206.

———1992c. Generic revision of the ant subfamily Dolichoderinae. *Sociobiology* **21:** 1–181.

———1993. Taxonomic catalog of the ant subfamilies Aneuretinae and Dolichoderinae. [Currently in press.]

Smith, D. R. 1979. Formicoidea. In K. V. Krombein, P. D. Hurd, Jr., D. R. Smith, and B. D. Burks, eds., *Catalog of Hymenoptera in America North of Mexico.* Vol. **2:** *Apocrita (Aculeata),* 1323–1467. [Apocrita, 1199–2209.] Washington, D.C.

Smith, M. R. 1951. Formicidae. In C. F. W. Muesebeck, K. V. Krombein, and H. K. Townes, eds., *Hymenoptera of America North of Mexico. Synoptic Catalog, United States Department of Agriculture. Agriculture Monograph* **2:** 778–875.

———1952. The correct name for the group of ants formerly known as *Pseudomyrma. Proceedings of the Entomological Society of Washington* **54:** 97–98.

———1953. A revision of the genus *Romblonella* W.M. Wheeler. *Proceedings of the Hawaiian Entomological Society* **15:** 75–80.

———1956a. A key to the workers of *Veromessor* Forel of the United States and the description of a new subspecies. *Pan-Pacific Entomologist* **32:** 36–38.

———1956b. A list of the species of *Romblonella* including two generic transfers. *Bulletin of the Brooklyn Entomological Society* **51:** 18.

———1957. Revision of the genus *Stenamma* Westwood in America north of Mexico. *American Midland Naturalist* **57:** 133–174.

———1958. Formicidae. In K. V. Krombein, ed., *Hymenoptera of America North of Mexico. Synoptic Catalog. First Supplement, United States Department of Agriculture. Agriculture Monograph* **2** (Supplement 1): 108–162.

———1961. A study of New Guinea ants of the genus *Aphaenogaster* Mayr. *Acta Hymenopterologica* **1:** 213–237.

———1962. A remarkable new *Stenamma* from Costa Rica, with pertinent facts on other Mexican and Central American species. *Journal of the New York Entomological Society* **70:** 33–38.

Snelling, R. R. 1967. Studies on California ants. 3. The taxonomic status of *Proceratium californicum* Cook. *Contributions in Science* **124:** 1–10.

———1973. Studies on California ants. 7. The genus *Stenamma. Contributions in Science* **245:** 1–38.

———1975. Descriptions of new Chilean ant taxa. *Contributions in Science* **274:** 1–19.

———1976. A revision of the honey ants, genus *Myrmecocystus. Natural History Museum of Los Angeles County Science Bulletin* **24:** 1–163.

———1979. *Aphomomyrmex* and a related new genus of arboreal African ants. *Contributions in Science* **316:** 3–8.

———1981. Systematics of social Hymenoptera. In H. R. Hermann, ed., *Social Insects* **2:** 369–453. New York. 491 pp.

———1982a. The taxonomy and distribution of some North American *Pogonomyrmex* and descriptions of two new species. *Bulletin of the Southern California Academy of Sciences* **80** (1981): 97–112.

———1982b. A revision of the honey ants, genus *Myrmecocystus,* first supplement. *Bulletin of the Southern California Academy of Sciences* **81:** 69–86.

———1988. Taxonomic notes on Nearctic species of *Camponotus,* subgenus *Myrmentoma.* In J. C. Trager, ed., *Advances in Myrmecology,* 55–78. New York. 551 pp.

Snelling, R. R., and W. R. Buren. 1985. Description of a new species of slave-making ant in the *Formica sanguinea* group. *Great Lakes Entomologist* **18:** 69–78.

Snelling, R. R., and J. H. Hunt, 1976. The ants of Chile. *Revista Chilena de Entomologia* **9** (1975): 63–129.

Snelling, R. R., and J. T. Longino. 1992. Revisionary notes on the fungus-growing ants of the genus *Cyphomyrmex, rimosus*-group. In D. Quintero and A. Aiello, eds., *Insects of Panama and Mesoamerica: Selected Studies,* 479–494. Oxford. 692 pp.

Spahr, U. 1987. Ergänzungen und Berichtigungen zu R. Keilbach's Bibliographie und Liste der Bernsteinfossilien—Ordnung Hymenoptera. *Stuttgarter Beiträge zur Naturkende,* ser. B (Geologie und Paläontologie) **127:** 1–121 (Formicidae, 41–64).

Sudd, J. H., and N. R. Franks. 1987. *The Behavioural Ecology of Ants.* Glasgow and London. 206 pp.

Tarbinsky, Y. S. 1976. *Murav'i Kirgizii.* Frunze. 217 pp.

Taylor, R. W. 1965a. The Australian ants of the genus *Pristomyrmex,* with a case of apparent character displacement. *Psyche* **72:** 35–54.

———1965b. A monographic revision of the rare tropicopolitan ant genus *Probolomyrmex* Mayr. *Transactions of the Royal Entomological Society of London* **117:** 345–365.

———1967. A monographic revision of the ant genus *Ponera* Latreille. *Pacific Insects Monograph* **13:** 1–112.

———1968a. Notes on the Indo-Australian basicerotine ants. *Australian Journal of Zoology* **16:** 333–348.

———1968b. A supplement to the revision of Australian *Pristomyrmex* species. *Journal of the Australian Entomological Society* **7:** 63–66.

———1968c. A new Malayan species of the ant genus *Epitritus,* and a related new genus from Singapore. *Journal of the Australian Entomological Society* **7:** 130–134.

———1970. Characterization of the Australian endemic ant genus *Peronomyrmex* Viehmeyer. *Journal of the Australian Entomological Society* **9:** 209–211.

———1973. Ants of the Australian genus *Mesostruma* Brown. *Journal of the Australian Entomological Society* **12:** 24–38.

———1977. New ants of the Australasian genus *Orectognathus,* with a key to the known species. *Australian Journal of Zoology* **25:** 581–612.

———1978a. *Nothomyrmecia macrops:* a living-fossil ant rediscovered. *Science* **201:** 979–985.

———1978b. A taxonomic guide to the ant genus *Orectognathus. CSIRO Division of Entomology,* report no. **3:** 1–11 (plus microfiche).

———1979. Melanesian ants of the genus *Amblyopone. Australian Journal of Zoology* **26** (1978): 823–839.

———1980a. New Australian ants of the genus *Orectognathus,* with summary description of the twenty-nine known species. *Australian Journal of Zoology* **27** (1979): 773- 788.

———1980b. Australian and Melanesian ants of the genus *Eurhopalothrix* Brown and Kempf—notes and new species. *Journal of the Australian Entomological Society* **19:** 229–239.

———1980c. Notes on the Russian endemic ant genus *Aulacopone* Arnoldi. *Psyche* **86** (1979): 353–361.

———1985. The ants of the Papuasian genus *Dacetinops.* In G. E. Ball, ed., *Taxonomy, Phylogeny and Zoogeography of Beetles and Ants,* 41–67. Dordrecht.

———1987a. A checklist of the ants of Australia, New Caledonia and New Zealand. *CSIRO Division of Entomology,* Report no. **41:** 1–92.

———1987b. A checklist of the ants of Australia, New Caledonia and New Zealand. (Supplement.) *CSIRO Division of Entomology,* Report no. **41** (supplement): 1–5.

———1989. Australian ants of the genus *Leptothorax* Mayr. *Memoirs of the Queensland Museum* **27:** 605–610.

———1990. New Asian ants of the tribe Basicerotini, with an on-line computer interactive key to the twenty-six known Indo-Australian species. *Invertebrate Taxonomy* **4:** 397–425.

———1991. Notes on the ant genera *Romblonella* and *Willowsiella,* with comments on their affinities, and the first description of Australian species. *Psyche* **97** (1990): 281–296.

Taylor, R. W., and D. R. Brown. 1985. *Zoological Catalogue of Australia.* Vol. **2:** *Hymenoptera: Formicoidea, Vespoidea and Sphecoidea.* Canberra. 381 pp.

Terayama, M. 1989. The ant tribe Amblyoponini of Taiwan, with description of a new species. *Japanese Journal of Entomology* **57:** 343–346.

Terayama, M., and S. Kubota. 1989. The ant tribe Dacetini of Taiwan, with descriptions of three new species. *Japanese Journal of Entomology* **57:** 778–792.

Terayama, M., and S. Yamane. 1989. The army ant genus *Aenictus* from Sumatra, with descriptions of three new species. *Japanese Journal of Entomology* **57:** 597–603.

Terron, G. 1974. Découverte au Cameroun de deux espèces nouvelles du genre *Prionopelta* Mayr. *Annales de la Faculté des Sciences du Cameroun* **17:** 105–119.

———1981. Deux nouvelles espèces éthiopiennes pour le genre *Proceratium. Annales de la Faculté des Sciences de Yaoundé* **28:** 95–103.

Thompson, C. R., and C. Johnson. 1989. Rediscovered species and revised key to the Florida thief ants. *Florida Entomologist* **72:** 697–698.

Tinaut, A. 1990. El género *Amblyopone* Erichson en la Península Ibérica. *Miscellania Zoologica* **12** (1988): 189–193.

Tohmé, G., and H. Tohmé. 1981. Les fourmis du genre *Messor* en Syrie. Position systématique. Description de quelques ailés et de formes nouvelles. Répartition géographique. *Ecologia Mediterranea* **7:** 139–153.

Trager, J. C. 1984. A revision of the genus *Paratrechina* of the continental United States. *Sociobiology* **9:** 51–162.

———1988. A revision of *Conomyrma* from the southeastern United States, especially Florida, with keys to the species. *Florida Entomologist* **71:** 11–29.

———1991. A revision of the fire ants, *Solenopsis geminata* group. *Journal of the New York Entomological Society* **99:** 141–198.

Wang, C., and J. Wu. 1991. Taxonomic studies on the genus *Polyrhachis* of China. *Forest Research* **4:** 596–601.

Wang, M., G. Xiao, and J. Wu. 1988. Taxonomic studies on the genus *Tetramorium* Mayr in China. *Forest Research* **1:** 264–274.

———1989a. Taxonomic studies on the genus *Camponotus* in China. *Forest Research* **2:** 221–228.

———1989b. Taxonomic studies on the genus *Camponotus* in China (conclusion). *Forest Research* **2:** 321- 328.

Ward, P. S. 1980. A systematic revision of the *Rhytidoponera impressa* group in Australia and New Guinea. *Australian Journal of Zoology* **28:** 475–498.

———1984. A revision of the ant genus *Rhytidoponera* in New Caledonia. *Australian Journal of Zoology* **32:** 131–175.
———1985. The Nearctic species of the genus *Pseudomyrmex. Quaestiones Entomologicae* **21:** 209–246.
———1988. Mesic elements in the western Nearctic ant fauna: taxonomic and biological notes on *Amblyopone, Proceratium* and *Smithistruma. Journal of the Kansas Entomological Society* **61:** 102–124.
———1989. Systematic studies on pseudomyrmecine ants: revision of the *Pseudomyrmex oculatus* and *P. subtilissimus* species groups, with taxonomic comments on other species. *Quaestiones Entomologicae* **25:** 393–468.
———1990. The ant subfamily Pseudomyrmecinae: generic revision and relationship to other formicids. *Systematic Entomology* **15:** 449–489.
Watkins, J. F., II. 1972. The taxonomy of *Neivamyrmex texanus* n. sp., *N. nigrescens* and *N. californicus,* with distribution map and keys to the species of *Neivamyrmex* of the United States. *Journal of the Kansas Entomological Society* **45:** 347–372.
———1976. *The Identification and Distribution of New World Army Ants.* Waco, Texas. 102 pp.
———1977. The species and subspecies of *Nomamyrmex. Journal of the Kansas Entomological Society* **50:** 203–214.
———1982. The army ants of Mexico. *Journal of the Kansas Entomological Society* **55:** 197–247.
Weber, N. A. 1944. The Neotropical coccid-tending ants of the genus *Acropyga* Roger. *Annals of the Entomological Society of America* **37:** 89–122.
———1947. A revision of the North American ants of the genus *Myrmica* Latreille with a synopsis of the Palaearctic species. 1. *Annals of the Entomological Society of America* **40:** 437–474.
———1948. A revision of the North American ants of the genus *Myrmica* Latreille, with a synopsis of the Palearctic species. 2. *Annals of the Entomological Society of America* **41:** 267–308.
———1950a. The African species of the genus *Oligomyrmex* Mayr. *American Museum Novitates* **1442:** 1–19.
———1950b. A revision of the North American ants of the genus *Myrmica* Latreille with a synopsis of the Palearctic species. 3. *Annals of the Entomological Society of America* **43:** 189–226.
———1950c. New Trinidad Myrmicinae, with a note on *Basiceros* Schulz. *American Museum Novitates* **1465:** 1–6.
———1952. Studies on African Myrmicinae, 1. *American Museum Novitates* **1548:** 1–32.
Wheeler, G. C., and J. Wheeler. 1985. A simplified conspectus of the Formicidae. *Transactions of the American Entomological Society* **111:** 255–264.
———1986. *The Ants of Nevada.* Lawrence, Kansas. 138 pp.
Wheeler, J. 1968. Male genitalia and the taxonomy of *Polyergus. Proceedings of the Entomological Society of Washington* **70:** 156–164.
Wheeler, W. M. 1902. An American *Cerapachys,* with remarks on the affinities of the Cerapachyinae. *Biological Bulletin* **3:** 181- 191.
———1918. The ants of the genus *Opisthopsis* Emery. *Bulletin of the Museum of Comparative Zoology at Harvard College* **62:** 341–362.
———1919. The ants of the genus *Metapone* Forel. *Annals of the Entomological Society of America* **12:** 173–191.
———1920. The subfamilies of Formicidae, and other taxonomic notes. *Psyche* **27:** 46–54.
———1922. The ants of the Belgian Congo. *Bulletin of the American Museum of Natural History* **45:** 1–1139.
———1924. Ants of Krakatau and other islands in the Sunda Straight. *Treubia* **5:** 239–258.
———1933. Three obscure genera of ponerine ants. *American Museum Novitates* **672:** 1–23.
———1934a. A second revision of the ants of the genus *Leptomyrmex* Mayr. *Bulletin of the Museum of Comparative Zoology at Harvard College* **77:** 69–118.
———1934b. Neotropical ants collected by Dr. Elisabeth Skwarra and others. *Bulletin of the Museum of Comparative Zoology at Harvard College* **77:** 159–240.
———1935. The Australian ant genus *Mayriella. Psyche* **42:** 151–160.
Willey, R. B., and W. L. Brown, Jr. 1983. New species of the ant genus *Myopias. Psyche* **90:** 249–285.
Wilson, E. O. 1955. A monographic revision of the ant genus *Lasius. Bulletin of the Museum of Comparative Zoology at Harvard College* **113:** 1–201.
———1957. The *tenuis* and *selenophora* groups of the ant genus *Ponera. Bulletin of the Museum of Comparative Zoology at Harvard College* **116:** 355–386.
———1958a. Studies on the ant fauna of Melanesia. 1. The tribe Leptogenyini. 2. The tribes Amblyoponini and Platythyreini. *Bulletin of the Museum of Comparative Zoology at Harvard College* **118:** 101–153.
———1958b. Studies on the ant fauna of Melanesia. 3. *Rhytidoponera* in western Melanesia and the Moluccas. 4. The tribe Ponerini. *Bulletin of the Museum of Comparative Zoology at Harvard College* **119:** 303–371.
———1959a. Studies on the ant fauna of Melanesia. 5. The tribe Odontomachini. *Bulletin of the Museum of Comparative Zoology at Harvard College* **120:** 483–510.
———1959b. Studies on the ant fauna of Melanesia. 6. The tribe Cerapachyini. *Pacific Insects* **1:** 39–57.
———1964. The true army ants of the Indo-Australian area. *Pacific Insects* **6:** 427–483.
———1971. *The Insect Societies.* Cambridge, Mass. 548 pp.
———1987. The earliest known ants: an analysis of the Cretaceous species and an inference concerning their social organisation. *Paleobiology* **13:** 44–53.
———1989. *Chimaeridris,* a new genus of hook-mandibled myrmicine ants from tropical Asia. *Insectes Sociaux* **36:** 62- 69.
Wilson, E. O., and W. L. Brown, Jr. 1955. Revisionary notes on the *sanguinea* and *neogagates* groups of the ant genus *Formica. Psyche* **62:** 108–129.
———1956. New parasitic ants of the genus *Kyidris,* with notes on ecology and behavior. *Insectes Sociaux* **3:** 439–454.
Wilson, E. O., T. Eisner, G. C. Wheeler, and J. Wheeler. 1956. *Aneuretus simoni* Emery, a major link in ant evolution. *Bulletin of the Museum of Comparative Zoology at Harvard College* **115:** 81–99.
Wilson, E. O., and R. W. Taylor. 1967. The ants of Polynesia. *Pacific Insects Monograph* **14:** 1–109.
Wing, W. M. 1968. Taxonomic revision of the Nearctic genus *Acanthomyops. Cornell University Agricultural Experiment Station Memoir* **405:** 1–173.
Wu, J. 1990. Taxonomic studies on the genus *Formica* of China. *Forest Research* **3:** 1–8.
Wu, J., and C. Wang. 1990. A taxonomic study on the genus *Tetraponera* Smith in China. *Scientia Silvae Sinicae* **26:** 515- 518.

Yamauchi, K. 1979. Taxonomical and ecological studies on the ant genus *Lasius* in Japan. 1. Taxonomy. *Science Reports of the Faculty of Education, Gifu University* (Nat. Sci.) **6:** 147- 181.

Yamauchi, K., and K. Hayashida. 1968. Taxonomic studies on the genus *Lasius* in Hokkaido, with ethological and ecological notes. 1. The subgenus *Dendrolasius* or jet black ants. *Journal of the Faculty of Science, Hokkaido University,* ser. 6, Zoology **16:** 396–412.

———1970. Taxonomic studies on the genus *Lasius* in Hokkaido, with ethological and ecological notes. 2. The subgenus *Lasius. Journal of the Faculty of Science, Hokkaido University,* ser. 6, Zoology **17:** 501–519.

Yasumatsu, K., and W. L. Brown, Jr. 1951. Revisional notes on *Camponotus herculeanus* Linné and close relatives in the Palaearctic regions. *Journal of the Faculty of Agriculture, Kyushu University* **10:** 29–44.

———1957. A second look at the ants of the *Camponotus herculeanus* group in eastern Asia. *Journal of the Faculty of Agriculture, Kyushu University* **11:** 45–51.

Yasumatsu, K., and Y. Murakami. 1960. A revision of the genus *Stenamma* of Japan. *Esakia* **1:** 27–31.

Index and Checklist

This section serves both as an index and a checklist of nominal taxa. As well as giving page references for subjects and taxonomic names, it also indicates the current status and rank of any taxon. The following convention is used.

(1) A name prefixed by * indicates an extinct taxon.
(2) A name prefixed by # indicates a taxon of subgenus rank.
(3) Currently recognized available names are printed in **bold**, with names in the genus group in ***bold italic*** and those in the family group in **bold roman**; main entries for subfamilies are in upper case.
(4) Names in synonymy, junior homonyms, and unavailable names in the genus group are in *italic;* those in the family group are in roman.
(5) Misspellings of names are omitted from this index but can be found in the synoptic classifications.